T0263492

DICTIONARY OF
ANIMAL HEALTH
TERMINOLOGY

DICTIONARY OF ANIMAL HEALTH TERMINOLOGY

in

English, French, Spanish, German and Latin

compiled by

OFFICE INTERNATIONAL DES EPIZOOTIES
Paris, France

Editor: Roy Mack FRCVS

ELSEVIER
Amsterdam – New York – Tokyo – Oxford 1992

ELSEVIER SCIENCE PUBLISHERS B.V.
Sara Burgerhartstraat 25
P.O. Box 211, 1000 AE Amsterdam, The Netherlands

Distributors for the United States and Canada:
ELSEVIER SCIENCE PUBLISHING COMPANY INC.
655 Avenue of the Americas
New York, N.Y. 10010, U.S.A.

Library of Congress Cataloging-in-Publication Data

Dictionary of animal health terminology : in English, French, Spanish,
  German, and Latin / compiled by Office international des epizooties,
  Paris, France ; editor, Roy Mack.
      p.   cm.
    Rev. ed. of the veterinary medicine portion of: Dictionary of
  animal production terminology. 2nd ed. 1979.
    ISBN 0-444-88085-2
    1. Animal health--Dictionaries--Polyglot. 2. Veterinary medicine-
  -Dictionaries--Polyglot. 3. Dictionaries, Polyglot.   I. Mack, Roy.
  II. International Office of Epizootics. III. Dictionary of animal
  production terminology.
  SF609.D55   1992
  636.089'03--dc20   92-9212
                                                                 CIP

ISBN: 9780444880857

© 1992 Elsevier Science Publishers B.V. All rights reserved.

No part of this publication may be reproduced, stored in a retrieval system or transmitted in any form or by any means, electronic, mechanical, photocopying, recording or otherwise, without the prior written permission of the publisher, Elsevier Science Publishers B.V., Copyright and Permissions Department, P.O. Box 521, 1000 AM Amsterdam, The Netherlands.

Special regulations for readers in the U.S.A. – This publication has been registered with the Copyright Clearance Center, Inc. (CCC), Salem, Massachusetts. Information can be obtained from the CCC about conditions under which photocopies of parts of this publication may be made in the U.S.A. All other copyright questions, including photocopying outside of the U.S.A., should be referred to the publisher, Elsevier Science Publishers B.V.

No responsibility is assumed by the publisher for any injury and/or damage to persons or property as a matter of products liability, negligence or otherwise, or from any use or operation of any methods, products, instructions or ideas contained in the material herein.

Transferred to digital print 2007

Printed and bound by CPI Antony Rowe, Eastbourne

PREFACE

The European Association for Animal Production (EAAP) was the first organization to publish, in collaboration with the Food and Agriculture Organization (FAO), a quadrilingual dictionary of zootechnical and veterinary terms in 1959. A second edition was published in 1985 under the title *Dictionary of Animal Production Terminology*.

During preparations for a third edition, Dr. J. Boyazoglu (Secretary General of EAAP), aware of the increasing specialization in animal production and of the associated terminology, judged that the time had come to divide the dictionary into two volumes, one devoted to animal husbandry terms and the other to veterinary terms.

Despite the size of the task of preparing a dictionary in four languages (English, French, German and Spanish - and also Latin), the Office International des Epizooties agreed to participate.

I am grateful to Dr. Boyazoglu for his confidence that the OIE, then run by Dr. Louis Blajan, was the appropriate organization to prepare a veterinary dictionary.

The division made between the animal husbandry and veterinary terms means that specialists in animal production and in animal health who have to work in many languages will need to consult not only the present volume, but also its companion volume devoted to animal production.

It gives me great pleasure to thank all those who have contributed to this dictionary, in particular Mr. Roy Mack, former Director of the Commonwealth Bureau of Animal Health, who made the initial selection of terms, prepared the linguistic equivalents and corrected the proofs.

I also wish to express my recognition of the assistance provided by those who revised the lists of terms, whose names are given on page vii.

I hope that the usefulness of this dictionary will be proportional to the efforts and time devoted to it by the contributors. I am sure that it will be a valuable reference tool for all those engaged in animal health who have to work in four languages.

Dr. J. Blancou
Secretary General OIE

LIST OF CONTRIBUTORS

Prof. Arturo Anadón
Departamento de Farmacología
Facultad de Medicina
Universidad Complutense
28040 Madrid, Spain

Prof. Dr. Miguel Cordero-del-Campillo
Catedrático de Parasitología y
Enfermedades Parasitarias
Universidad de León, Spain

Prof. J. Euzeby
Ecole Nationale Vétérinaire de Lyon
B.P. 31 Marcy l'Etoile
F-69752 Charbonnières, France

Prof. Dr. B. Hoffmann
Justus-Liebig-Universität
Frankfurter Strasse 106
D-6300 Giessen, Germany

Prof. G. Milhaud
Service de Pharmacie et Toxicologie
Ecole Nationale Vétérinaire d'Alfort
7, Avenue du Général de Gaulle
F-94704 Maisons-Alfort, France

Prof. Dr. H. Ludwig
Institut für Virologie
Fachbereich Veterinärmedizin
Nordufer 20
D-1000 Berlin 65, Germany

Prof. C. Pavaux
Laboratoire d'Anatomie
Ecole Nationale Vétérinaire de Toulouse
23, Chemin des Capelles
F-31076 Toulouse, France

Prof. Elias Fernando Rodriguez Ferri
Decano de la Facultad de Veterinaria
Campus Universitario de Vegazana
24071 León, Spain

Prof. Dr. M. Rommel
Institut für Parasitologie
Tierärztliche Hochschule
Bünteweg 17
D-3000 Hannover 71, Germany

Prof. J. Sandoval
Departamento de Anatomía y Embriología
Facultad de Veterinaria
24071 León, Spain

Prof. Dr. H. Waibl
Institut für Tieranatomie
Veterinärstrasse 13
D-8000 München 22, Germany

Basic Table

A

1 abdomen; belly
f abdomen
e abdomen; vientre
d Abdomen; Bauch
l abdomen

2 abdominal air sac
f sac aérien abdominal
e saco aéreo abdominal
d Bauchluftsack
l saccus abdominis

3 abdominal aorta
f aorte abdominale
e aorta abdominal
d Bauchaorta
l aorta abdominalis

4 abdominal aortic plexus
f plexus aortique abdominal
e plexo aórtico abdominal
d Plexus aorticus abdominalis
l plexus aorticus abdominalis

5 abdominal cavity
f cavité abdominale
e cavidad abdominal
d Bauchhöhle
l cavum abdominis

6 abdominal fissure
f fissure abdominale
e cisura abdominal
d Bauchspalte
l fissura abdominalis

7 abdominal hernia
f hernie abdominale
e hernia abdominal
d Eingeweidebruch
l hernia abdominalis

8 abdominal mammary gland
f mamelle abdominale
e mama abdominal
d abdominale Milchdrüse
l mamma abdominalis

9 abdominal muscles
f muscles de l'abdomen
e músculos del abdomen
d Bauchmuskeln
l musculi abdominis

10 abdominal part of cutaneous muscle
f muscle cutané (peaucier) du tronc;
   pannicule charnu
e músculo cutáneo del tronco
d Rumpfhautmuskel
l musculus cutaneus trunci

11 abducent nerve
f nerf abducens; nerf moteur oculaire
   externe
e nervio abducente
d Nervus abducens
l nervus abducens

12 abductor muscle of second digit
f muscle abducteur de l'index (doigt II)
e músculo abductor del dedo 2°
d Musculus abductor digiti II
l musculus abductor digiti II

13 aberrant ductules
f canalicules aberrants
e conductillos aberrantes
d Ductuli aberrantes
l ductuli aberrantes

* abnormal behaviour → 425

14 abnormality
f anomalie
e anomalía
d Anomalie; Abweichung

15 abomasal groove
f sillon abomasal
e surco del abomaso
d Labmagenfurche
l sulcus abomasi

16 abomasitis; inflammation of abomasum
f abomasite; inflammation de la caillette
e abomasitis
d Abomasitis; Labmagenentzündung

17 abomasum
f abomasum; caillette
e abomoso; cuajar
d Labmagen; Drüsenmagen
l abomasum

18 abortion
f avortement
e aborto
d Abort; Fehlgeburt; Verwerfen

19 **abscess**
f abcès
e absceso
d Abszeß
l abscessus

* **absence of an organ** → 100

* **absence of limbs** → 138

* **acanthocephalid infestation** → 21

20 **acanthocephalids; thorny-headed worms**
f acanthocéphales
e acantocéfalos
d Akanthozephale; Kratzer
l Acanthocephala

21 **acanthocephalosis; acanthocephalid infestation**
f acanthocéphalose; infestation par acanthocéphales
e acantocefalosis
d Akanthozephalose; Befall mit Kratzern

22 **acanthosis**
f acanthose
e acantosis
d Akanthose

23 **acarapisosis; acariasis of bees; Acarapis woodii infestation**
f acariose trachéale des abeilles; infestation par Acarapis woodii
e acariosis de las abejas
d Akarapidose; Tracheenmilbenkrankheit der Biene

* **Acarapis woodii infestation** → 23

24 **acardia; congenital absence of heart**
f acardie; absence congénitale du cœur
e acardia
d Akardie; Fehlen des Herzens

* **acariasis of bees** → 23

25 **acaricide**
f acaricide
e acaricida
d Akarizid

* **acaudia** → 4434

26 **accessory carpal bone**
f os accessoire du carpe; os pisiforme
e hueso accesorio del carpo; hueso pisiforme

d Os carpi accessorium; Erbsenbein; Anhangsbein
l os carpi accessorium; os pisiforme

27 **accessory cephalic vein**
f veine céphalique accessoire
e vena cefálica accesoria
d Vena cephalica accessoria
l vena cephalica accessoria

* **accessory digit** → 1276

28 **accessory ligament; check ligament**
f ligament accessoire
e ligamento accesorio
d Ligamentum accessorium; Unterstützungsband
l ligamentum accessorium

29 **accessory lobe of lung**
f lobe accessoire du poumon; lobe azygos
e lóbulo accesorio
d Anhangslappen der Lunge
l lobus accessorius pulmonis dextra

30 **accessory nerve; spinal nerve**
f nerf accessoire; nerf spinal
e nervio accesorio; nervio espinal
d Nervus accessorius
l nervus accessorius

31 **accessory organs of eye**
f organes accessoires de l'œil
e órganos accesorios del ojo
d Nebenorgane des Auges
l organa oculi accessoria

32 **accessory pancreatic duct**
f conduit pancréatique accessoire; canal de Santorini
e conducto pancreático accesorio; conducto de Santorini
d Ductus Santorini; Nebengang des exokrinen Pankreas
l ductus pancreaticus accessorius

33 **accessory pillar of rumen**
f pilier accessoire du rumen
e pilar accesorio del rumen
d Nebenpfeiler des Rumens

34 **accessory sex glands**
f glandes génitales accessoires
e glándulas genitales accessorias
d akzessorische Geschlechtsdrüsen
l glandulae genitales accessoriae

35 accompanying veins
f veines accompagnantes
e venas comitantes
d Begleitvene
l venae comitantes

36 acetabular fossa
f fosse acétabulaire
e fosa del acetábulo
d Fossa acetabuli
l fossa acetabuli

37 acetabular incisure
f incisure de l'acétabulum
e escotadura del acetábulo
d Incisura acetabuli; Pfannenausschnitt
l incisura acetabuli

38 acetabular lip
f bourrelet acétabulaire
e labro acetabular
d Labrum acetabulare
l labrum acetabulare

39 acetabulum
f acétabulum
e acetábulo
d Acetabulum; Hüftpfanne
l acetabulum

40 acetonaemia
f acétonémie; cétonémie
e acetonemia; quetonemia
d Azetonämie; Ketonämie

41 acetonuria; ketonuria
f acétonurie; cétonurie
e acetonuria; quetonuria
d Azetonurie; Ketonurie

42 achalasia
f achalasie
e acalasia; espasmo de un esfínter
d Achalasie

43 acheilia
f acheilie
e aqueilia
d Acheilie

* **Achilles tendon** → 1004

44 achondroplasia
f achondroplasie
e acondroplasia
d Achondroplasie

45 acid-base equilibrium
f équilibre acido-basique
e equilibrio ácido-básico
d Säure-Basengleichgewicht

46 acid-fast bacterium
f bactérie acido-résistante
e bacteria ácido-resistente
d säurefeste Bakterie

47 acidosis
f acidose
e acidosis
d Acidose; Azidose

* **Ackerknecht's organ** → 3163

48 acorn poisoning
f intoxication par les glands
e intoxicación por bellotas
d Eichelvergiftung

* **acoustic area** → 2184, 4332

* **acral lick dermatitis** → 49

49 acrodermatitis; acral lick dermatitis
f acrodermatite
e acrodermatitis
d Akrodermatite; Hautenzündung an Akren

50 acromion
f acromion
e acromion
d Akromion; Gräteneck
l acromion

* **ACTH** → 1096

51 actinobacillosis; wooden tongue of cattle; Actinobacillus infection
f actinobacillose
e actinobacilosis
d Aktinbazillose

* **Actinobacillus equuli infection** → 4074

* **Actinobacillus infection** → 51

* **Actinobacillus pleuropneumoniae infection** → 3559

* **Actinomyces bovis infection** → 53

52 actinomycetes
f actinomycètes
e actinomicetos
d Aktinomyzeten
l Actinomycetales

53 **actinomycosis; lumpy jaw of cattle;
Actinomyces bovis infection**
f actinomycose; infection à Actinomyces
bovis
e actinomicosis
d Aktinomykose; Actinomyces-bovis-
Infektion

54 **active immunity**
f immunité active
e inmunidad activa
d aktive Immunität

55 **active ingredient; active principle**
f principe actif
e ingrediente activo; principio activo
d Wirkstoff

* **active principle** → 55

* **Acuaria infestation** → 56

56 **acuariosis; Acuaria infestation**
f acuariose
e acuariosis
d Acuariose

57 **acupuncture**
f acupuncture
e acupuntura
d Akupunktur

58 **acute**
f aigu
e agudo
d akut

59 **acute disease**
f maladie aiguë
e enfermedad aguda
d akute Krankheit

60 **acute toxicity**
f toxicité aiguë
e toxicidad aguda
d akute Toxizität

* **Adam's apple** → 2429

61 **adductor muscle**
f muscle adducteur
e músculo aductor
d Einwärtszieher
l musculus adductor

62 **adductor muscle of fifth digit**
f muscle adducteur du doigt V
e músculo aductor del dedo 5°
d Musculus adductor digiti V
l musculus adductor digiti V

63 **adductor muscle of first digit**
f muscle adducteur du doigt I
e músculo aductor del dedo 1°
d Musculus adductor digiti I
l musculus adductor digiti I

64 **adductor muscle of rectrices**
f muscle adducteur des rectrices
e músculo aductor de las rectrices
d Musculus adductor rectricius
l musculus adductor rectricius

65 **adductor muscle of second digit**
f muscle adducteur du doigt II
e músculo aductor del dedo 2°
d Musculus adductor digiti II
l musculus adductor digiti II

66 **adenitis**
f adénite
e adenitis
d Adenitis; Drüsenentzündung

67 **adenocarcinoma**
f adénocarcinome
e adenocarcinoma
d Adenokarzinoma; Drüsenepithelkrebs

68 **adenohypophysis; anterior lobe of pituitary
gland**
f adénohypophyse; lobe antérieur de la
glande pituitaire
e adenohipófisis; lóbulo anterior de la
hipófisis
d Adenohypophysis;
Hypophysenvorderlappen
l adenohypophysis; lobus anterior
hypophyseos

69 **adenoma**
f adénome
e adenoma
d Adenom

70 **adenomatosis**
f adénomatose
e adenomatosis
d Adenomatose

71 **adenovirus**
f adénovirus
e adenovirus
d Adenovirus

72 **adenovirus infection**
f adénovirose
e adenovirosis
d Adenovirus-Infektion

73 **adhesion**
 f adhérence
 e adherencia
 d Adhäsion; Verklebung
 l adhesio

74 **adipose capsule of kidney; kidney fat**
 f capsule adipeuse du rein; graisse de
   rognon
 e cápsula adiposa del riñón; grasa del riñón
 d Capsula adiposa renis; Nierenfett
 l capsula adiposa renis

75 **adipose tissue**
 f tissu adipeux
 e tejido adiposo
 d Fettgewebe

76 **adrenal cortex**
 f cortex de la glande surrénale; glande
   cortico-surrénale
 e corteza adrenal; córticosuprarrenal
 d Nebennierenrinde
 l cortex glandulae suprarenalis

77 **adrenal gland**
 f glande surrénale
 e glándula suprarrenal
 d Nebenniere
 l glandula suprarenalis

 * **adrenaline** → 1502

78 **adrenal medulla**
 f médullosurrénale
 e médula suprarrenal
 d Nebennierenmark
 l medulla glandulae suprarenalis

79 **adventitia**
 f adventice
 e túnica adventicia
 d Adventitia
 l tunica adventitia

80 **adventitious coat of oesophagus**
 f tunique adventice de l'œsophage
 e túnica adventicia del esófago
 d äußere Schicht der Speiseröhre
 l tunica adventitia (o)esophagi

81 **adverse effect**
 f effet secondaire; réaction indésirable
 e efecto adverso
 d Nebenwirkung; Nebenerscheinung;
   Schadwirkung

 * **Aegyptianella pullorum infection** → 82

82 **aegyptianellosis; Aegyptianella pullorum
   infection**
 f ægyptianellose aviaire
 e egipcianelosis de las aves
 d Aegyptianellose

83 **aerobe; aerobic bacterium**
 f aérobie; bactérie aérobie
 e aerobia; bacteria aerobia
 d Aerobier; aerobe Bakterie

 * **aerobic bacterium** → 83

84 **aerocystitis; swim bladder inflammation**
 f aérocystite
 e aerocistitis
 d Aerozystitis; Schwimmblasenentzündung

 * **Aeromonas salmonicida infection** → 85

85 **aeromoniasis of Salmonidae; furunculosis
   of Salmonidae; Aeromonas salmonicida
   infection**
 f furonculose des salmonidés; aéromonose
   par Aeromonas salmonicida
 e furunculosis de los salmónidos;
   aeromonosis por Aeromonas salmonicida
 d Aeromonas-salmonicida-Infektion

86 **aerophagia; air swallowing; wind sucking**
 f aérophagie; déglutition d'air; tic
   aérophagique
 e aerofagia
 d Aerophagie; Luftschlucken des Pferdes;
   Windschnappen

87 **aetiology; etiology**
 f étiologie
 e etiología
 d Ätiologie

88 **afebrile**
 f afébrile; apyrétique
 e afebril
 d fieberlos

89 **aflatoxicosis; aflatoxin poisoning**
 f aflatoxicose; intoxication par l'aflatoxine
 e aflatoxicosis
 d Aflatoxikose; Vergiftung durch Aflatoxin

90 **aflatoxin; Aspergillus flavus toxin**
 f aflatoxine
 e aflatoxina
 d Aflatoxin

 * **aflatoxin poisoning** → 89

91 African horse sickness
f peste équine
e peste equina
d afrikanische Pferdesterbe
l pestis equorum

92 African horse sickness virus
f virus de la peste équine
e virus de la peste equina
d afrikanisches Pferdesterbe-Virus

93 African swine fever
f peste porcine africaine
e peste porcina africana
d afrikanische Schweinepest
l pestis suum africana

94 African swine fever virus
f virus de la peste porcine africaine
e virus de la peste porcina africana
d afrikanisches Schweinepest-Virus

95 African trypanosomosis; nagana
f trypanosomoses africaines; nagana
e tripanosomosis africana
d afrikanische Trypanosomose; Nagana

96 afterbirth
f délivre
e secundinas
d Nachgeburt; Secundinae

97 afterfeather
f hypopenne
e hipopluma
d Nebenschaft
l hypopenna

98 agalactia
f agalactie; agalaxie
e agalaxia
d Agalaktie; Milchmangel

99 agammaglobulinaemia
f agammaglobulinémie
e agammaglobulinemia
d Agammaglobulinämie

100 agenesis; absence of an organ
f agénésie; absence de développement
e agenesia
d Agenesie

101 agglutination
f agglutination
e aglutinación
d Agglutination; Verklumpung

102 agglutination test
f réaction d'agglutination
e prueba de aglutinación
d Agglutinationsreaktion

103 aggregated lymphatic nodules; Peyer's patches
f nodules lymphatiques agglomérés; plaques de Peyer
e nodulillos linfáticos agregados; placas de Peyer
d aggregierte Lymphknötchen; Peyersche Platten
l lymphonoduli aggregati

104 agnathia
f agnathie
e agnacia
d Agnathie
l Agnathia

105 agranulocytosis
f agranulocytose; absence de granulocytes
e agranulocitosis
d Agranulozytose

106 airborne infection
f infection aérogène
e infección aerógena
d aerogene Infektion

* air capillaries → 3521

* airsacculitis → 107

107 air sac inflammation; airsacculitis
f aérosacculite; inflammation des sacs aériens
e inflamación de los sacos aéreos
d Luftsackentzündung

108 air sacs
f sacs aériens
e sacos aéreos
d Luftsäcke
l sacci pneumatici

* air swallowing → 86

109 alar foramen
f foramen alaire; trou alaire
e agujero alar
d Flügelloch des Atlas
l foramen alare

110 alar ligaments
f ligaments alaires
e ligamentos alares
d Ligamenta alaria
l ligamenta alaria

* alar membrane → 3632

111 albendazole
f albendazole
e albendazol
d Albendazol
l albendazolum

112 albinism
f albinisme
e albinismo
d Albinismus

113 albuminuria
f albuminurie
e albuminuria
d Albuminurie

* alcelaphine herpesvirus type 1 → 2650

114 aldosterone
f aldostérone
e aldosterona
d Aldosteron
l aldosteronum

115 Aleutian disease; plasmacytosis of mink
f maladie de l'Aléoutien
e enfermedad aleutiana del visón
d Aleutenkrankheit des Nerzes;
  Plasmozytose der Nerze

116 alfaprostol
f alfaprostol
e alfaprostol
d Alfaprostol
l alfaprostolum

117 alfaxalone; alphaxalone
f alfaxalone
e alfaxalona
d Alfaxalon
l alfaxalonum

118 alimentary canal
f tractus alimentaire; tube digestif
e canal alimentario; tubo digestivo
d Verdauungskanal
l canalis alimentarius; tubus digestorius

119 alkaloid
f alcaloïde
e alcaloide
d Alkaloid

120 alkalosis
f alcalose
e alcalosis
d Alkalose

* allantochorion → 885

121 allantois
f allantoïde
e alantoides
d Allantois
l allantois

122 allergen
f allergène
e alergeno
d Allergen

123 allergenic
f allergisant
e alergénico
d allergenisch

124 allergic reaction
f réaction allergique
e reacción alérgica
d allergische Reaktion

125 allergy
f allergie
e alergia
d Allergie

126 allocortex; olfactory cortex
f allocortex; cortex hétérogénétique
e alocórtex
d Allokortex
l allocortex

127 alopecia; hair loss
f alopécie; pelade
e alopecia
d Alopezie; Haarschwund

128 alphavirus
f alphavirus
e alphavirus
d Alphavirus

* alphaxalone → 117

129 altitude sickness
f mal d'altitude
e mal de altura
d Höhenkrankheit

130 altrenogest
f altrénogest
e altrenogest
d Altrenogest
l altrenogestum

131 **alveolar ductules**
  f canalicules alvéolaires
  e conductillos alveolares
  d Alveolengänge
  l ductuli alveolares

132 **alveolar foramina of maxilla**
  f foramens alvéolaires
  e agujeros alveolares
  d Foramina alveolares
  l foramina alveolares maxillaris

133 **alveolar process of maxilla**
  f processus alvéolaire maxillaire
  e apófisis alveolar
  d Alveolarfortsatz des Oberkiefers
  l processus alveolaris maxillae

134 **alveolar saccules**
  f sacs alvéolaires
  e sáculos alveolares
  d Alveolensäckchen
  l sacculi alveolares

135 **alveus of hippocampus**
  f alvéus de l'hippocampe
  e canal del hipocampo
  d Muldenblatt
  l alveus hippocampi

136 **amantadine**
  f amantadine
  e amantadina
  d Amantadin
  l amantadinum

137 **ambiguous nucleus**
  f noyau ambigu
  e núcleo ambiguo
  d Nucleus ambiguus
  l nucleus ambiguus

138 **amelia; absence of limbs**
  f amélie; absence des membres
  e amelia
  d Amelie; Fehlen der Extremitäten

139 **American foulbrood; malignant foulbrood;
    Bacillus larvae infection**
  f loque américaine; loque maligne; infection
    à Bacillus larvae
  e loque americana
  d bösartige Faulbrut; Bacillus-larvae-
    Infektion

140 **amicarbalide**
  f amicarbalide
  e amicarbalida
  d Amicarbalid
  l amicarbalidum

141 **amidostomosis; Amidostomum infestation**
  f amidostomose
  e amidostomosis
  d Amidostomose

* **Amidostomum infestation** → 141

142 **amikacin**
  f amikacine
  e amikacina
  d Amikacin
  l amikacinum

143 **amitraz**
  f amitraz
  e amitraz
  d Amitraz
  l amitrazum

* **Ammon's horn** → 2014

144 **amnion**
  f amnios
  e amnios
  d Amnion
  l amnion

145 **amoxicillin; amoxycillin**
  f amoxicilline
  e amoxicilina
  d Amoxicillin
  l amoxicillinum

* **amoxycillin** → 145

146 **amphiarthrosis; cartilaginous joint**
  f amphiarthrose; jointure cartilagineuse
    semi-mobile
  e anfiartrosis; articulación cartilaginosa
  d Amphiarthrose; Knorpelverbindung
  l junctura cartilaginea

147 **amphotericin B**
  f amphotéricine B
  e anfotericina B
  d Amphotericin B
  l amphotericinum B

148 **ampicillin**
  f ampicilline
  e ampicilina
  d Ampicillin
  l ampicillinum

149 **amprolium**
  f amprolium
  e amprolio
  d Amprolium
  l amprolium

150 ampulla of deferent duct
  f ampoule du canal déférent
  e ampolla del conducto deferente
  d Samenleiterampulle
  l ampulla ductus deferentis

151 ampulla of uterine tube
  f ampoule de la trompe utérine
  e ampolla de la trompa uterina
  d Ampulle des Eileiters
  l ampulla tuba uterina

152 ampullar nerve
  f nerf ampullaire
  e nervio ampular
  d Nervus ampullaris
  l nervus ampullaris

153 amygdaloid body
  f corps amygdalien
  e cuerpo amigdaloide
  d Mandelkörper
  l corpus amygdaloideum

154 amyloidosis
  f amyloïdose; amylose
  e amiloidosis
  d Amyloidose

155 amyotrophy; muscular atrophy
  f amyotrophie; atrophie musculaire
  e amiotrofia; atrofia muscular
  d Myatrophie; Muskelatrophie

156 anaemia; anemia
  f anémie
  e anemia
  d Anämie; Blutarmut

157 anaemic; anemic
  f anémique
  e anémico
  d anämisch

158 anaerobe; anaerobic bacterium
  f anaérobie; bactérie anaérobie
  e anaerobia; bacteria anaerobia
  d Anaerobier; anaerobe Bakterie

* anaerobic bacterium → 158

159 anaesthesia; anesthesia
  f anesthésie
  e anestesia
  d Anästhesie; Narkose

160 anaesthetic; anesthetic
  f anesthésique
  e anestésico
  d Anästhetikum; Narkotikum

161 anal atresia
  f atrésie anale
  e atresia anal
  d Analatresie
  l atresia ani

162 anal canal
  f canal anal
  e conducto anal
  d Canalis analis
  l canalis analis

163 analeptic
  f analeptique
  e analéptico
  d Analeptikum

164 analgesic
  f analgésique
  e analgésico
  d Analgetikum

165 anal glands
  f glandes anales
  e glándulas anales
  d Analdrüsen
  l glandulae anales

166 anal sac; paranal sinus
  f sac anal
  e seno paranal
  d Analbeutel
  l sinus paranalis

167 anal sphincter
  f sphincter de l'anus
  e esfínter del ano
  d Afterschließmuskel
  l musculus sphincter ani

* anamnesis → 699

168 anaphylactic reaction
  f réaction anaphylactique
  e reacción anafiláctica
  d anaphylaktische Reaktion

169 anaphylaxis
  f anaphylaxie
  e anafilaxia
  d Anaphylaxie

* Anaplasma infection → 170

170 anaplasmosis; Anaplasma infection
  f anaplasmose
  e anaplasmosis; infección por Anaplasma spp.
  d Anaplasmose; Anaplasma-Infektion

171 **anastomosis**
f anastomose
e anastomosis
d Anastomose
l anastomosis

172 **anastomotic vessel**
f vaisseau anastomotique
e vaso anastomótico
d Vas anastomoticum
l vas anastomoticum

173 **anatomy**
f anatomie
e anatomía
d Anatomie
l anatomia

174 **anconeal process; beak of olecranon**
f processus anconé; bec de l'olécrâne
e apófisis ancónea
d Processus anconeus
l processus anconeus

175 **anconeus muscle**
f muscle anconé
e músculo ancóneo
d Ellbogenhöckermuskel
l musculus ancon(a)eus

* **Ancylostoma infestation** → 176

* **ancylostome** → 2029

176 **ancylostomosis; Ancylostoma infestation**
f ankylostomose
e anquilostomosis; ancilostomosis
d Ankylostomiase

177 **androgen; androgenic hormone**
f hormone androgène
e hormona androgénica
d Androgen; männliches Sexualhormon

* **androgenic hormone** → 177

178 **androstanolone; stanolone**
f androstanolone
e androstanolona
d Androstanolon
l androstanolonum

* **anemia** → 156

* **anemic** → 157

179 **anencephaly**
f anencéphalie; encéphalo-araphie
e anencefalía
d Anenzephalie

* **anesthesia** → 159

* **anesthetic** → 160

* **anestrus** → 192

180 **aneurysm**
f anévrysme
e aneurisma
d Aneurysma

181 **angle of jaw; angle of mandible**
f angle de la mandibule
e ángulo de la mandíbula
d Unterkieferwinkel
l angulus mandibulae

* **angle of mandible** → 181

182 **angle of mouth**
f angle de la bouche
e ángulo de la boca
d Mundwinkel
l angulus oris

* **angle of pelvis** → 2163

183 **angle of rib**
f angle costal
e ángulo de la costilla
d Rippenwinkel
l angulus costae

184 **angular bone**
f os angulaire
e hueso angular
d Os angulare
l os angulare

185 **angular vein of eye**
f veine angulaire de l'œil
e vena angular del ojo
d Vena angularis oculi
l vena angularis oculi

186 **anhidrosis**
f anhidrose
e anhidrosis
d Anhidrose

187 **animal clinic**
f clinique pour animaux
e clínica para animales
d Tierklinik

188 **animal disease**
f maladie des animaux
e enfermedad animal
d Tierkrankheit

189 animal health
 f santé animale
 e sanidad animal
 d Tiergesundheit

 * **ankle bone → 4574**

 * **ankle joint → 2022**

190 ankylosis
 f ankylose
 e anquilosis
 d Ankylose

191 annular ligament of radius
 f ligament annulaire du radius
 e ligamento anular del radio
 d Ligamentum anulare radii
 l ligamentum anulare radii

192 anoestrus; anestrus
 f anœstrus
 e anestro
 d Anöstrie; Brunstlosigkeit

 * **anoplurosis → 2578**

193 anorectal atresia
 f atrésie ano-rectale
 e atresia anorectal
 d Afterverschluß ohne Rektum
 l atresia ani et recti

194 anorexia
 f anorexie
 e anorexia
 d Anorexie

195 anotia
 f anotie
 e anotia
 d Anotie

196 anoxia
 f anoxie
 e anoxia
 d Anoxie; Sauerstoffmangel

197 antebrachial fascia; fascia of forearm
 f fascia antébrachial
 e fascia del antebrazo
 d Fascia antebrachii
 l fascia antebrachii

198 antebrachiocarpal articulation; radiocarpal joint
 f articulation antébrachiocarpienne
 e articulación antebraquiocarpiana
 d Arm-Vorderfußwurzelgelenk
 l articulatio antebrachiocarpea

 * **antemetic → 217**

199 anterior chamber of eye
 f chambre antérieure de l'œil
 e cámara anterior del globo
 d vordere Augenkammer
 l camera anterior bulbi

200 anterior limiting lamina; Bowman's membrane
 f lame limitante externe; membrane de Bowman
 e lámina limitante anterior
 d vordere Grenzschicht der Hornhaut
 l lamina limitans anterior

 * **anterior lobe of pituitary gland → 68**

201 anthelmintic; vermifuge
 f anthelminthique
 e antihelmíntico; vermífugo
 d Anthelminthikum; Vermifugum

202 anthrax; Bacillus anthracis infection
 f charbon bactéridien; fièvre charbonneuse
 e carbunco bacteridiano
 d Milzbrand; Bacillus-anthracis-Infektion

203 anthrax bacillus
 f bacille charbonneux; bactéridie charbonneuse
 e bacilo carbuncoso; bacteridia carbuncosa
 d Milzbrandbazillus
 l Bacillus anthracis

204 antibacterial agent
 f agent antibactérien
 e fármaco antibacteriano
 d antibakterielles Mittel; Bakteriostatikum

205 antibiotic
 f antibiotique
 e antibiótico
 d Antibiotikum

206 antibiotic resistance
 f antibiorésistance
 e resistencia antibiótica
 d Antibiotikaresistenz

207 antibiotic therapy
 f antibiothérapie; traitement aux antibiotiques
 e terapia antibiótica
 d Antibiotikatherapie

208 **antibloat agent**
 f agent antimétéorisant; météorifuge
 e fármaco contra timpanitis
 d Mittel gegen Tympanie

209 **antibody**
 f anticorps
 e anticuerpo
 d Antikörper

210 **antibody formation**
 f formation des anticorps
 e formación de anticuerpos
 d Antikörperbildung

211 **anticoagulant**
 f anticoagulant
 e anticoagulante
 d Antikoagulans

212 **anticoccidal drug; coccidiostat**
 f anticoccidien; coccidiostatique
 e fármaco anticoccidiósico; coccidiostático
 d Antikokzidium; Kokzidiostatikum

213 **anticonvulsant**
 f anticonvulsivant
 e fármaco anticonvulsivante
 d Antikonvulsivum

214 **antidiarrhoea agent**
 f antidiarrhéique
 e fármaco antidiarréico
 d Antidiarrhöikum

215 **antidiuretic hormone; vasopressin**
 f hormone antidiurétique; vasopressine
 e vasopresina
 d antidiuretisches Hormon; Vasopressin

216 **antidote**
 f antidote
 e antídoto
 d Antidot; Gegengift

217 **antiemetic; antemetic**
 f antiémétique
 e antiemético
 d Ant(i)emetikum

218 **antifungal agent**
 f agent antifongique; antimycosique
 e fármaco antifúngico
 d Antimykotikum; Fungistatikum

219 **antigen**
 f antigène
 e antígeno
 d Antigen

220 **antigen-antibody reaction**
 f réaction antigène-anticorps
 e reacción antígeno-anticuerpo
 d Antigen-Antikörper-Reaktion

221 **antigenicity**
 f antigénicité
 e antigenicidad
 d Antigenität

222 **antihistamine; histamine antagonist**
 f antihistaminique
 e fármaco antialérgico; antihistamínico
 d Antihistaminikum; Antiallergikum

223 **anti-infective agent; chemotherapeutic drug**
 f anti-infectieux
 e fármaco antiinfeccioso; fármaco quimioterápico
 d Chemotherapeutikum; Antiinfektiosum

224 **anti-inflammatory agent**
 f anti-inflammatoire
 e fármaco antiinflamatorio
 d Antiphlogistikum

225 **antineoplastic agent; cytostatic agent**
 f médicament cytostatique; antinéoplasique
 e fármaco antineoplásico
 d Zytostatikum

226 **antiparasitic agent**
 f antiparasitaire
 e fármaco antiparasitario
 d Antiparasitikum

227 **antiprotozoal agent**
 f antiprotozoaire
 e fármaco antiprotozoario
 d Mittel gegen Protozoen

228 **antiseptic**
 f antiseptique
 e antiséptico
 d Antiseptikum

* **antiserum** → 2139

229 **antitoxin**
 f antitoxine
 e antitoxina
 d Antitoxin

230 **antitragus**
 f antitragus
 e antitrago
 d Antitragus
 l antitragus

231 antivenin
f sérum antivenimeux
e contraveneno
d Antidot gegen Schlangengift

232 antiviral agent
f agent antiviral; virucide
e fármaco antivírico; viricida
d Virostatikum

233 anus
f anus
e ano
d Anus; After
l anus

* anvil → 2167

234 aorta
f aorte
e aorta
d Aorta; Hauptschlagader
l aorta

235 aortic aneurysm
f anévrysme de l'aorte
e aneurisma de la aorta
d Aneurysma aortae

236 aortic arch
f arc aortique
e arco de la aorta
d Aortenbogen
l arcus aortae

237 aortic hiatus of diaphragm
f hiatus aortique
e hiato aórtico
d Aortenschlitz
l hiatus aorticus

238 aortic isthmus
f isthme aortique
e istmo de la aorta
d Isthmus aortae; Aortenverengung
l isthmus aortae

239 aortic opening
f orifice aortique du cœur
e orificio de la aorta
d Aortenöffnung
l ostium aortae

240 aorticorenal ganglia
f ganglions aortico-rénaux
e ganglios aorticorrenales
d Ganglia aorticorenalia
l ganglia aorticorenalia

241 aortic rupture
f rupture de l'aorte
e ruptura aórtica
d Aortenruptur

* aortic valve → 4816

242 apex of cochlea; cupula of cochlea
f coupole de la cochlée
e cúpula del cóclea
d Cupula cochleae; Schneckenspitze
l cupula cochleae

243 apex of frog
f apex cunéal; pointe de la fourchette
e vértice de la cuña
d Strahlspitze
l apex cunei

244 apex of heart
f apex du cœur; pointe du cœur
e vértice del corazón
d Herzspitze
l apex cordis

245 apex of nose; tip of nose
f pointe du nez; extrémité du nez
e vértice de la nariz; punta de la nariz
d Nasenspitze
l apex nasi

* aphosphorosis → 3456

246 aphtha
f aphte
e afta
d Aphthe

247 apical lobe of lung; cranial lobe
f lobe crânial du poumon
e lóbulo craneal
d Spitzenlappen der Lunge
l lobus cranialis pulmonis

248 aplasia
f aplasie
e aplasia
d Aplasie

249 aponeurosis
f aponévrose
e aponeurosis
d Aponeurose; Sehnenblatt
l aponeurosis

250 apophysis
f apophyse
e apófisis
d Apophyse
l apophysis

251 **appetite disorder**
f trouble de l'appétit
e trastorno del apetito
d Appetitstörung

* **application to the skin** → 1195

252 **apramycin**
f apramycine
e apramicina
d Apramycin
l apramycinum

253 **apterial muscles**
f muscles aptériaux
e músculos de las apterias
d Musculi apteriales
l musculi apteriales

254 **aqueduct of mesencephalon**
f aqueduc mésencéphalique (de Sylvius)
e acueducto del mesencéfalo (de Silvio)
d Mittelhirnkanal; Sylvius' Aquädukt
l aqueductus mesencephali

255 **aqueous humour**
f humeur aqueuse
e humor acuoso
d Kammerwasser
l humor aquosus

256 **arachnoidal villi; pacchionian granulations**
f granulations arachnoïdiennes (de Pacchioni)
e granulaciones aracnoideas
d Pacchionische Granulationen; Arachnoidealzotten
l granulationes arachnoideales

257 **arachnoid of brain; cranial arachnoid**
f arachnoïde encéphalique
e aracnoides del encéfalo
d Spinngewebshaut des Gehirns
l arachnoidea encephali

258 **arachnoid of spinal cord; spinal arachnoid**
f arachnoïde spinale
e aracnoides espinal
d Spinngewebshaut des Rückenmarks
l arachnoidea spinalis

259 **arbovirus; arthropod-borne virus**
f arbovirus
e arbovirus
d Arbovirus

260 **arcuate artery**
f artère arquée
e arteria arqueada
d Bogenarterie
l arteria arcuata

261 **arcuate nerve fibres of cerebrum; association fibres**
f fibres arquées du cerveau
e fibras arqueadas del cerebro
d bogenförmige Nervenfasern des Großhirns; Assoziationsfasern
l fibrae arcuatae cerebri

262 **arcuate nucleus**
f noyau arqué
e núcleo arqueado
d Nucleus arcuatus
l nucleus arcuatus

263 **arcuate veins**
f veines arquées
e venas arqueadas
d Venae arcuatae
l venae arcuatae

264 **arenavirus**
f arénavirus
e arenavirus
d Arenavirus

* **argasidosis** → 4581

265 **argasid tick; soft tick**
f argasidés
e garrapata blanda; argas
d Lederzecke
l Argasidae

266 **Arizona infection**
f arizonose; infection à Arizona
e arizonosis
d Arizona-Infektion

* **armed tapeworm** → 3474

* **armpit** → 369

* **Arnold's ganglion** → 3183

267 **arprinocid**
f arprinocide
e arprinocida
d Arprinocid
l arprinocidum

268 **arrector muscles; erector muscles of hairs**
f muscles érecteurs des poils; muscles pilomoteurs

e músculos erectores de los pelos
d Haarbalgmuskeln; Aufrichtmuskeln der
  Haare
l musculi arrectores pilorum

269 arrhythmia
f arythmie
e arritmia
d Arrhythmie; Herzrhythmusstörung

270 arterial arch
f arcade artérielle
e arco arterioso
d Arterienring
l arcus arteriosus

271 arterial cone
f cône artériel
e cono arterioso
d Conus arteriosus
l conus arteriosus

272 arterial duct
f canal artériel (de Botal)
e conducto arterioso
d Ductus arteriosus
l ductus arteriosus

273 arterial network
f réseau artériel
e red arteriosa
d arterielles Gefäßnetz
l rete arteriosum

274 arteriole
f artériole
e arteriola
d Arteriole
l arteriola

275 arteriosclerosis
f artériosclérose
e arterioesclerosis
d Arteriosklerose

276 arteriovenous anastomosis
f anastomose artério-veineuse
e anastomosis arteriovenosa
d arteriovenöse Anastomose
l anastomosis arteriovenosa

277 arteritis
f artérite
e arteritis
d Arteriitis; Arterienentzündung

278 artery
f artère

e arteria
d Arterie
l arteria

279 arthritis
f arthrite
e artritis
d Arthritis; Gelenkentzündung

280 arthrocentesis
f arthrocentèse; ponction articulaire
e punción articular
d Arthrozentese; Gelenkpunktion

281 arthrodesis
f arthrodèse
e artrodesis
d Arthrodese; Gelenkverödung

* arthrodia → 3494

282 arthrogryposis
f arthrogrypose
e artrogriposis
d Arthrogryposis

283 arthropathy; joint disease
f arthropathie; affection articulaire
e artropatía; enfermedad articular
d Arthropathie; Gelenkkrankheit

* arthropod-borne virus → 259

284 arthropod parasite
f arthropode parasite
e artrópodo parásito
d parasitischer Arthropode

285 articular bone
f os articulaire
e hueso articular
d Os articulare
l os articulare

* articular capsule → 2358

286 articular cartilage
f cartilage articulaire
e cartílago articular
d Gelenkknorpel
l cartilago articularis

287 articular disk
f disque articulaire
e disco articular
d Gelenkdiskus
l discus articularis

288 **articular fovea**
 f fossette articulaire
 e fóvea articular
 d Gelenkgrube
 l fovea articularis

289 **articular meniscus**
 f ménisque articulaire
 e menisco articular
 d Gelenkmeniskus
 l meniscus articularis

290 **articular muscle**
 f muscle articulaire
 e músculo articular
 d Gelenkmuskel
 l musculus articularis

291 **articular muscle of hip joint**
 f muscle articulaire de la hanche
 e músculo articular de la cadera
 d Kapselmuskel des Hüftgelenkes
 l musculus articularis coxae

292 **articular muscle of shoulder joint**
 f muscle articulaire de l'épaule
 e músculo articular del hombro
 d Schultergelenkmuskel
 l musculus articularis humeri

293 **articular muscle of stifle**
 f muscle articulaire du genou
 e músculo articular de la rodilla
 d Kniegelenkmuskel
 l musculus articularis genus

294 **articular process**
 f processus articulaire; apophyse articulaire
 e apófisis articular
 d Gelenkfortsatz
 l processus articularis

295 **articular surface**
 f facette articulaire
 e cara articular
 d Gelenkfläche
 l facies articularis

296 **articular tubercle of temporal bone**
 f tubercule articulaire de l'os temporal
 e tubérculo articular del hueso temporal
 d Gelenkhöcker des Jochbeinfortsatzes
 l tuberculum articulare ossis temporalis

297 **articulation; joint**
 f articulation
 e articulación
 d Gelenk
 l articulatio; junctura ossium

298 **articulation of head of rib; capitular articulation**
 f articulation de la tête costale
 e articulación de la cabeza de la costilla
 d Rippenköpfchengelenk
 l articulatio capitis costae

299 **arytenoid cartilage**
 f cartilage aryténoïde
 e cartílago aritenoideo
 d Gießkannenknorpel; Stellknorpel
 l cartilago arytenoidea

* **ascariasis** → 302

300 **ascarid**
 f ascaride
 e ascárido
 d Spulwurm
 l Ascaridoidea

* **Ascaridia infestation** → 301

* **ascaridiasis** → 301

301 **ascaridiosis; ascaridiasis; Ascaridia infestation**
 f ascaridiose; infestation par Ascaridia spp.
 e ascaridiosis
 d Askaridiose; Spulwurmbefall des Geflügels

302 **ascariosis; ascariasis; Ascaris infestation**
 f ascaridose; infestation par Ascaris spp.
 e ascaridosis
 d Askaridose; Spulwurmbefall des Schweins

* **Ascaris infestation** → 302

303 **ascending aorta**
 f aorte ascendante
 e aorta ascendente
 d aufsteigende Aorta
 l aorta ascendens

304 **ascending colon**
 f côlon ascendant
 e colon ascendente
 d aufsteigendes Kolon
 l colon ascendens

305 **ascites**
 f ascite
 e ascitis
 d Aszites; Bauchwassersucht

* **ascorbic acid** → 4936

* **Ascosphaera apis infection** → 845

306 **aspergillosis; Aspergillus infection**
  f aspergillose
  e aspergilosis; infección por Aspergillus spp.
  d Aspergillose; Aspergillus-Infektion

  * **Aspergillus flavus toxin** → 90

  * **Aspergillus infection** → 306

307 **aspermia**
  f aspermie; aspermatisme
  e aspermia
  d Aspermie

308 **asphyxia**
  f asphyxie
  e asfixia
  d Asphyxie

  * **association fibres** → 261

309 **asternal rib; false rib**
  f côte asternale
  e costilla asternal; costilla falsa
  d falsche Rippe
  l costa asternalis; costa spuria

310 **asthenia**
  f asthénie
  e astenia
  d Asthenie

  * **astragalus** → 4574

311 **ataxia**
  f ataxie
  e ataxia
  d Ataxie

312 **atelectasis**
  f atélectasie
  e atelectasia
  d Atelektase

313 **atherosclerosis**
  f athérosclérose
  e aterosclerosis
  d Atherosklerose

314 **atlantal fossa; fossa of atlas wing**
  f fosse atloïdienne
  e fosa del atlas
  d Flügelgrube des Atlas
  l fossa atlantis

315 **atlanto-axial articulation**
  f articulation atloïdo-axoïdienne
  e articulación atlantoaxial
  d Atlas-Axis-Gelenk; zweites Kopfgelenk
  l articulatio atlantoaxialis

316 **atlanto-occipital articulation**
  f articulation atlanto-occipitale
  e articulación atlantooccipital
  d Atlas-Hinterhauptsbein-Gelenk; erstes
     Kopfgelenk
  l articulatio atlantooccipitalis

317 **atlas; first cervical vertebra**
  f atlas; première vertèbre cervicale
  e atlas; primera vértebra cervical
  d Atlas; Kopfträger; erster Halswirbel
  l atlas

318 **atony**
  f atonie
  e atonía
  d Atonie; Schlaffheit

319 **atopy**
  f atopie
  e atopia
  d Atopie

320 **atresia**
  f atrésie
  e atresia
  d Atresie

321 **atrichia; hairlessness**
  f atrichie; absence de poils
  e atriquia; atricosis
  d Atrichie; Haarlosigkeit

322 **atrioventricular bundle; bundle of His**
  f faisceau atrio-ventriculaire de His
  e fascículo atrioventricular
  d atrioventrikuläres Bündel; His-Bündel
  l fasciculus atrioventricularis

323 **atrioventricular septum**
  f septum atrio-ventriculaire; cloison
     interauriculo-ventriculaire
  e septo aurículoventricular
  d Septum atrioventricularis
  l septum atrioventricularis

324 **atrium of heart**
  f atrium du cœur; oreillette du cœur
  e atrio del corazón
  d Herzvorkammer
  l atrium cordis

325 **atrophic rhinitis**
  f rhinite atrophique
  e rinitis atrófica
  d atrophische Rhinitis
  l rhinitis atrophica

326 atrophy
f atrophie
e atrofia
d Atrophie; Schwund

327 attenuated virus
f virus atténué
e virus atenuado
d attenuiertes Virus

328 auditory ossicles
f osselets
e huesecillos del oído; osículos auditivos
d Gehörknöchelchen
l ossicula auditus

329 auditory teeth of Huschke
f dents acoustiques/auditives
e dientes acústicos
d Gehörzähne
l dentes acustici

330 auditory tube; eustachian tube
f trompe auditive (d'Eustache)
e trompa auditiva (de Eustaquio); tuba acústica
d Ohrtrompete; Eustachische Röhre
l tuba auditiva

* **Auerbach's plexus** → 2907

331 Aujeszky's disease; pseudorabies
f maladie d'Aujeszky
e enfermedad de Aujeszky; seudorrabia
d Aujeszkysche Krankheit; Pseudowut
l paralysis bulbaris infectiosa

332 Aujeszky virus; pseudorabies virus; porcine herpesvirus type 1
f virus de la maladie d'Aujeszky; herpèsvirus porcin type 1
e virus de la enfermedad de Aujeszky; virus de la seudorrabia; herpesvirus porcino tipo 1
d Aujeszky-Virus; Schweineherpesvirus Typ 1; SHV-1

333 auricle of atrium; auricular appendage
f auricule de l'atrium
e aurícula del atrio
d Herzohr
l auricula atrii

* **auricle of ear** → 3478

* **auricular appendage** → 333

334 auricular arteries
f artères auriculaires
e arterias auriculares
d Ohrmuschelarterien
l arteriae auriculares

* **auricular cartilage** → 1021

335 auricular concha; concha of auricle
f conque auriculaire
e concha de la oreja
d Muschel des äußeren Ohres
l concha auriculae

336 auricular gland
f glande auriculaire
e glándula auricular
d Ohrendrüse
l glandula auricularis

* **auricular muscles** → 1417

337 auricular nerves
f nerfs auriculaires
e nervios auriculares
d Ohrmuschelnerven
l nervi auriculares

* **auricular veins** → 1421

338 auriculopalpebral nerve
f nerf auriculo-palpébral
e nervio auriculopalpebral
d Nervus auriculopalpebralis
l nervus auriculopalpebralis

339 auriculotemporal nerve
f nerf auriculo-temporal
e nervio auriculotemporal
d Nervus auriculotemporalis
l nervus auriculotemporalis

340 autoantibody
f auto-anticorps
e autoanticuerpo
d Autoantikörper

341 autogenous vaccine
f autovaccin
e autovacuna
d Autovakzine

342 autoimmune disease
f maladie auto-immune
e enfermedad autoinmune
d Autoimmunerkrankung

343 autonomic ganglia
f ganglions autonomes
e ganglios autónomos
d autonome Ganglien
l ganglia autonomica

344 autonomic nervous system
f système nerveux autonome
e sistema nervioso autónomo; sistema
  nervioso vegetativo
d autonomes Nervensystem; vegetatives
  Nervensystem
l systema nervosum autonomicum

345 autonomic plexus
f plexus autonome
e plexo autónomo
d autonomer Plexus
l plexus autonomicus

* autopsy → 3569

346 aviadenovirus; avian adenovirus
f adénovirus aviaire
e adenovirus aviar
d aviäres Adenovirus

* avian adenovirus → 346

347 avian coccidiosis
f coccidiose aviaire
e coccidiosis aviar
d Geflügelkokzidiose

348 avian encephalomyelitis; epidemic tremor
f encéphalomyélite infectieuse aviaire
e encefalomielitis aviar
d aviäre Enzephalomyelitis; ansteckende
  Küken-Enzephalomyelitis
l encephalomyelitis avium

349 avian encephalomyelitis virus
f virus de l'encéphalomyélite aviaire
e virus de la encefalomielitis aviar
d aviäres Enzephalomyelitis-Virus

350 avian erythroblastosis
f érythroblastose aviaire
e eritroblastosis aviar
d aviäre Erythroblastose

351 avian herpesvirus
f herpèsvirus aviaire
e herpesvirus aviar
d Geflügelherpesvirus; aviäres Herpesvirus

352 avian infectious bronchitis
f bronchite infectieuse aviaire

e bronquitis infecciosa aviar
d infektiöse Bronchitis des Huhnes
l bronchitis infectiosa avium

* avian infectious coryza → 2175

353 avian infectious laryngotracheitis; ILT
f laryngotrachéite infectieuse aviaire
e laringotraqueitis infecciosa
d infektiöse Laryngotracheitis der Hühner;
  ILT
l laryngotracheitis infectiosa avium

354 avian infectious synovitis; Mycoplasma synoviae infection
f synovite infectieuse aviaire; infection à
  Mycoplasma synoviae
e sinovitis infecciosa aviar; infección por
  Mycoplasma synoviae
d infektiöse Synovitis des Geflügels;
  Mycoplasma-synoviae-Infektion

355 avian influenza; classical fowl plague
f grippe aviaire; peste aviaire
e influenza aviar; peste aviar
d Vogelgrippe; klassische Geflügelpest
l pestis avium

356 avian influenzavirus
f influenzavirus aviaire; virus de la grippe
  aviaire
e virus de la influenza aviar
d aviäres Influenzavirus

357 avian laryngotracheitis herpesvirus; ILT virus
f herpèsvirus de la laryngotrachéite
  infectieuse aviaire
e virus de la laringo-traqueitis infecciosa
  aviar
d infektiöses Laryngotracheitis-Herpesvirus

358 avian leukosis
f leucose aviaire
e leucosis aviar
d aviäre Leukose; Geflügelleukose
l leucosis gallinarum

359 avian myeloblastosis
f myéloblastose aviaire
e mieloblastosis aviar
d aviäre Myeloblastose

* avian neurolymphomatosis → 2683

360 avian paramyxovirus
f paramyxovirus aviaire
e paramyxovirus aviar
d aviäres Paramyxovirus

* avian paramyxovirus-1 → 3006

* avian pasteurellosis → 1754

* avian poxvirus → 365

361 avian reovirus
f réovirus aviaire
e reovirus aviar
d aviäres Reovirus

362 avian spirochaetosis; Borrelia gallinarum infection
f spirochétose aviaire; infection à Borrelia gallinarum
e espiroquetosis aviar; treponematosis aviar; infección por Borrelia gallinarum
d Spirochätose des Geflügels; Borrelia-gallinarum-Infektion

363 avian tubercle bacillus
f bacille de la tuberculose aviaire
e bacilo de la tuberculosis aviar
d aviärer Tuberkelbazillus
l Mycobacterium avium

364 avian tuberculosis; Mycobacterium avium infection
f tuberculose aviaire; infection à Mycobacterium avium
e tuberculosis aviar; infección por Mycobacterium avium
d Geflügeltuberkulose; Mycobacterium-avium-Infektion

365 avipoxvirus; avian poxvirus
f poxvirus aviaire
e poxvirus aviar
d Geflügelpoxvirus

366 avirulent
f avirulent; dépourvu de virulence
e avirulento
d avirulent

367 avitaminosis; vitamin deficiency
f avitaminose; carence en vitamines
e avitaminosis; carencia vitamínica
d Avitaminose; Vitaminmangelkrankheit

368 avoparcin
f avoparcine
e avoparcina
d Avoparcin
l avoparcinum

369 axilla; armpit
f aisselle

e axila
d Achsel
l axilla

370 axillary artery
f artère axillaire
e arteria axilar
d Arteria axillaris
l arteria axillaris

371 axillary fascia
f fascia axillaire
e fascia axilar
d Fascia axillaris
l fascia axillaris

372 axillary fold
f pli axillaire
e pliegue axilar
d Achselfalte
l plica axillaris

373 axillary fossa
f fosse axillaire
e fosa axilar
d Fossa axillaris; Achselhöhle
l fossa axillaris

374 axillary loop
f anse axillaire
e asa axilar
d Ansa axillaris
l ansa axillaris

375 axillary lymph nodes
f nœuds lymphatiques axillaires
e nódulos linfáticos axilares
d axillare Lymphknoten
l lymphonodi axillares

376 axillary lymphocentre
f lymphocentre axillaire
e linfocentro axilar
d Achsellymphzentrum
l lymphocentrum axillare

377 axillary nerve; circumflex nerve
f nerf axillaire; nerf circonflexe
e nervio axilar; nervio circunflejo
d Nervus axillaris
l nervus axillaris

378 axillary region
f région axillaire
e región axilar
d Achselgegend
l regio axillaris

379 axillary vein
  f veine axillaire
  e vena axilar
  d Vena axillaris; Achselvene
  l vena axillaris

380 axis; second cervical vertebra
  f axis; deuxième vertèbre cervicale
  e eje; segunda vértebra cervical
  d Axis; zweiter Halswirbel
  l axis

381 azaperone
  f azapérone

  e azaperona
  d Azaperon
  l azaperonum

382 azoospermia
  f azoospermie
  e azoospermia
  d Azoospermie

383 azoturia
  f azoturie
  e azoturia
  d Azoturie

B

* **Babesia infection** → 384

* **babesiasis** → 384

384 babesiosis; babesiasis; Babesia infection
f babésiose; infection par Babesia spp.
e babesiosis
d Babesiose; Babesieninfektion

385 bacillary haemoglobinuria; Clostridium haemolyticum infection
f hémoglobinurie bacillaire; infection à Clostridium haemolyticum
e hemoglobinuria bacilar; infección por Clostridium haemolyticum
d bazilläre Hämoglobinurie; Clostridium-haemolyticum-Infektion

386 bacillus
f bacille
e bacilo
d Bazillus
l Bacillus

* **Bacillus anthracis infection** → 202

* **Bacillus larvae infection** → 139

* **Bacillus piliformis infection** → 4720

387 bacitracin
f bacitracine
e bacitracina
d Bacitracin
l bacitracinum

388 back
f dos
e dorso
d Rücken
l dorsum

* **backbone** → 4875

389 back muscles
f muscles du dos
e músculos del dorso
d Rückenmuskeln
l musculi dorsi

* **back of head** → 3068

390 back of knee
f creux du genou; creux poplité
e poples; corva
d Kniekehle; Kniebeuge

391 back of the hand
f dos de la main
e dorso de la mano
d Handrücken
l dorsum manus

392 back regions
f régions du dos
e regiones del dorso
d Rückengegenden
l regiones dorsi

393 bacteraemia; bacteremia
f bactériémie
e bacteremia; bacteriemia
d Bakteriämie

* **bacteremia** → 393

394 bacterial antigen
f antigène bactérien
e antígeno bacteriano
d Bakterienantigen

395 bacterial disease
f maladie bactérienne; bactériose
e enfermedad bacteriana
d bakterielle Krankheit

396 bacterial kidney disease; BKD; Renibacterium salmoninarum infection
f rénibactériose; infection à Renibacterium salmoninarum
e renibacteriosis; infección por Renibacterium salmoninarum
d bakterielle Nierenkrankheit der Salmoniden; Renibacterium-salmoninarum-Infektion

397 bactericidal
f bactéricide
e bactericida
d bakterizid

398 bacteriophage; phage
f bactériophage
e bacteriófago
d Bakteriophage

399 bacterium
f bactérie
e bacteria
d Bakterium
l Bacterium

* Bacteroides nodosus infection → 2177

400 balanitis
 f balanite
 e balanitis
 d Balanitis

401 balanoposthitis
 f balanoposthite
 e balanopostitis
 d Balanoposthitis; Präputialkatarrh

402 balantidiosis; Balantidium infection
 f balantidiose
 e balantidiosis
 d Balantidiose

* **Balantidium infection → 402**

403 ball-and-socket joint; enarthrosis
 f articulation sphéroïde; énarthrose
 e articulación esferoidal
 d Sphäroidgelenk; Enarthrose
 l articulatio spheroidea

* **ball of sole → 3496**

404 bambermycin
 f bambermycine
 e bambermicina; flavomicina
 d Bambermycin; Flavomycin
 l bambermycinum

405 barbiturate
 f barbiturate
 e barbiturato
 d Barbiturat

406 barb of feather
 f barbe
 e barba de pluma
 d Federast
 l barba

407 barbule
 f barbule
 e barbícula
 d Barbula
 l barbula

408 bar of hoof; reflected part of hoof wall
 f barre; arc-boutant; partie réfléchie
 e porción inflexa
 d Eckstrebe; Abbiegung der Hornwand des
    Hufes
 l pars inflexa

409 bar of mouth
 f barre
 e barra
 d Zahnlücke

* **Bartonella infection → 410**

410 bartonellosis; Bartonella infection
 f bartonellose; infection par Bartonella spp.
 e bartonelosis; infección por Bartonella spp.
 d Bartonellose; Bartonella-Infektion

411 basal lamina; basal plate
 f lame basale
 e lámina basal
 d Lamina basalis
 l lamina basalis

412 basal layer of epidermis; germinative layer
 f stratum germinativum; couche basale de
    l'épiderme
 e estrato basal
 d Stratum basale; Keimschicht
 l stratum basale epidermidis

* **basal plate → 411**

413 base of heart
 f base du cœur
 e base del corazón
 d Herzbasis
 l basis cordis

414 base of horn
 f base de la corne
 e base del cuerno
 d Hornbasis
 l basis cornus

* **base of tail region → 3911**

415 basihyoid; body of hyoid bone
 f basihyal; corps de l'os hyoïde
 e basihioidea
 d Basihyoid
 l basihyoideum

416 basilar artery
 f artère basilaire
 e arteria basilar
 d Arteria basilaris
 l arteria basilaris

417 basilar membrane of cochlear duct
 f membrane basilaire du conduit cochléaire
 e lámina basilar del conducto coclear
 d Basilarmembran des Schneckengangs
 l lamina basilaris ductus cochlearis

418 basioccipital bone
 f os basi-occipital
 e hueso basioccipital
 d Os basioccipital
 l os basioccipitale

419 basipodium
 f basipode
 e basípodo
 d Basipodium
 l basipodium

420 basisphenoid bone
 f os basi-sphénoïde
 e hueso basiesfenoides
 d hinteres Keilbein
 l os basisphenoidale

 * **B cell** → 498

421 beak; bill
 f rostre; bec
 e pico
 d Schnabel
 l rostrum

 * **beak of olecranon** → 174

422 bee disease
 f maladie des abeilles
 e enfermedad de las abejas
 d Bienenkrankheit

423 beef tapeworm; unarmed tapeworm
 f ténia du bœuf; ténia inerme
 e tenia inerme
 d Rinderbandwurm
 l Taenia saginata

424 bee louse
 f pou des abeilles
 e piojo de las abejas
 d Bienenlaus
 l Braula coeca

425 behavioural disorder; abnormal behaviour
 f trouble du comportement
 e anomalía del comportamiento
 d Verhaltungsstörung; Fehlverhalten

 * **belly** → 1

426 benign course
 f évolution bénigne
 e curso benigno
 d gutartiger Verlauf

427 benzalkonium chloride
 f chlorure de benzalkonium
 e cloruro de benzalconio
 d Benzalkoniumchlorid
 l benzalkonii chloridum

428 benzathine benzylpenicillin; benzathine penicillin
 f benzathine benzylpénicilline
 e bencilpenicilina benzatina; penicilina benzatina
 d Benzathin-Benzylpenicillin
 l benzathini benzylpenicillinum

 * **benzathine penicillin** → 428

429 benzimidazole
 f benzimidazole
 e benzimidazol
 d Benzimidazol

430 benzocaine
 f benzocaïne
 e benzocaína
 d Benzocain
 l benzocainum

431 benzylpenicillin; penicillin G
 f benzylpénicilline
 e bencilpenicilina
 d Benzylpenicillin
 l benzylpenicillinum

432 bephenium hydroxynaphthoate
 f hydroxynaphtoate de béphénium
 e hidroxinaftoato de befenio
 d Bepheniumhydroxynaphthoat
 l bephenii hydroxynaphthoas

 * **Besnoitia infection** → 433

433 besnoitiosis; Besnoitia infection
 f besnoïtiose; infection par Besnoitia spp.
 e besnoitiosis
 d Besnoitiose

434 beta-adrenergic receptor
 f récepteur bêta-adrénergique
 e receptor betaadrenérgico
 d Beta-Rezeptor

 * **beta-adrenergic receptor blockader** → 435

435 beta-blocker; beta-adrenergic receptor blockader
 f bêta-bloquant
 e betabloqueador
 d Beta-Rezeptoren-Blocker

436 beta-haemolytic
 f bêta-hémolytique
 e betahemolítico
 d beta-hämolytisch

437 betamethasone
 f bêtaméthasone
 e betametasona
 d Betamethason
 l betamethasonum

 * **beta-propiolactone** → 3636

438 bezoar
 f bézoard; égagropile
 e bezoar
 d Bezoar

439 bicarotid trunk
 f tronc bicarotidien
 e tronco bicarotídeo
 d Truncus bicaroticus
 l truncus bicaroticus

440 bicellular corpuscle; Grandry's corpuscle
 f corpuscule bicellulaire (de Grandry)
 e corpúsculo bicelular
 d Grandrysches Körperchen
 l corpusculum bicellulare

441 biceps brachii muscle
 f muscle biceps du bras
 e músculo bíceps del brazo
 d zweiköpfiger Oberarmmuskel
 l musculus biceps brachii

442 biceps femoris muscle
 f muscle biceps fémoral
 e músculo bíceps del muslo
 d zweiköpfiger Oberschenkelmuskel
 l musculus biceps femoris

443 bicipital artery
 f artère bicipitale
 e arteria bicipital
 d Arteria bicipitalis
 l arteria bicipitalis

444 bicipital vein
 f veine bicipitale
 e vena bicipital
 d Vena bicipitalis
 l vena bicipitalis

 * **bicuspid valve** → 2473

445 bile
 f bile; fiel
 e bilis; hiel
 d Galle
 l fel

446 bile pigment
 f pigment biliaire
 e pigmento biliar
 d Gallenfarbstoff

 * **biliary calculus** → 1795

447 biliary ductules
 f canalicules biliaires
 e conductillos bilíferos
 d Gallengänge
 l ductuli biliferi

448 bilirubin
 f bilirubine
 e bilirrubina
 d Bilirubin

 * **bill** → 421

 * **Billroth's cords** → 4613

449 bioassay; biological assay
 f titrage biologique; analyse biologique
 e ensayo biológico
 d biologische Analyse; biologische Prüfung

450 biocenosis
 f biocénose
 e biocenosis
 d Biozoenose

451 biochemical polymorphism
 f polymorphisme biochimique
 e polimorfismo bioquímico
 d biochemischer Polymorphismus

 * **biological assay** → 449

452 biopsy
 f biopsie
 e biopsia
 d Biopsie

453 biotechnology
 f biotechnologie
 e biotecnología
 d Biotechnologie; Biotechnik

454 bipennate muscle
 f muscle bipenne
 e músculo bipenniforme
 d doppelt gefiederter Muskel
 l musculus bipennatus

455 birnavirus
 f birnavirus
 e birnavirus
 d Birnavirus

456 bite
 f morsure
 e mordedura
 d Bißwunde

457 bithionol
 f bithionol
 e bitionol
 d Bithionol
 l bithionolum

458 biting louse
 f mallophage
 e piojo masticador; malófago
 d Federling; Haarling
 l Mallophaga

459 biting louse of dogs
 f trichodecte du chien
 e piojo grande del perro
 d Hundehaarling
 l Trichodectes canis

 * BKD → 396

460 black disease of sheep; necrotic hepatitis; Clostridium novyi infection
 f hépatite infectieuse nécrosante du mouton; infection à Clostridium novyi
 e hepatitis necrótica; enfermedad negra de los ovinos
 d deutscher Bradsot; Clostridium-novyi-Infektion
 l hepatitis infectiosa necrotica

461 black fly
 f simulie
 e simulio; mosca negra
 d Kriebelmücke
 l Simulium

 * blackhead of turkeys → 2018

462 blackleg; Clostridium chauvoei infection; Clostridium septicum infection
 f charbon symptomatique; infection à Clostridium chauvoei; infection à Clostridium septicum
 e carbunco sintomático; pata negra; infección por Clostridium chauvoei; infección por Clostridium septicum
 d Rauschbrand; Clostridium-chauvoei-Infektion; Clostridium-septicum-Infektion

463 bladder; urinary bladder
 f vessie
 e vejiga
 d Harnblase
 l vesica urinaria

 * bladder worm → 1209

 * Blastomyces dermatitidis infection → 3025

 * bleeding → 1936

464 blepharitis
 f blépharite
 e blefaritis; tarsitis
 d Blepharitis; Lidentzündung

465 blindness
 f cécité
 e ceguedad
 d Blindheit

466 blind part of retina
 f partie aveugle de la rétine
 e porción ciega de la retina
 d blinder Teil der Netzhaut
 l pars c(a)eca retinae

467 blind sac of stomach (*horse*)
 f cul-de-sac de l'estomac
 e saco ciego del estómago
 d Blindsack des Magens (*Pferd*)
 l saccus c(a)ecus ventriculi

 * bloat → 4718

 * blockage → 3055

468 blood
 f sang
 e sangre
 d Blut
 l sanguis

469 blood-brain barrier
 f barrière hémato-encéphalique
 e barrera sangre-cerebro
 d Blut-Hirn-Schranke

470 blood cell
 f cellule sanguine
 e glóbulo de la sangre
 d Blutzelle

471 blood chemistry; blood composition
 f constituants chimiques du sang
 e química del sangre
 d Blutchemie; Blutzusammensetzung

472 blood clot
 f caillot
 e coágulo sanguíneo
 d Blutgerinnsel

473 blood coagulation
f coagulation sanguine
e coagulación sanguínea
d Blutgerinnung

* **blood composition** → 471

474 blood disorder
f trouble sanguin; hémopathie
e trastorno del sangre
d Hämopathie; hämatologische Störung

475 blood flow
f débit sanguin
e corriente de la sangre
d Blutfluß; Durchblutung

476 blood formation; haematopoiesis
f hématopoïèse
e hemopoyesis; hematosis
d Blutbildung; Hämatopoese

477 blood group
f groupe sanguin
e grupo sanguíneo
d Blutgruppe

478 blood plasma
f plasma sanguin
e plasma sanguíneo
d Blutplasma

479 blood platelet; thrombocyte
f thrombocyte; plaquette sanguine
e trombocito; plaqueta sanguínea
d Blutplättchen; Thrombozyt

480 blood pressure
f pression sanguine
e presión sanguínea
d Blutdruck

481 blood protein
f protéine sanguine
e proteína sanguínea
d Bluteiweiß

482 blood protein disorder
f trouble des protéines sanguines
e trastorno de las proteínas sanguíneas
d Bluteiweiß-Störung

483 blood sample
f échantillon de sang
e muestra sanguínea
d Blutprobe

484 blood serum
f sérum sanguin
e suero sanguíneo
d Blutserum

485 blood smear
f frottis de sang
e embadurnamiento de sangre
d Blutausstrich

486 blood stream
f courant sanguin
e corriente sanguíneo
d Blutstrom

487 blood sucker
f hématophage
e sanguijuela
d Blutsauger

488 blood sugar
f glucose sanguin; glycémie
e glucosa sanguínea
d Blutzucker

489 blood supply
f apport sanguin; irrigation sanguine
e irrigación sanguínea
d Blutversorgung; Durchblutung

490 blood transfusion
f transfusion sanguine
e transfusión sanguínea
d Bluttransfusion

491 blood vessel
f vaisseau sanguin
e vaso sanguíneo
d Blutgefäß
l vas sanguina

492 blood volume
f volume sanguin
e volumen sanguíneo
d Blutvolumen

493 blowfly; bluebottle fly
f calliphore
e califórido; moscón azul
d blaue Schmeißfliege
l Calliphora

* **bluebottle fly** → 493

494 bluecomb disease
f maladie de la crête bleue; coronavirose aviaire
e enfermedad de la cresta azul; coronavirosis aviar
d Blaukammkrankheit

495 **blue-green algae; cyanobacteria**
 f cyanophycées
 e algas azul-verde; cianobacteria
 d blaugrüne Algen; Cyanobakterien

496 **bluetongue; ovine orbivirus infection**
 f fièvre catarrhale du mouton
 e lengua azul
 d Bluetongue
 l febris catarrhalis ovium

497 **bluetongue orbivirus**
 f virus de la fièvre catarrhale du mouton
 e virus de la lengua azul
 d Bluetongue-Orbivirus

498 **B lymphocyte; B cell**
 f lymphocyte B
 e linfocito B
 d B-Lymphozyt

499 **body feather; covert feather**
 f tectrice
 e tectríz; pluma de revestimiento
 d Deckfeder
 l tectrix

500 **body fluid**
 f liquide biologique
 e líquido del organismo
 d Körperflussigkeit

501 **body of basisphenoid bone**
 f corps de l'os basisphénoïde
 e cuerpo del hueso basiesfenoides
 d Keilbeinkörper
 l corpus ossis basisphenoidale

502 **body of epididymis**
 f corps de l'épididyme
 e cuerpo del epidídimo
 d Mittelteil des Nebenhodens
 l corpus epididymidis

* **body of hyoid bone** → 415

503 **body of maxilla**
 f corps de l'os maxillaire
 e cuerpo del maxilar
 d Oberkieferkörper
 l corpus maxillae

504 **body of penis**
 f corps du pénis
 e cuerpo del pene
 d Peniskörper
 l corpus penis

505 **body of tongue**
 f corps de la langue
 e cuerpo de la lengua
 d Zungenkörper
 l corpus linguae

506 **body of uterus**
 f corps utérin
 e cuerpo del útero
 d Gebärmutterkörper
 l corpus uteri

507 **body regions**
 f régions du corps
 e regiones corporales
 d Körpergegenden
 l regiones corporis

508 **body temperature**
 f température corporelle
 e temperatura corporal
 d Körpertemperatur

509 **boldenone**
 f boldénone
 e boldenona
 d Boldenon
 l boldenonum

510 **bone**
 f os
 e hueso
 d Knochen
 l os; ossis

* **bone cell** → 3173

511 **bone disease**
 f ostéopathie
 e osteopatía
 d Knochenerkrankung; Osteopathie

512 **bone formation; osteogenesis**
 f ostéogenèse
 e osteogénesis
 d Knochenbildung; Osteogenese

513 **bone marrow**
 f moelle osseuse
 e médula ósea
 d Knochenmark
 l medulla ossium

514 **bone marrow disorder**
 f trouble de la moelle osseuse
 e trastorno de la médula ósea
 d Knochenmarkstörung

* **bones of skull** → 1126

* **bone softening** → 3176

515 bony crest
 f crête osseuse; casque; écusson
 e cresta ósea
 d knöcherner Kamm
 l crista ossea

* **bony labyrinth** → 3167

* **bony outgrowth** → 3179

516 bony palate
 f palais osseux
 e paladar óseo
 d knöcherner Gaumen
 l palatum osseum

517 bony part of auditory tube
 f partie osseuse de la trompe auditive
 e porción ósea del la trompa auditiva
 d knöcherner Teil der Ohrtrompete
 l pars ossea tubae auditivae

518 bony spiral lamina; spiral plate of cochlea
 f lame spirale osseuse
 e lámina espiral ósea
 d Lamina spiralis ossea
 l lamina spiralis ossea

519 booster dose
 f dose de rappel
 e dosis reforzador
 d Boosterdosis; Auffrischungsdosis

520 border disease of sheep; ovine pestivirus infection
 f maladie de la frontière; pestivirose ovine
 e enfermedad de la frontera
 d ovine Pestivirose

521 Borna disease
 f maladie de Borna
 e enfermedad de Borna
 d Bornasche Krankheit

522 Borna disease virus
 f virus de la maladie de Borna
 e virus de la enfermedad de Borna
 d Borna-Virus

* **Borrelia burgdorferi infection** → 2611

* **Borrelia gallinarum infection** → 362

523 borreliosis
 f borréliose; infection à Borrelia
 e borreliosis; infección por Borrelia spp.
 d Borreliose; Borrelia-Infektion

524 botulism; Clostridium botulinum intoxication
 f botulisme; intoxication à Clostridium botulinum
 e botulismo; intoxicación por Clostridium botulinum
 d Botulismus; Clostridium-botulinum-Intoxikation

525 bovine adenovirus
 f adénovirus bovin
 e adenovirus bovino
 d bovines Adenovirus

526 bovine babesiosis; piroplasmosis in cattle
 f babésiose bovine
 e babesiosis bovina
 d Babesiose des Rindes

527 bovine brucellosis; contagious abortion; Brucella abortus infection
 f brucellose bovine; avortement contagieux
 e brucelosis bovina; aborto contagioso; infección por Brucella abortus
 d Rinderbrucellose; seuchenhaftes Verwerfen

528 bovine diarrhoea pestivirus; BVD virus; mucosal disease virus
 f pestivirus de la diarrhée virale bovine; virus de la maladie des muqueuses
 e virus de la diarrea vírica bovina; virus BVD
 d bovines Diarrhöe-Pestivirus; BVD-Virus

529 bovine enterovirus
 f entérovirus bovin
 e enterovirus bovino
 d bovines Enterovirus

530 bovine ephemeral fever virus
 f virus de la fièvre éphémère bovine
 e virus de la fiebre efímera bovina
 d bovines Ephemeralfieber-Virus

531 bovine farcy; Mycobacterium farcinogenes infection
 f infection à Mycobacterium farcinogenes
 e infección por Mycobacterium farcinogenes
 d Mycobacterium-farcinogenes-Infektion

532 bovine herpesvirus
 f herpèsvirus bovin
 e herpesvirus bovino
 d bovines Herpesvirus

* **bovine herpesvirus type 1** → 2101

* **bovine herpesvirus type 2** → 537

533 **bovine infectious keratitis; Moraxella bovis infection**
f kératite infectieuse bovine; infection à Moraxella bovis
e queratitis infecciosa bovina; infección por Moraxella bovis
d bovine infektiöse Keratitis; Rinderkeratitis; Moraxella-bovis-Infektion

* **bovine infectious petechial fever** → 3121

534 **bovine leukosis; enzootic bovine leukosis; EBL**
f leucose bovine enzootique
e leucosis bovina enzoótica
d Rinderleukose

535 **bovine leukosis oncovirus; EBL virus**
f virus de la leucose bovine enzootique
e virus de la leucosis bovina enzoótica
d Rinderleukose-Virus

536 **bovine malignant catarrh; malignant catarrhal fever**
f coryza gangréneux des bovins
e fiebre catarral maligna
d bösartiges Katarrhalfieber
l coryza gangraenosa bovum

537 **bovine mammillitis herpesvirus; bovine herpesvirus type 2**
f virus de la thélite bovine; herpèsvirus bovin type 2
e herpesvirus de la mamilitis bovina; herpesvirus bovino tipo 2
d bovines Herpesvirus Typ 2; Mamillitis-Virus; BHV-2

538 **bovine mastitis**
f mammite bovine
e mastitis bovina
d Rindermastitis

539 **bovine papillomatosis; cattle warts**
f papillomatose bovine
e verrucosis bovina; papilomatosis bovina
d Rinderpapillomatose

540 **bovine papillomavirus**
f papillomavirus bovin
e papillomavirus bovino; virus del papiloma de los bovinos
d bovines Papillomavirus

541 **bovine papular stomatitis**
f stomatite papuleuse bovine
e estomatitis papular bovina
d Stomatitis papulosa
l stomatitis papulosa

542 **bovine papular stomatitis orthopoxvirus**
f orthopoxvirus de la stomatite bovine papuleuse
e virus de la estomatitis papular bovina
d Stomatitis-papulosa-Orthopoxvirus

543 **bovine parainfluenza virus**
f virus du parainfluenza bovin
e virus de la parainfluenza bovina
d bovines Parainfluenzavirus

* **bovine pestivirus infection** → 553

544 **bovine pustular stomatitis parapoxvirus**
f parapoxvirus de la stomatite pustuleuse bovine
e parapoxvirus de la estomatitis pustular bovina
d Stomatitis-pustulosa-Parapoxvirus

545 **bovine respiratory syncytial pneumovirus; BRSV**
f virus syncytial respiratoire bovin
e virus sincitial respiratorio bovino
d bovines respiratorisches Synzytialvirus; BRSV

546 **bovine somatotropin; BST**
f hormone somatotrope bovine
e somatotropina bovina
d bovines Somatotropin

547 **bovine spongiform encephalopathy; BSE**
f encéphalopathie spongiforme bovine
e encefalopatía espongiforme bovina
d bovine spongiforme Enzephalopathie

548 **bovine syncytial virus**
f virus syncytial bovin
e virus sincitial bovino
d bovines Synzytialvirus

* **bovine theileriosis** → 1422

549 **bovine trichomonosis; tritrichomonosis**
f trichomonose bovine
e tricomonosis bovina
d Trichomonadenseuche des Rindes; Tritrichomonose

550 **bovine tubercle bacillus**
f bacille de la tuberculose bovine
e bacilo de la tuberculosis bovina
d boviner Tuberkelbazillus
l Mycobacterium bovis

551 **bovine tuberculosis; Mycobacterium bovis infection**
f tuberculose bovine; infection à Mycobacterium bovis

e tuberculosis bovina; infección por
Mycobacterium bovis
d Rindertuberkulose; Mycobacterium-
bovis-Infektion

552 bovine ulcerative mammillitis
f mamillite ulcérative bovine
e mamilitis ulcerativa bovina
d bovine Herpesmamillitis

**553 bovine virus diarrhoea; BVD; bovine
pestivirus infection**
f diarrhée virale bovine
e diarrea vírica de los bovinos; BVD
d bovine Virusdiarrhöe; BVD

* **Bowman's capsule** → 1853

* **Bowman's glands** → 3095

* **Bowman's membrane** → 200

554 brachial artery
f artère brachiale
e arteria braquial
d Arteria brachialis; Armarterie
l arteria brachialis

555 brachial fascia
f fascia brachial
e fascia del brazo
d Fascia brachii
l fascia brachii

556 brachial muscle
f muscle brachial
e músculo braquial
d Armmuskel
l musculus brachialis

557 brachial plexus
f plexus brachial
e plexo braquial
d Armgeflecht
l plexus brachialis

558 brachial region
f région brachiale
e región del brazo
d Oberarmgegend
l regio brachii

559 brachial vein
f veine brachiale
e vena braquial
d Vena brachialis
l vena brachialis

560 brachiocephalic muscle
f muscle brachio-céphalique
e músculo braquiocefálico
d Arm-Kopfmuskel
l musculus brachiocephalicus

561 brachiocephalic trunk
f tronc brachio-céphalique
e tronco braquiocefálico
d Truncus brachiocephalicus
l truncus brachiocephalicus

562 brachiocephalic vein
f veine brachio-céphalique
e vena braquicefálica
d Vena brachiocephalica
l vena brachiocephalica

563 brachioradial muscle
f muscle brachio-radial
e músculo braquiorradial
d Oberarm-Speichenmuskel
l musculus brachioradialis

564 brachygnathia
f brachygnathie
e braquignatia
d Brachygnathia

565 bracken poisoning
f ptéridisme
e intoxicación por helecho
d Adlerfarnvergiftung; Farnkrautvergiftung

* **brain** → 1458

566 brain disease; encephalopathy
f encéphalopathie
e encefalopatía
d Enzephalopathie; Gehirnerkrankung

* **Branchiomyces infection** → 567

567 branchiomycosis; Branchiomyces infection
f branchiomycose; infection à
Branchiomyces
e branquiomicosis; infección por
Branchiomyces spp.
d Branchiomykose; Kiemenfäule

**568 braxy; Clostridium septicum
enterotoxaemia**
f entérotoxémie à Clostridium septicum;
œdème malin de la caillette
e bradsot de los ovinos; enterotoxemia por
Clostridium septicum
d Labmagenpararauschbrand; nordischer
Bradsot

569 breast
  f poitrine; poitrail
  e pecho
  d Brust
  l pectus

570 breast blister of poultry; keel bursitis
  f ampoule du bréchet; bursite sternale
  e ampolla esternal
  d Brustbeule des Geflügels; Entzündung der Bursa sternalis

* **breastbone** → 4220

* **breathing** → 3829

571 bridge of nose
  f chanfrein
  e dorso de la nariz
  d Nasenrücken
  l dorsum nasi

572 bristle
  f soie de porc
  e cerda
  d Borste
  l seta

573 broadest muscle of back
  f muscle grand dorsal
  e músculo latísimo del dorso
  d breiter Rückenmuskel
  l musculus latissimus dorsi

574 broad fish tapeworm
  f Diphyllobothrium latum
  e difilobotrio; botriocéfalo
  d Fischbandwurm des Menschen; Grubenkopfbandwurm
  l Diphyllobothrium latum

* **broad ligament of liver** → 1618

575 broad ligament of uterus
  f ligament large de l'utérus
  e ligamento ancho del útero
  d breites Gebärmutterband
  l ligamentum latum uteri

576 bromociclen; bromocyclen
  f bromociclène
  e bromocicleno
  d Bromociclen
  l bromociclenum

577 bromocriptine
  f bromocriptine
  e bromocriptina
  d Bromocriptin
  l bromocriptinum

* **bromocyclen** → 576

578 bromofos; bromophos
  f bromophos
  e bromofos
  d Bromofos
  l bromofosum

* **bromophos** → 578

579 bronchial glands
  f glandes bronchiques
  e glándulas bronquiales
  d Bronchialdrüsen
  l glandulae bronchales .

580 bronchial lymphocentre
  f lymphocentre bronchique
  e linfocentro bronquial
  d Bronchiallymphzentrum
  l lymphocentrum bronchale

581 bronchial spasm; bronchospasm
  f bronchospasme
  e broncoespasmo
  d Bronchospasmus

582 bronchial tree
  f arbre bronchique
  e árbol bronquio
  d Bronchialbaum
  l arbor bronchalis

583 bronchial veins
  f veines bronchiques
  e venas bronquiales
  d Venae bronchales
  l venae bronchales

584 bronchiectasis
  f bronchestasie
  e bronquiectasia
  d Bronchiektasie

585 bronchitis
  f bronchite
  e bronquitis
  d Bronchitis

586 bronchoesophageal artery
  f artère broncho-œsophagienne
  e arteria broncoesofágica
  d Arteria bronchoesophagea
  l arteria bronchoesophagea

587 bronchoesophageal muscle
  f muscle broncho-œsophagien
  e músculo broncoesofágico
  d Musculus bronchoesophageus
  l musculus bronchoesophageus

588 bronchopneumonia
 f broncho-pneumonie
 e bronconeumonía
 d Bronchopneumonie

589 bronchopulmonary segments
 f segments bronchopulmonaires
 e segmentos broncopulmonares
 d Lungensegmente
 l segmenta bronchopulmonalia

 * **bronchospasm** → 581

590 bronchus
 f bronche
 e bronquio
 d Bronchus; Luftröhrenast
 l bronchus

 * **brood patch** → 2166

591 brotianide
 f brotianide
 e brotianida
 d Brotianid
 l brotianidum

592 brown dog tick
 f tique brune du chien
 e garrapata morena del perro
 d braune Hundezccke
 l Rhipicephalus sanguineus

 * **BRSV** → 545

 * **Brucella abortus infection** → 527

 * **Brucella infection** → 594

 * **Brucella suis infection** → 3552

593 brucellin
 f brucelline
 e brucelina
 d Brucellin

594 brucellosis; Brucella infection
 f brucellose
 e brucelosis; infección por Brucella spp.
 d Brucellose

 * **BSE** → 547

 * **BST** → 546

595 buccal artery
 f artère buccale
 e arteria bucal
 d Arteria buccalis
 l arteria buccalis

596 buccal glands
 f glandes buccales
 e glándulas bucales
 d Backendrüsen
 l glandulae buccales

597 buccal nerve
 f nerf buccal
 e nervio bucal
 d Nervus buccalis
 l nervus buccalis

598 buccal papillae (*of ruminants*)
 f papilles buccales
 e papilas bucales
 d Backenpapillen
 l papillae buccales

599 buccal region
 f région buccale
 e región bucal
 d Backengegend
 l regio buccalis

600 buccinator muscle
 f muscle buccinateur
 e músculo buccinador
 d Backenmuskel; Trompetermuskel
 l musculus buccinator

601 buffalo louse
 f pou du buffle
 e piojo del búfalo
 d Büffellaus
 l Haematopinus tuberculatus

602 buiatrics
 f buiatrie
 e buiatría
 d Buiatrik

603 bulbar conjunctiva
 f conjonctive oculaire
 e túnica conjuntiva del globo
 d Bindehaut des Augapfels
 l tunica conjunctiva bulbi

 * **bulbar fascia** → 4065

604 bulbocavernous muscle
 f muscle bulbo-spongieux; muscle bulbo-
    caverneux
 e músculo bulboesponjoso; músculo
    bulbocavernoso
 d Musculus bulbospongiosus; Musculus
    bulbocavernosus
 l musculus bulbospongiosus; musculus
    bulbocavernosus

* bulb of eye → 1603

* bulb of Krause → 605

605 **bulboid capsule; bulb of Krause**
 f corpuscule bulboïde (de Krause)
 e corpúsculo bulboideo (de Krause)
 d Krausesche Endkolben
 l corpusculum bulboideum

606 **bulbourethral gland; Cowper's gland**
 f glande bulbo-urétrale (de Cowper)
 e glándula bulbouretral
 d Harnröhrenzwiebeldrüse; Cowpersche
   Drüse
 l glandula bulbourethralis

607 **bunamidine**
 f bunamidine
 e bunamidina
 d Bunamidin
 l bunamidinum

* bundle of His → 322

608 **bunostomosis; Bunostomum infestation**
 f bunostomose
 e bunostomosis
 d Bunostomose; Befall mit Bunostomum

* Bunostomum infestation → 608

609 **bunyavirus**
 f bunyavirus
 e bunyavirus
 d Bunyavirus

610 **buparvaquone**
 f buparvaquone
 e buparvacuona
 d Buparvaquon
 l buparvaquonum

611 **bupivacaine**
 f bupivacaïne
 e bupivacaina
 d Bupivacain
 l bupivacainum

612 **buquinolate**
 f buquinolate
 e buquinolato
 d Buquinolat
 l buquinolatum

* Burdach's column → 1622

* bursa of Fabricius → 950

613 **bursectomy**
 f bursectomie
 e bursectomía
 d Bursektomie

614 **bursitis**
 f bursite
 e bursitis
 d Bursitis; Schleimbeutelentzündung

615 **buserelin**
 f buséréline
 e buserelina
 d Buserelin
 l buserelinum

616 **buttocks**
 f fesse
 e nalgas
 d Gesäß
 l nates; clunis

* BVD → 553

* BVD virus → 528

C

* cachexia → 1444

617 caecal artery; cecal artery
f artère cæcale
e arteria cecal
d Arteria caecalis; Blinddarmarterie
l arteria c(a)ecalis

618 caecal sacculations; cecal sacculations
f haustrations du cæcum; bosselures du cæcum
e saculaciones del ciego
d Haustra caeci; Aussackungen des Blindarmes
l haustra c(a)eci

619 caecal veins; cecal veins
f veines cæcales
e venas cecales
d Venae caecales
l venae c(a)ecales

620 caecocolic orifice; cecocolic opening
f ostium cæco-colique; orifice cæco-colique
e orificio cecocólico
d Blinddarm-Dickdarmöffnung
l ostium c(a)ecocolicum

621 caecum; cecum
f cæcum
e ciego
d Zäkum; Blinddarm
l caecum; cecum

622 caesarean section; cesarean section
f opération césarienne
e histerotomía
d Kaiserschnitt; Schnittentbindung
l sectio caesarea

* CAE virus → 654

623 cage layer fatigue
f fatigue de la pondeuse en cage
e fatiga de la gallina ponedora
d Käfigmüdigkeit

624 calamus; quill
f calamus; hampe
e cálamo; cañón de pluma

d Calamus; Spule
l calamus

625 calcaneal tuberosity; point of hock
f tubérosité du calcaneus; pointe du jarret
e tuberosidad del calcáneo
d Fersenhöcker
l tuber calcanei

* calcaneocuboid joint → 626

626 calcaneoquartal joint; calcaneocuboid joint
f articulation calcanéo-cuboïdienne
e articulación calcanocuartal
d Articulatio calcaneoquartalis
l articulatio calcaneoquartalis; articulatio calcaneocuboidea

* calcaneum → 1692

627 calcinosis
f calcinose
e calcinosis
d Kalzinose; Kalkablagerung

628 calculus
f calcul
e cálculo
d Calculus; Steinchen
l calculus

629 calf diarrhoea; calf scours
f diarrhée des veaux
e diarrea de los terneros
d Kälberdiarrhöe; Kälberdurchfall

630 calf disease
f maladie des veaux
e enfermedad de los terneros
d Kälberkrankheit

631 calf of leg
f mollet
e pantorilla
d Wade
l sura

632 calf pneumonia
f pneumonie des veaux
e neumonía de los terneros
d Kälberpneumonie

* calf scours → 629

633 calicivirus
f calicivirus
e calicivirus
d Calicivirus

* **calliphoridosis** → 2908

* **callosal gyrus** → 919

* **CAM** → 884

634 **cambendazole**
 f cambendazole
 e cambendazol
 d Cambendazol
 l cambendazolum

* **Campylobacter infection** → 635

635 **campylobacteriosis; Campylobacter
 infection**
 f campylobactériose; infection à
 Campylobacter
 e campilobacteriosis; infección por
 Campylobacter spp.
 d Kampylobakteriose; Campylobacter-
 Infektion

636 **canalicule**
 f canalicule
 e canalículo
 d Kanälchen
 l canaliculus

* **Candida infection** → 637

637 **candidiasis; candidosis; Candida infection**
 f candidose; infection à Candida
 e candidosis; candidiasis; infección por
 Candida spp.
 d Kandidamykose; Candida-Infektion

* **candidosis** → 637

638 **canine adenovirus**
 f adénovirus canin
 e adenovirus canino
 d canines Adenovirus

639 **canine distemper; distemper**
 f maladie de Carré
 e moquillo; enfermedad de Carré
 d Staupe; Hundestaupe
 l febris catarrhalis et nervosum caninum

* **canine distemper virus** → 1342

640 **canine hepatitis virus**
 f virus de l'hépatite contagieuse du chien
 e virus de la hepatitis canina infecciosa
 d HCC-Virus

641 **canine herpesvirus**
 f herpèsvirus canin
 e herpesvirus canino
 d canines Herpesvirus

642 **canine muscle**
 f muscle canin
 e músculo canino
 d Eckzahnmuskel
 l musculus caninus

643 **canine parainfluenza virus**
 f virus du parainfluenza canin
 e virus de la parainfluenza canina
 d canines Parainfluenzavirus

644 **canine parvoviral infection**
 f parvovirose canine
 e parvovirosis canina
 d canine Parvovirose

645 **canine teeth**
 f dents canines
 e dientes caninas
 d Hundszähne; Hakenzähne; Fangzähne
 l dentes canini

646 **canine viral hepatitis**
 f hépatite contagieuse du chien; maladie de
 Rubarth
 e hepatitis canina infecciosa; enfermedad de
 Rubarth
 d Hepatitis contagiosa canis; HCC;
 ansteckende Leberentzündung
 l hepatitis contagiosa canis

647 **cannibalism**
 f cannibalisme
 e canibalismo
 d Kannibalismus

* **cannon bone** → 4527, 4528

648 **capillariosis; hairworm infestation**
 f capillariose; infestation par Capillaria spp.
 e capilariosis
 d Kapillariose; Haarwurmbefall

649 **capillary vessel**
 f vaisseau capillaire
 e vaso capilar
 d Kapillare; Haargefäß
 l vas capillare

* **capitular articulation** → 298

650 **caponization**
 f chaponnage
 e caponización
 d Kapaunisierung

651 capped elbow; elbow hygroma
 f hygroma du coude
 e higroma del codo
 d Ellbogenbeule; Stollbeule

652 capped hock; tarsal hygroma
 f capelet; hygroma du tarse
 e higroma del tarso
 d Tarsalhygrom; Gelenkgalle des Tarsus

653 caprine arthritis-encephalitis
 f arthrite-encéphalite caprine
 e artritis-encéfalitis caprina
 d Arthritis und Enzephalitis der Ziege

654 caprine arthritis-encephalitis virus; CAE virus
 f virus de l'arthrite-encéphalite caprine
 e virus de la artritis-encefalitis caprina
 d Ziegenarthritis und Enzephalitis-Virus

 * **caprine morbillivirus** → 3493

655 capripoxvirus
 f capripoxvirus
 e capripoxvirus
 d Capripoxvirus

656 capture of animals
 f capture des animaux
 e captura de los animales
 d Tierfang

657 carazolol
 f carazolol
 e carazolol
 d Carazolol
 l carazololum

658 carbachol
 f carbachol
 e carbacol
 d Carbachol
 l carbacholum

659 carbadox
 f carbadox
 e carbadox
 d Carbadox
 l carbadoxum

660 carbaril; carbaryl; sevin
 f carbaryl
 e carbarilo
 d Carbaryl
 l carbarilum

 * **carbaryl** → 660

661 carbohydrate metabolism disorder
 f trouble du métabolisme glucidique
 e trastorno del metabolismo de carbohidrato
 d Störung des Kohlenhydratstoffwechsel

662 carcass condemnation
 f saisie des carcasses
 e condenación de las canales
 d Untauglichkeit

663 carcass disposal
 f élimination des carcasses
 e eliminación de las canales
 d Tierkörperbeseitigung

664 carcinogen
 f substance cancérogène
 e sustancia cancerígena
 d kanzerogene Substanz

665 carcinogenicity
 f pouvoir cancérogène
 e potencia cancerígena
 d Kanzerogenität; Karzinogenität

666 carcinoma
 f carcinome
 e carcinoma
 d Karzinom

667 cardia; cardiac part of stomach
 f cardia; partie cardiale de l'estomac
 e porción cardíaca
 d Kardia des Magens
 l pars cardiaca ventriculi

668 cardiac ganglia
 f ganglions cardiaques
 e ganglios cardíacos
 d Ganglia cardiaca
 l ganglia cardiaca

669 cardiac hypertrophy
 f hypertrophie cardiaque
 e hipertrofia cardíaca
 d Herzhypertrophie

670 cardiac impression
 f empreinte cardiaque
 e impresión cardíaca
 d Impressio cardiaca; Herzeindruck
 l impressio cardiaca

671 cardiac insufficiency
 f insuffisance cardiaque
 e insuficiencia cardíaca
 d Herzinsuffizienz

672 **cardiac notch of lung**
  f incisure cardiaque du poumon
  e escotadura cardíaca del pulmón
  d Incisura cardiaca pulmonis; Herzeinschnitt
  l incisura cardiaca pulmonis

  * **cardiac part of stomach** → 667

673 **cardiac plexus**
  f plexus cardiaque
  e plexo cardíaco
  d Herzgeflecht
  l plexus cardiacus

674 **cardiac veins**
  f veines du cœur
  e venas del corazón
  d Venen der Herzwand
  l venae cordis

675 **cardiomegaly; enlargement of heart**
  f cardiomégalie; gros cœur
  e cardiomegalía
  d Kardiomegalie; Herzvergrößerung

676 **cardiovascular agent**
  f modificateur cardiovasculaire
  e agente cardiovascular
  d Kreislaufmittel

677 **cardiovascular disease**
  f maladie cardiovasculaire
  e enfermedad cardiovascular
  d kardiovaskuläre Krankheit

678 **cardiovirus; encephalomyocarditis virus**
  f cardiovirus; virus de
    l'encéphalomyocardite
  e cardiovirus; virus de la encefalomiocarditis
  d Cardiovirus; Enzephalomyocarditis-Virus

  * **cariniform cartilage** → 2681

679 **carnassial tooth; sectorial tooth**
  f dent carnassière
  e diente cortante
  d Reißzahn
  l dens sectorius

680 **caroticotympanic nerves**
  f nerfs carotico-tympaniques
  e nervios caroticotimpánicos
  d Nervi caroticotympanici
  l nervi caroticotympanici

681 **carotid body**
  f glomus carotidien
  e glomo carotídeo
  d Glomus caroticum; Karotiskörper
  l glomus caroticum

682 **carotid plexus**
  f plexus carotidien
  e plexo carotídeo
  d Plexus caroticus
  l plexus caroticus

  * **carpal bone II** → 4028

  * **carpal bone III** → 4525

683 **carpal bones**
  f os du carpe
  e huesos del carpo
  d Karpalknochen
  l ossa carpi

  * **carpal bones II & III** → 4026

684 **carpal glands** (*pig*)
  f glandes carpiennes (*du porc*)
  e glándulas carpianas
  d Karpaldrüsen (*Schwein*)
  l glandulae carpae

685 **carpal hairs of cat**
  f poils carpaux
  e pelos carpianos
  d Karpaltasthaare; Karpalvibrissen
  l pili carpales

686 **carpal joint**
  f articulation du carpe
  e articulación del carpo
  d Karpalgelenk; Vorderfußwurzelgelenk
  l articulatio carpi

687 **carpal pad**
  f torus carpien; coussinet carpien
  e almohadilla carpiana
  d Karpalballen
  l torus carpeus

688 **carpal region**
  f région du carpe
  e región del carpo
  d Karpalregion
  l regio carpi

689 **carpometacarpal joints**
  f articulations carpo-métacarpiennes
  e articulaciones carpometacarpianas
  d Karpal-Metakarpal-Gelenke
  l articulationes carpometacarpeae

690 **carpometacarpal ligaments**
  f ligaments carpo-métacarpiens
  e ligamentos carpometacarpianos
  d Ligamenta carpometacarpea
  l ligamenta carpometacarpea

691 carpometacarpus
f carpo-métacarpe
e carpometacarpo
d Carpometacarpus
l carpometacarpus

692 carpus; wrist
f carpe
e carpo
d Carpus; Vorderfußwurzel
l carpus

693 carrier (*of a pathogen*)
f porteur asymptomatique
e portador asintomático
d Keimträger; Überträger

694 carrier state
f portage latent
e estado portador
d Keimträgerzustand

695 cartilage of heart
f cartilage du cœur
e cartílago del corazón
d Herzknorpel
l cartilago cordis

696 cartilage of third phalanx
f cartilage unguéal; cartilage de la troisième
phalange
e cartílago ungular; cartílago de la tercera
falange
d Hufknorpel
l cartilago ungularis

* **cartilaginous joint** → 146

697 cartilaginous part of auditory tube
f partie cartilagineuse de la trompe auditive
e porción cartilaginosa de la trompa auditiva
d knorpeliger Teil der Ohrtrompete
l pars cartilaginea tubae auditivae

698 caruncle
f caroncule
e carúncula
d Karunkel
l caruncula

699 case history; anamnesis
f antécédents; commémoratifs
e anamnesis; historia clínica
d Anamnese; Vorbericht

**700 caseous lymphadenitis; Corynebacterium
ovis infection**
f lymphadénite caséeuse; infection à
Corynebacterium ovis
e linfadenitis caseosa ovina; infección por
Corynebacterium ovis
d Pseudotuberkulose des Schafes; käsige
Lymphadenitis
l lymphadenitis caseosa

701 castor-bean tick
f Ixodes ricinus
e garrapata ricina
d gemeiner Holzbock
l Ixodes ricinus

702 castrate *v*
f castrer; châtrer
e castrar
d kastrieren

703 castrated
f castré; châtré
e castrado
d kastriert

704 castration; orchidectomy
f castration; orchidectomie
e castración
d Kastration; Orchitomie
l castratio

705 casualty slaughter; emergency slaughter
f abattage d'urgence
e sacrificio de urgencia
d Notschlachtung

706 cataract
f cataracte
e catarata
d Katarakt; Linsentrübung
l cataracta

707 catarrh
f catarrhe
e catarro
d Katarrh
l catarrhus

708 catarrhal inflammation
f inflammation catarrhale
e inflamación catarral
d katarrhalische Entzündung; Katarrh

709 cat disease
f maladie des chats
e enfermedad de los gatos
d Katzenkrankheit

710 catecholamine
f catécholamine
e catecolamina
d Katecholamin

711 cat flea
  f puce du chat
  e pulga del gato
  d Katzenfloh
  l Ctenocephalides felis

712 catheter
  f cathéter; sonde
  e catéter
  d Katheter

713 cattle biting louse
  f Bovicola bovis
  e piojo del pelo de los bovinos
  d Rinderhaarling
  l Bovicola bovis

714 cattle disease
  f maladie des bovidés
  e enfermedad del ganado
  d Rinderkrankheit

  * **cattle plague** → 3898

  * **cattle warts** → 539

715 caudal abdominal artery
  f artère abdominale caudale
  e arteria abdominal caudal
  d Arteria abdominalis caudalis
  l arteria abdominalis caudalis

716 caudal abdominal region
  f région abdominale caudale
  e región caudal del abdomen
  d kaudale Bauchregion; Unterbauchgegend
  l regio abdominis caudalis

717 caudal abdominal vein
  f veine abdominale caudale
  e vena abdominal caudal
  d Vena abdominalis caudalis
  l vena abdominalis caudalis

718 caudal abductor cruris muscle
  f muscle abducteur caudal de la jambe
  e músculo abductor caudal de la pierna
  d hinterer Auswärtsführer des Unterschenkels
  l musculus abductor cruris caudalis

  * **caudal aperture of pelvis** → 3367

719 caudal arteries; coccygeal arteries
  f artères caudales; artères coccygiennes
  e arterias caudales
  d Arteriae caudales
  l arteriae caudales

720 caudal clunial nerves
  f nerfs cluniaux caudaux
  e nervios caudales de las nalgas
  d Nervi clunium caudales
  l nervi clunium caudales

721 caudal colliculus; quadrigeminal body
  f colliculus caudal; tubercule quadrijumeau antérieur
  e colículo caudal; tubérculo cuadrigémino
  d Colliculus caudalis; hintere Vierhügel
  l colliculus caudalis; corpora quadrigemina

722 caudal communicating artery
  f artère communicante caudale
  e arteria comunicante caudal
  d kaudale Verbindungsarterie
  l arteria communicans caudalis

723 caudal cutaneous nerve of thigh
  f nerf cutané fémoral caudal
  e nervio cutáneo caudal del muslo
  d Nervus cutaneus femoris caudalis
  l nervus cutaneus femoris caudalis

724 caudal epigastric vein
  f veine épigastrique caudale
  e vena epigástrica caudal
  d Vena epigastrica caudalis
  l vena epigastrica caudalis

725 caudal extremity of vertebra; vertebral fossa
  f extrémité caudale de la vertèbre; fosse vertébrale
  e extremidad caudal; fosa de la vértebra
  d Wirbelpfanne
  l extremitas caudalis vertebrae

726 caudal femoral arteries
  f artères fémorales caudales
  e arterias caudales del muslo
  d Arteriae caudales femoris
  l arteriae caudales femoris

727 caudal femoral veins
  f veines fémorales caudales
  e venas caudales del muslo
  d Venae caudales femoris
  l venae caudales femoris

728 caudal ganglia; coccygeal ganglia
  f ganglions caudaux; ganglions coccygiens
  e ganglios caudales; ganglios cocígeos
  d Schwanzganglien
  l ganglia caudalia; ganglia coccygea

729 caudal glands; coccygeal glands (*of dog*)
  f glandes coccygiennes (*chien*); glandes de la queue

e glándulas de la cola
d Schwanzdrüsen
l glandulae caudae; glandulae coccygis

730 caudal gluteal artery
f artère glutéale caudale
e arteria glútea caudal
d Arteria glutea caudalis
l arteria glutea caudalis

731 caudal gluteal nerve
f nerf glutéal caudal
e nervio glúteo caudal
d kaudaler Gesäßnerv
l nervus gluteus caudalis

732 caudal gluteal vein
f veine glutéale caudale
e vena glútea caudal
d Vena glutea caudalis
l vena glutea caudalis

733 caudal groove of rumen
f sillon caudal du rumen
e surco caudal del rumen
d kaudale Pansenfurche
l sulcus caudalis (ruminis)

734 caudal laryngeal nerve
f nerf laryngé caudal
e nervio laríngeo caudal
d Nervus laryngeus caudalis
l nervus laryngeus caudalis

735 caudal ligament
f ligament caudal
e ligamento caudal
d Ligamentum caudale
l ligamentum caudale

736 caudal lobe of cerebellum
f lobe caudal du cervelet
e lóbulo caudal del cerebelo
d Lobus caudalis cerebelli
l lobus caudalis cerebelli

737 caudal mandibular adductor muscle
f muscle adducteur caudal de la mandibule
e músculo aductor caudal de la mandíbula
d Musculus adductor mandibulae caudalis
l musculus adductor mandibulae caudalis

738 caudal medullary velum
f voile médullaire caudal
e velo medular caudal
d Velum medullare caudale; hinteres Marksegel
l velum medullare caudale

739 caudal mesenteric artery
f artère mésentérique caudale
e arteria mesentérica caudal
d Arteria mesenterica caudalis
l arteria mesenterica caudalis

740 caudal mesenteric ganglion
f ganglion mésentérique caudal
e ganglio mesentérico caudal
d Ganglion mesentericum caudale
l ganglion mesentericum caudale

741 caudal mesenteric lymph nodes
f nœuds lymphatiques mésentériques caudaux
e nódulos linfáticos mesentéricos caudales
d hintere Mesenteriallymphknoten
l lymphonodi mesenterici caudales

742 caudal mesenteric lymphocentre
f lymphocentre mésentérique caudal
e linfocentro mesentérico caudal
d Lymphocentrum mesentericum caudale
l lymphocentrum mesentericum caudale

743 caudal mesenteric vein
f veine mésentérique caudale
e vena mesentérica caudal
d Vena mesenterica caudalis
l vena mesenterica caudalis

744 caudal nasal nerve
f nerf nasal caudal
e nervio nasal caudal
d Nervus nasalis caudalis
l nervus nasalis caudalis

745 caudal nerves; tail nerves
f nerfs caudaux/coccygiens
e nervios caudales/coccígeos
d Schwanznerven
l nervi caudales/coccygei

746 caudal oblique muscle of head
f muscle oblique caudal de la tête
e músculo oblicuo caudal de la cabeza
d hinterer schiefer Kopfmuskel
l musculus obliquus capitis caudalis

747 caudal omental recess
f récessus omental caudal
e receso caudal del omento
d kaudale Netzbeutelhöhle
l recessus caudalis omentalis

748 caudal pancreaticoduodenal artery
f artère pancréatico-duodénale caudale
e arteria pancreaticoduodenal caudal
d Arteria pancreaticoduodenalis caudalis
l arteria pancreaticoduodenalis caudalis

749 **caudal phrenic artery**
 f artère phrénique caudale
 e arteria frénica caudal
 d kaudale Zwerchfellarterie
 l arteria phrenica caudalis

750 **caudal phrenic vein**
 f veine phrénique caudale
 e vena frénica caudal
 d Vena phrenica caudalis
 l vena phrenica caudalis

751 **caudal pillar of rumen**
 f pilier caudal du rumen
 e pilar caudal del rumen
 d kaudaler Pansenpfeiler
 l pila caudalis ruminis

752 **caudal rectal nerves**
 f nerfs rectaux caudaux
 e nervios rectales caudales
 d hintere Mastdarmnerven
 l nervi rectales caudales

753 **caudal scapulohumeral muscle**
 f muscle scapulo-huméral caudal
 e músculo escapulohumeral caudal
 d Musculus scapulohumeralis caudalis
 l musculus scapulohumeralis caudalis

754 **caudal stylopharyngeal muscle**
 f muscle stylo-pharyngien caudal
 e músculo estilofaríngeo caudal
 d Musculus stylopharyngeus caudalis
 l musculus stylopharyngeus caudalis

755 **caudal thoracic air sac**
 f sac aérien thoracique caudal
 e saco aéreo torácico caudal
 d kaudaler Brustluftsack
 l saccus thoracicus caudalis

756 **caudal thyroid artery**
 f artère thyroïdienne caudale
 e arteria tiroidea caudal
 d kaudale Schilddrüsenarterie
 l arteria thyroidea caudalis

757 **caudal tibial artery**
 f artère tibiale caudale
 e arteria tibial caudal
 d kaudale Schienbeinarterie
 l arteria tibialis caudalis

758 **caudal tibial muscle**
 f muscle tibial caudal
 e músculo tibial caudal
 d kaudaler Schienbeinmuskel
 l musculus tibialis caudalis

759 **caudal tibial vein**
 f veine tibiale caudale
 e vena tibial caudal
 d hintere Tibialvene
 l vena tibialis caudalis

* **caudal veins** → 4439

760 **caudal vena cava**
 f veine cave caudale
 e vena cava caudal
 d Vena cava caudalis; hintere
   Körperhohlvene
 l vena cava caudalis

761 **caudal vertebra; coccygeal vertebra**
 f vertèbre caudale; vertèbre coccygienne
 e vértebra caudal; vértebra coccígea
 d Schwanzwirbel
 l vertebra caudalis

* **caudal wing web** → 3574

762 **caudate lobe of liver**
 f lobe caudé du foie
 e lóbulo caudado del hígado
 d geschwänzter Leberlappen
 l lobus caudatus hepatis

763 **caudate nucleus**
 f noyau caudé
 e núcleo caudado
 d Nucleus caudatus
 l nucleus caudatus

764 **caudate process of liver**
 f processus caudé du foie
 e apófisis caudada del hígado
 d geschwänzter Fortsatz der Leber
 l processus caudatus hepatis

765 **caudodorsal blind sac of rumen**
 f cul-de-sac caudodorsal du rumen
 e saco ciego caudodorsal
 d dorsaler Endblindsack
 l saccus c(a)ecus caudodorsalis

766 **caudofemoral muscle**
 f muscle caudo-fémoral
 e músculo caudofemoral
 d Musculus caudofemoralis
 l musculus caudofemoralis

767 **caudo-ilio-femoral muscle**
 f muscle caudo-ilio-fémoral
 e músculo caudoiliofemoral
 d Musculus caudo-ilio-femoralis
 l musculus caudo-ilio-femoralis

768 **caudoventral blind sac of rumen**
  f cul-de-sac caudoventral du rumen
  e saco ciego caudoventral
  d ventraler Endblindsack
  l saccus c(a)ecus caudoventralis

769 **causal factor**
  f facteur causal
  e factor causal
  d kausaler Faktor

770 **cautery**
  f cautère; cautérisation
  e cauterio
  d Brennen; Kauterisation

771 **cavernous body of penis**
  f corps caverneux du pénis
  e cuerpo cavernoso del pene
  d Penisschwellkörper
  l corpus cavernosum penis

772 **cavernous sinus**
  f sinus caverneux
  e seno cavernoso
  d Sinus cavernosus
  l sinus cavernosus

773 **cavernous veins of penis**
  f veines caverneuses du pénis
  e venas cavernosas del pene
  d Venae cavernosae penis
  l venae cavernosae penis

774 **cavity of pituitary stalk**
  f cavité de la tige pituitaire
  e porción hueca del infundíbulo
  d Infundibularhöhle
  l pars cava infundibuli

 * **CBPP** → 1041

 * **CCPP** → 1042

 * **cecal artery** → 617

 * **cecal sacculations** → 618

 * **cecal veins** → 619

 * **cecocolic opening** → 620

 * **cecum** → 621

775 **cefalexin; cephalexin**
  f céfaléxine
  e cefalexina
  d Cefalexin
  l cefalexinum

776 **cefaloridine**
  f céfaloridine
  e cefaloridina
  d Cefaloridin
  l cefaloridinum

777 **cefalotin; cephalothin**
  f céfalotine
  e cefalotina
  d Cefalotin
  l cefalotinum

778 **cefotaxime**
  f céfotaxime
  e cefotaxima
  d Cefotaxim
  l cefotaximum

 * **celiac artery** → 971

 * **celiac ganglia** → 972

 * **celiac lymphocentre** → 973

 * **celiac plexus** → 974

779 **cell**
  f cellule
  e célula
  d Zell
  l cellula

780 **cell count**
  f numération cellulaire; nombre de cellules
  e numeración celular
  d Zellzahl; Zellgehalt

781 **cell culture**
  f culture cellulaire
  e cultivo celular
  d Zellkultur

782 **cell division**
  f division cellulaire
  e división celular
  d Zellteilung

783 **cell line**
  f lignée cellulaire
  e línea celular
  d Zellinie

784 **cell-mediated immunity**
  f immunité à médiation cellulaire
  e inmunidad celular
  d zellvermittelte Immunität

785 cell membrane
 f membrane cellulaire
 e membrana celular
 d Zellmembran

786 cell nucleus
 f noyau cellulaire
 e núcleo celular
 d Zellkern; Nukleus

787 cells of reticulum
 f cellules réticulaires
 e celdillas del retículo
 d Feldern der Haube
 l cellulae reticuli

788 cell ultrastructure
 f ultrastructure cellulaire
 e ultraestructura celular
 d Ultrastruktur der Zelle

789 cell wall
 f paroi cellulaire
 e pared celular
 d Zellwand

790 CELO virus; chick embryo lethal orphan virus
 f virus CELO
 e virus CELO
 d CELO-Virus

* **CEM → 1046**

791 central artery of retina
 f artère centrale de la rétine
 e arteria central de la retina
 d Arteria centralis retinae
 l arteria centralis retinae

792 central canal of spinal cord
 f canal central de la moelle épinière; canal de l'épendyme
 e canal central del epéndimo
 d Zentralkanal im Rückenmark
 l canalis centralis medullae spinalis

793 central fovea of retina
 f fovea centralis de la rétine
 e fóvea central de la retina
 d Fovea centralis retinae
 l fovea centralis retinae

794 central groove of frog
 f sillon cunéal central; lacune médiane de la fourchette
 e surco cuneal central
 d mittlere Strahlfurche
 l sulcus cunealis centralis

795 central nervous system; CNS
 f système nerveux central; SNC
 e sistema nervioso central
 d zentrales Nervensystem; ZNS
 l systema nervosum centrale

796 central nucleus of thalamus
 f noyau central du thalamus
 e núcleo central del tálamo
 d Zentralkern des Thalamus
 l nucleus centralis thalami

* **central pillar of cochlea → 2848**

797 central tarsal bone; navicular bone
 f os central du tarse; os naviculaire
 e hueso central del tarso; hueso navicular
 d Os tarsi centrale; Kahnbein
 l os tarsi centrale; os naviculare

798 central tendon of perineum; perineal body
 f centre tendineux du périnée
 e centro tendinoso del periné; cuerpo perineal
 d sehniger Teil des Perineums
 l centrum tendineum perinei; corpus perineale

799 central veins of liver
 f veines centrales du foie
 e venas centrales del hígado
 d Venae centrales hepatis
 l venae centrales hepatis

800 centrodistal joint; cuneonavicular joint
 f articulation cunéo-naviculaire
 e articulación centrodistal
 d Articulatio centrodistalis
 l articulatio centrodistalis; articulatio cuneonavicularis

801 centroquartal bone
 f os naviculo-cuboïde
 e hueso centrocuartal; hueso naviculocuboideo
 d Os centroquartale
 l os centroquartale; os naviculocuboideum

* **cephalexin → 775**

802 cephalic vein
 f veine céphalique
 e vena cefálica
 d Vena cephalica
 l vena cephalica

803 cephalosporin
 f céphalosporine
 e cefalosporina
 d Cephalosporin

* **cephalothin** → 777

804 ceratohyoid
f cératohyal
e queratohioideo
d Keratohyoid
l ceratohyoideum

805 ceratohyoid muscle
f muscle cérato-hyoïdien
e músculo queratohioideo
d Musculus ceratohyoideus
l musculus ceratohyoideus

806 cere
f cire
e cera
d Wachshaut
l cera

807 cerebellar arteries
f artères cérébelleuses
e arterias del cerebelo
d Arteriae cerebelli; Kleinhirnarterien
l arteriae cerebelli

808 cerebellar cortex; cortex of cerebellum
f cortex cérébelleux
e corteza del cerebelo
d Kleinhirnrinde
l cortex cerebelli

809 cerebellar peduncles
f pédoncules cérébelleux
e pedúnculos del cerebelo
d Kleinhirnstiele
l pedunculi cerebellares

810 cerebellomedullary cistern; great cistern
f citerne cérébello-médullaire
e cisterna cerebelomedular; cisterna magna
d Cisterna cerebellomedullaris; Cisterna magna
l cisterna cerebellomedullaris; cisterna magna

811 cerebellum
f cervelet
e cerebelo
d Zerebellum; Kleinhirn
l cerebellum

812 cerebral arteries
f artères cérébrales
e arterias del cerebro
d Arteriae cerebri; Großhirnarterien
l arteriae cerebri

813 cerebral convolutions
f circonvolutions cérébrales
e giros del cerebro; circonvoluciones cerebrales
d Gehirnwindungen
l gyri cerebri

814 cerebral cortex
f cortex cérébral
e corteza del cerebro
d Großhirnrinde
l cortex cerebri

815 cerebral hemisphere
f hémisphère cérébral
e hemisferio
d Gehirnhälfte
l hemispherium

816 cerebral peduncle
f pédoncule cérébral
e pilar del cerebro; pedúnculo cerebral
d Großhirnstiel; Gehirnschenkel
l crus cerebri; pedunculus cerebri

817 cerebral pia mater
f pie-mère de l'encéphale
e piamadre del encéfalo
d weiche Hirnhaut
l pia mater encephali

818 cerebral veins
f veines cérébrales
e venas del cerebro
d Venae cerebri
l venae cerebri

819 cerebrocortical necrosis; polioencephalomalacia
f nécrose du cortex cérébral; polioencéphalomalacie
e necrosis cerebrocortical; polioencefalomalacia
d Zerebrokortikalnekrose; Hirnrindennekrose; Polioenzephalomalazie

820 cerebrospinal artery
f artère cérébro-spinale
e arteria cerebrospinal
d Arteria cerebrospinalis
l arteria cerebrospinalis

821 cerebrospinal fluid
f liquide cérébro-spinal; liquide céphalo-rachidien
e líquido cerebroespinal; líquido cefalorraquídeo
d Liquor; Zerebrospinalflüssigkeit
l liquor cerebrospinalis

822 **cerebrovascular disorder**
  f trouble cérébrovasculaire
  e trastorno cerebrovascular
  d Gehirngefäßstörung

823 **cerebrum**
  f cerveau
  e cerebro
  d Zerebrum; Großhirn
  l cerebrum

824 **ceruminous glands**
  f glandes cérumineuses
  e glándulas ceruminosas
  d Ohrenschmalzdrüsen
  l glandulae ceruminosae

825 **cervical air sac**
  f sac aérien cervical
  e saco aéreo cervical
  d Halsluftsack
  l saccus cervicalis

826 **cervical appendages**
  f appendices cervicaux
  e apéndices del cuello
  d Berlocke; Glöckchen
  l appendices colli

827 **cervical arteries**
  f artères cervicales
  e arterias cervicales
  d Arteriae cervicales
  l arteriae cervicales

828 **cervical beard** (*of male turkey*)
  f barbe cervicale (*du dindon*)
  e barba cervical
  d Halsbart (*des Truthahns*)
  l barba cervicalis

829 **cervical canal**
  f canal cervical
  e canal del cuello uterino
  d Zervikalkanal
  l canalis cervicis uteri

830 **cervical cardiac nerves**
  f nerfs cardiaques cervicaux
  e nervios cardíacos cervicales
  d Nervi cardiaci cervicales
  l nervi cardiaci cervicales

831 **cervical enlargement of spinal cord**
  f intumescence cervicale
  e intumescencia cervical
  d Intumescentia cervicalis;
    Halsanschwellung
  l intumescentia cervicalis

832 **cervical fascia; fascia of neck**
  f fascia cervical
  e fascia cervical
  d Halsfaszie
  l fascia cervicalis

833 **cervical mucus**
  f glaire cervicale
  e moco cervical
  d Zervikalschleim

834 **cervical nerves**
  f nerfs cervicaux
  e nervios cervicales
  d Halsnerven
  l nervi cervicales

835 **cervical part of cutaneous muscle**
  f muscle cutané (peaucier) du cou
  e músculo cutáneo del cuello
  d Halshautmuskel
  l musculus cutaneus colli

836 **cervical plexus**
  f plexus cervical
  e plexo cervical
  d Halsnervengeflecht
  l plexus cervicalis

837 **cervical vertebra**
  f vertèbre cervicale
  e vértebra cervical
  d Halswirbel
  l vertebra cervicalis

838 **cervicitis**
  f cervicite
  e cervicitis
  d Zervizitis; Cervicitis

839 **cervicothoracic ganglion; stellate ganglion**
  f ganglion cervico-thoracique; ganglion
    étoilé/stellaire
  e ganglio cervicotorácico; ganglio estrellado
  d sternförmiges Ganglion
  l ganglion cervicothoracicum

840 **cervix; uterine cervix; neck of uterus**
  f col utérin
  e cuello del útero
  d Zervix; Gebärmutterhals
  l cervix uteri

* **cesarean section** → 622

841 **cestode; tapeworm**
  f cestode; ténia
  e cestodo; tenia
  d Zestode; Bandwurm
  l Cestoda

* cestode infestation → 842

842 cestodosis; cestode infestation; tapeworm
    infestation
  f cestodose
  e cestodosis
  d Bandwurmbefall

843 cetrimide
  f cétrimide
  e cetrimida
  d Cetrimid
  l cetrimidum

844 cetylpyridinium chloride
  f chlorure de cétylpyridinium
  e cloruro de cetilpiridinio
  d Cetylpyridiniumchlorid
  l cetylpyridinii chloridum

* CF test → 1018

* CFU → 995

845 chalk brood of bees; Ascosphaera apis
    infection
  f infection des abeilles à Ascosphaera apis
  e cría encalada; infección de las abejas por
      Ascosphaera apis
  d Kalkbrut; Ascosphaera-apis-Infektion

846 challenge infection
  f infection d'épreuve; épreuve virulente
  e infección de prueba
  d Belastungsinfektion

* check ligament → 28

847 Chediak-Higashi syndrome
  f syndrome de Chediak-Higashi
  e síndrome de Chediak-Higashi
  d Chediak-Higashi-Syndrom

848 cheek
  f joue
  e mejilla
  d Backe; Wange
  l bucca

849 cheek hairs
  f poils jugaux; poils de la joue
  e pelos bucales
  d Backentasthaare
  l pili buccales

850 chelating agent
  f chélateur
  e agente de quelación
  d Chelatbildner

851 chemoprophylaxis
  f chimioprophylaxie
  e quimioprofilaxis
  d Chemoprophylaxe

852 chemotaxis
  f chimiotactisme
  e quimiotaxis
  d Chemotaxis

853 chemotherapeutic
  f chimiothérapeutique
  e quimioterapéutico
  d chemotherapeutisch

* chemotherapeutic drug → 223

854 chemotherapeutic index
  f index chimiothérapeutique
  e índice quimioterapéutico
  d chemotherapeutischer Index;
      therapeutische Breite

855 chemotherapy
  f chimiothérapie
  e quimioterapia
  d Chemotherapie

* chest → 4547

* chest muscles → 4540

856 chestnut; tarsal pad
  f coussinet tarsien
  e almohadilla tarsiana
  d Tarsalballen; Kastanie (*Pferd*)
  l torus tarseus

* chick embryo lethal orphan virus → 790

857 chicken flea
  f puce des volailles
  e pulga de las gallinas
  d Hühnerfloh
  l Ceratophyllus gallinae

* chigoe → 2357

858 chin
  f menton
  e mentón
  d Kinn
  l mentum

859 chin gland; mental organ
  f glande mentonnière
  e glándula mentoniana
  d Kinndrüse; Mentalbüschel
  l glandula mentalis

860 **chin region**
 f région mentonnière
 e región mentoniana
 d Kinngegend
 l regio mentalis

861 **chlamydia**
 f chlamydie
 e chlamydia
 d Chlamydie
 l Chlamydiales

 * **Chlamydia infection → 862**

 * **chlamydial abortion of ewes → 1484**

 * **Chlamydia psittaci infection → 3162**

862 **chlamydiosis; Chlamydia infection**
 f chlamydiose; infection à Chlamydia
 e clamidiosis; infección por Chlamydia spp.
 d Chlamydiose; Chlamydia-Infektion

863 **chloral hydrate**
 f hydrate de chloral
 e hidrato de cloral
 d Chloralhydrat

864 **chloralose**
 f chloralose
 e cloralosa
 d Chloralose

865 **chloramphenicol**
 f chloramphénicol
 e cloranfenicol
 d Chloramphenicol
 l chloramphenicolum

 * **chlorfenvinphos → 954**

866 **chlorhexidine**
 f chlorhexidine
 e clorhexidina
 d Chlorhexidin
 l chlorhexidinum

867 **chlormadinone**
 f chlormadinone
 e clormadinona
 d Chlormadinon
 l chlormadinonum

868 **chloroform**
 f chloroforme
 e cloroformo
 d Chloroform

869 **chloroquine**
 f chloroquine
 e cloroquina
 d Chloroquin
 l chloroquinum

870 **chloroxylenol**
 f chloroxylénol
 e cloroxilenol
 d Chloroxylenol
 l chloroxylenolum

871 **chlorpromazine**
 f chlorpromazine
 e clorpromazina
 d Chlorpromazin
 l chlorpromazinum

872 **chlortetracycline**
 f chlortétracycline
 e clortetraciclina
 d Chlortetracyclin
 l chlortetracyclinum

873 **cholangiectasis**
 f cholangiectasie
 e colangiectasia
 d Cholangiektasie; Gallengangserweiterung

874 **cholangitis**
 f cholangite
 e colangitis
 d Cholangitis

875 **cholecalciferol; colecalciferol; vitamin D3**
 f cholécalciférol; vitamine D3
 e colecalciferol; vitamina D3
 d Cholekalziferol; Colecalciferol; Vitamin D3
 l colecalciferolum

876 **cholecystitis**
 f cholécystite
 e colecistitis
 d Cholezystitis; Gallenblasenentzündung

 * **choledochous duct → 1003**

 * **cholelith → 1795**

877 **cholesterol**
 f cholestérol
 e colesterol
 d Cholesterin

878 **cholinergic**
 f cholinergique
 e colinérgico
 d cholinergisch

879 **cholinesterase**
  f cholinestérase
  e colinesterasa
  d Cholinesterase

880 **cholinesterase inhibitor**
  f anticholinestérique
  e inhibidor de colinesterasa
  d Cholinesterase-Hemmstoff

881 **chondritis**
  f chondrite
  e condritis
  d Chondritis

882 **chondrodystrophy**
  f chondrodystrophie
  e condrodistrofia
  d Chondrodystrophie

883 **chondrosesamoid ligaments**
  f ligaments chondro-sésamoïdiens
  e ligamentos condrosesamoideos
  d Hufknorpel-Strahlbeinbänder
  l ligamenta chondrosesamoidea

884 **chorioallantoic membrane; CAM**
  f membrane chorio-allantoïdienne; allanto-
  chorion
  e membrana corio-alantoica
  d Chorioallantoismembran; CAM

885 **chorioallantois; allantochorion**
  f chorioallantois; allanto-chorion
  e alantocorión
  d Chorioallantois; Allantochorion

* **choriocapillar lamina** → 891

886 **chorion**
  f chorion
  e corión
  d Chorion
  l chorion

887 **chorionic gonadotrophin; HCG**
  f gonadotrophine chorionique
  e gonadotrofina coriónica
  d Choriongonadotrophin
  l gonadotrophinum chorionicum

888 **chorionic villi**
  f villosités choriales
  e villosidades del corión
  d Chorionzotten
  l villi chorii

889 **chorioptic mange; chorioptosis**
  f gale chorioptique
  e sarna corióptica
  d Chorioptesräude

* **chorioptosis** → 889

890 **choroid**
  f choroïde
  e coroidea
  d Aderhaut
  l choroidea; chorioidea

891 **choroidocapillar lamina; choriocapillar
  lamina**
  f lame chorio-capillaire
  e lámina coroidocapilar
  d Lamina choroidocapillaris
  l lamina choroidocapillaris

892 **choroid plexus**
  f plexus choroïde
  e plexo coroideo
  d Plexus choroideus; Adergeflecht
  l plexus choroideus

893 **chromatin**
  f chromatine
  e cromatina
  d Chromatin

894 **chromatography**
  f chromatographie
  e cromatografía
  d Chromatographie

* **chromosomal anomaly** → 896

895 **chromosome**
  f chromosome
  e cromosoma
  d Chromosom

896 **chromosome abnormality; chromosomal
  anomaly**
  f anomalie chromosomique
  e anomalía cromosomal
  d Chromosomenanomalie

897 **chronic**
  f chronique
  e crónico
  d chronisch

898 **chronic disease**
  f maladie chronique
  e enfermedad crónica
  d chronische Krankheit

899 **chronic respiratory disease of poultry**
  f maladie respiratoire chronique;
    mycoplasmose aviaire
  e enfermedad crónica respiratoria;
    micoplasmosis respiratoria aviar; infección
    por Mycoplasma gallisepticum
  d chronische Atmungskrankheit;
    Geflügelmykoplasmose

900 **chronic toxicity**
  f toxicité chronique
  e toxicidad crónica
  d chronische Toxizität

  * **Chrysosporium infection** → 1958

901 **chyle**
  f chyle
  e quilo
  d Chylus
  l chylus

902 **chyle cistern; chylocyst; cistern of Pecquet**
  f citerne du chyle; citerne de Pecquet
  e cisterna del quilo (de Pecquet)
  d Chyluszisterne
  l cisterna chyli

  * **chylocyst** → 902

903 **chylothorax**
  f chylothorax
  e quilotorax
  d Chylothorax

904 **chyme**
  f chyme
  e quimo
  d Chymus
  l chymus

905 **chymosin**
  f chymosine
  e quimosina
  d Chymosin

906 **chymotrypsin**
  f chymotrypsine
  e quimotripsina
  d Chymotrypsin
  l chymotrypsinum

907 **cicatrix**
  f cicatrice
  e cicatriz
  d Cicatrix; Narbe

908 **cicatrization**
  f cicatrisation
  e cicatrización
  d Narbenbildung

909 **cilia**
  f cils
  e cilia
  d Zilien; Wimpern

910 **ciliary arteries**
  f artères ciliaires
  e arterias ciliares
  d Arteriae ciliares
  l arteriae ciliares

911 **ciliary body**
  f corps ciliaire
  e cuerpo ciliar
  d Ziliarkörper; Strahlenkörper
  l corpus ciliare

912 **ciliary ganglion**
  f ganglion ciliaire
  e ganglio ciliar
  d Ziliarganglion
  l ganglion ciliare

913 **ciliary glands** (*of conjunctiva*)
  f glandes ciliaires
  e glándulas ciliares
  d Ziliardrüsen
  l glandulae ciliares

914 **ciliary muscle**
  f muscle ciliaire.
  e músculo ciliar
  d Ziliarmuskel
  l musculus ciliaris

915 **ciliary nerves**
  f nerfs ciliaires
  e nervios ciliares
  d Ziliarnerven
  l nervi ciliares

916 **ciliary veins** (*of eye*)
  f veines ciliaires
  e venas ciliares
  d Venae ciliares
  l venae ciliares

917 **ciliary zonule; Zinn's membrane**
  f zonule ciliaire de Zinn
  e zónula ciliar
  d Zonula ciliaris
  l zonula ciliaris

918 ciliates
f ciliés
e ciliados
d Ziliaten; Wimpertierchen
l Ciliata

919 cingulate gyrus; callosal gyrus
f gyrus du cingulum
e giro del cíngulo
d Gyrus cinguli
l gyrus cinguli

920 ciprofloxacin
f ciprofloxacine
e ciprofloxacino
d Ciprofloxacin
l ciprofloxacinum

921 circulation
f circulation
e circulación
d Kreislauf

922 circulatory disorder
f trouble circulatoire
e trastorno circulatorio
d Kreislaufstörung

923 circulatory system
f appareil circulatoire
e aparato circulatorio
d Kreislaufsystem

924 circumanal glands
f glandes circumanales
e glándulas circumanales
d Zirkumanaldrüsen
l glandulae circumanales

925 circumflex arteries
f artères circonflexes
e arterias circumflejas
d Arteriae circumflexae
l arteriae circumflexae

* **circumflex nerve** → 377

926 circumflex veins
f veines circonflexes
e venas circumflejas
d Venae circumflexae
l venae circumflexae

* **circumvallate papillae** → 4814

* **cistern of Pecquet** → 902

* **classical fowl plague** → 355

* **classical swine fever** → 4385

927 claudication
f claudication
e claudicación
d Klaudikation
l claudicatio

* **Claviceps paspali intoxication** → 3337

* **Claviceps purpurea intoxication** → 1537

* **Claviceps purpurea toxin** → 1536

928 clavicle; wish bone
f clavicule; os furculaire
e clavícula
d Schlüsselbein
l clavicula; furcula

929 clavicular air sac
f sac aérien claviculaire
e saco aéreo clavicular
d Schlüsselbeinluftsack
l saccus clavicularis

930 claw
f griffe; onglon
e unguícula
d Kralle; Klaue
l unguicula

* **claw region** → 2027

931 clazuril
f clazuril
e clazurilo
d Clazuril
l clazurilum

932 clearance
f clairance
e aclaramiento
d Ausräumen

933 clear layer of epidermis
f stratum lucidum de l'épiderme
e estrato lúcido
d Stratum lucidum; durchscheinende
  Zellschicht
l stratum lucidum epidermidis

934 cleft palate; palatoschisis
f fissure de la voûte palatine; palatoschizis
e cisura del paladar; palatosquisis
d Gaumenspalte; Palatoschisis
l palatum fissum

* cleg → 2045

935 cleidobrachial muscle
f muscle cléido-brachial
e músculo cleidobraquial
d Musculus cleidobrachialis
l musculus cleidobrachialis

936 cleidocephalic muscle
f muscle cléido-céphalique
e músculo cleidocefálico
d Musculus cleidocephalicus
l musculus cleidocephalicus

937 cleidocervical muscle
f muscle cléido-cervical
e músculo cleidocervical
d Musculus cleidocervicalis
l musculus cleidocervicalis

938 cleidomastoid muscle
f muscle cléido-mastoïde
e músculo cleidomastoideo
d Musculus cleidomastoideus
l musculus cleidomastoideus

939 cleido-occipital muscle
f muscle cléido-occipital
e músculo cleidooccipital
d Musculus cleidooccipitalis
l musculus cleidooccipitalis

940 cleidotracheal muscle
f muscle cléido-trachéal
e músculo cleidotraqueal
d Musculus cleidotrachealis
l musculus cleidotrachealis

941 clenbuterol
f clenbutérol
e clenbuterol
d Clenbuterol
l clenbuterolum

942 clinic
f clinique
e clínica
d Klinik

943 clinical examination
f examen clinique
e examen clínico
d klinische Untersuchung

944 clinical features
f manifestations cliniques
e manifestaciones clínicas
d Klinik; klinisches Erscheinungsbild

945 clioquinol
f clioquinol
e clioquinol
d Clioquinol
l clioquinolum

946 clioxanide
f clioxanide
e clioxanida
d Clioxanid
l clioxanidum

947 clitoral fossa
f fosse clitoridienne
e fosa del clítoris
d Fossa clitoridis
l fossa clitoridis

948 clitoris
f clitoris
e clítoris
d Klitoris; Kitzler
l clitoris

949 cloaca
f cloaque
e cloaca
d Kloake
l cloaca

950 cloacal bursa; bursa of Fabricius
f bourse cloacale (de Fabricius)
e bolsa de la cloaca
d Kloakenbursa; Bursa fabricii
l bursa cloacalis

* cloacal gland → 4843

951 cloacal promontory
f promontoire du cloaque
e promontorio de la cloaca
d Kloakenwulst; Mündung der Samenleiter
l promontorium cloacale

952 cloacal sphincter muscle
f muscle sphincter du cloaque
e músculo esfínter de la cloaca
d Kloakenschließmuskel
l musculus sphincter cloacae

953 cloacitis
f cloacite
e cloacitis
d Kloazitis; Kloakenentzündung

* clofenotane → 1216

954 clofenvinfos; chlorfenvinphos
  f chlorfenvinphos
  e clofenvinfos
  d Clofenvinfos
  l clofenvinfosum

955 clopidol
  f clopidol
  e clopidol
  d Clopidol
  l clopidolum

956 cloprostenol
  f cloprosténol
  e cloprostenol
  d Cloprostenol
  l cloprostenolum

957 clorsulon
  f clorsulone
  e clorsulon
  d Clorsulon
  l clorsulonum

958 closantel
  f closantel
  e closantel
  d Closantel
  l closantelum

959 clostridial enterotoxaemia; Clostridium perfringens infection
  f entérotoxémie clostridienne; infection à Clostridium perfringens
  e enterotoxemia ovina; infección por Clostridium perfringens
  d Klostridien-Enterotoxämie; Clostridium-perfringens-Infektion

960 clostridial infection
  f clostridiose
  e clostridiosis
  d Klostridieninfektion

  * **Clostridium botulinum intoxication** → 524

  * **Clostridium chauvoei infection** → 462

  * **Clostridium haemolyticum infection** → 385

  * **Clostridium novyi infection** → 460, 2652

  * **Clostridium perfringens infection** → 959

  * **Clostridium perfringens type A infection** → 1805

  * **Clostridium perfringens type B infection** → 2407

  * **Clostridium perfringens type C infection** → 3207

  * **Clostridium perfringens type D infection** → 3698

  * **Clostridium septicum enterotoxaemia** → 568

  * **Clostridium septicum infection** → 462

  * **Clostridium tetani intoxication** → 4503

961 cloxacillin
  f cloxacilline
  e cloxacilina
  d Cloxacillin
  l cloxacillinum

  * **CNS** → 795

962 coagulation disorder; coagulopathy
  f coagulopathie; trouble de la coagulation sanguine
  e trastorno de la coagulación sanguínea
  d Koagulopathie; Blutgerinnungsstörung

  * **coagulopathy** → 962

963 coccidia
  f coccidies
  e coccidios
  d Kokzidien

  * **Coccidioides immitis infection** → 964

964 coccidioidomycosis; Coccidioides immitis infection
  f coccidioïdomycose; infection à Coccidioides immitis
  e coccidioidomicosis; infección por Coccidioides immitis
  d Kokzidioidomykose; Coccidioides-immitis-Infektion

965 coccidiosis
  f coccidiose
  e coccidiosis
  d Kokzidiose

  * **coccidiostat** → 212

  * **coccygeal arteries** → 719

  * **coccygeal ganglia** → 728

* coccygeal glands → 729

966 coccygeal muscle
 f muscle coccygien
 e músculo coxígeo
 d langer Seitwärtszieher des Schwanzes
 l musculus coccygeus

* coccygeal veins → 4439

* coccygeal vertebra → 761

* coccyx → 4432

967 cochlea
 f cochlée
 e cóclea
 d Cochlea; Schnecke
 l cochlea

968 cochlear duct
 f conduit cochléaire
 e conducto coclear
 d Schneckengang
 l ductus cochlearis

969 cochlear recess; Reichert's recess
 f récessus cochléaire
 e receso coclear
 d Recessus cochlearis
 l recessus cochlearis

970 cochlear root of vestibulocochlear nerve
 f racine cochléaire du nerf vestibulo-
 cochléaire
 e raíz coclear del nervio vestibulococlear
 d Schneckenwurzel des
 Gleichgewichtshörnervs
 l radix cochlearis nervi vestibulocochlearis

971 coeliac artery; celiac artery
 f artère cœliaque
 e arteria celíaca
 d Arteria coeliaca
 l arteria c(o)eliaca

972 coeliac ganglia; celiac ganglia
 f ganglions cœliaques
 e ganglios celíacos
 d Ganglia coeliaca
 l ganglia c(o)eliaca

973 coeliac lymphocentre; celiac lymphocentre
 f lymphocentre cœliaque
 e linfocentro celíaco
 d Lymphocentrum coeliacum
 l lymphocentrum c(o)eliacum

974 coeliac plexus; celiac plexus
 f plexus cœliaque
 e plexo celíaco
 d Sonnengeflecht
 l plexus c(o)eliacus

975 coenurosis; gid of sheep
 f cœnurose du mouton
 e cenurosis
 d Drehkrankheit der Schafe; Zönurose

976 coenurus; gid worm
 f cénure; cœnure
 e cenuro
 d Zönurus; Gehirnblasenwurm
 l Coenurus cerebralis

* coffin bone → 1340

* coffin joint → 1339

* colecalciferol → 875

* colibacillosis → 977

977 colibacteriosis; colibacillosis; Escherichia
 coli infection
 f colibacillose; infection à Escherichia coli
 e colibacilosis; infección por Escherichia coli
 d Kolibazillose; Escherichia-coli-Infektion

978 coli bacterium
 f colibacille
 e colibacilo
 d Kolibazillus
 l Escherichia coli

979 colic
 f colique
 e cólico
 d Kolik

980 colic arteries
 f artères coliques
 e arterias cólicas
 d Arteriae colicae
 l arteriae colicae

981 colic impression on liver
 f empreinte colique du foie
 e impresión cólica
 d Impressio colica; Koloneindruck
 l impressio colica

982 colic veins
 f veines coliques
 e venas cólicas
 d Venae colicae
 l venae colicae

983 **coliform bacterium**
  f bacille coliforme
  e bacilo coliforme
  d koliforme Bakterie

984 **coliform mastitis**
  f mammite coliforme
  e mastitis coliforma
  d Koli-Mastitis

985 **colistin**
  f colistine
  e colistina
  d Colistin
  l colistinum

986 **colitis**
  f colite
  e colitis
  d Colitis; Dickdarmentzündung

987 **collateral artery**
  f artère collatérale
  e arteria colateral
  d Kollateralarterie
  l arteria collateralis

988 **collateral groove of frog**
  f sillon paracunéal; lacune latérale de la
     fourchette
  e surco paracuneal lateral/medial
  d seitliche Strahlfurche
  l sulcus paracunealis lateralis/medialis

989 **collateral ligament**
  f ligament collateral
  e ligamento colateral
  d Seitenband
  l ligamentum collaterale

990 **collateral veins**
  f veines collatérales
  e venas colaterales
  d Kollateralvene
  l venae collaterales

991 **collateral vessel**
  f vaisseau collatéral
  e vaso colateral
  d Kollateralgefäß
  l vas collaterale

992 **coloboma**
  f colobome
  e coloboma
  d Kolobom

993 **colon**
  f côlon
  e colon
  d Kolon; Grimmdarm
  l colon

* **colonic mesentery** → 2776

994 **colony count**
  f dénombrement des colonies
  e recuento de colonias microbianas
  d Koloniezahl; Mikrobenzahl

995 **colony-forming unit; CFU**
  f unité formant colonies; UFC
  e unidad formando las colonias
  d koloniebildende Einheit; KBE

996 **colostral immunity**
  f immunité colostrale
  e inmunidad calostral
  d Kolostralimmunität; kolostrale Immunität

997 **colostrum**
  f colostrum
  e calostro
  d Kolostrum; Kolostralmilch

998 **colostrum-deprived**
  f privé de colostrum
  e privado de calostro
  d kolostrumfrei aufgezogen

999 **colostrum-deprived disease; Cytophaga columnaris
    infection**
  f infection à Cytophaga columnaris
  e infección por Cytophaga columnaris
  d Cytophaga-columnaris-Infektion

* **comb** → 1718

1000 **combined vaccine; polyvalent vaccine**
  f vaccin associé; vaccin polyvalent
  e vacuna polivalente
  d Kombinationsvakzine; assoziierte Vakzine;
     Mehrfachvakzine

1001 **commissure of eyelids**
  f commissure des paupières
  e comisura de los párpados
  d Augenlidkommisur
  l commissura palpebrarum

1002 **commissure of lips of mouth**
  f commissure labiale
  e comisura de los labios
  d Lippenkommissur
  l commissura labiorum oris

1003 **common bile duct; choledochous duct**
  f canal cholédoque
  e conducto colédoco
  d Hauptgallengang
  l ductus choledochus

1004 **common calcaneal tendon; Achilles tendon; hamstring tendon**
  f tendon calcanéen commun; tendon d'Achille; corde du jarret
  e tendón calcáneo común; tendón de Aquiles
  d Fersensehnenstrang; Achillessehne
  l tendo calcaneus communis

1005 **common carotid artery**
  f artère carotide commune
  e arteria carótida común
  d Arteria carotis communis
  l arteria carotis communis

1006 **common digital extensor muscle**
  f muscle extenseur commun des doigts
  e músculo extensor digital común
  d gemeinsamer Zehenstrecker
  l musculus extensor digitorum communis

1007 **common hepatic duct**
  f conduit hépatique commun; canal hépatique commun
  e conducto hepático común
  d gemeinsamer Gallengang
  l ductus hepaticus communis

1008 **common iliac artery**
  f artère iliaque commune
  e arteria ilíaca común
  d Arteria iliaca communis
  l arteria iliaca communis

1009 **common iliac vein**
  f veine iliaque commune
  e vena ilíaca común
  d Vena iliaca communis; gemeinsame Hüftvene
  l vena iliaca communis

1010 **common integument**
  f tégument commun; appareil tégumentaire
  e tegumento común
  d allgemeine Körperdecke
  l integumentum commune

1011 **common liver fluke**
  f douve hépatique; grande douve du foie
  e distoma hepático; gran duela del hígado
  d großer Leberegel
  l Fasciola hepatica

1012 **common ox warble fly**
  f varron; hypoderme du bœuf
  e hipoderma bovina; barro
  d große Dasselfliege
  l Hypoderma bovis

1013 **common peroneal nerve**
  f nerf péronier commun
  e nervio peroneo común
  d Nervus peroneus communis
  l nervus peroneus communis

1014 **common synovial sheath of flexor muscles**
  f gaine synoviale des muscles fléchisseurs
  e vaina sinovial común de los músculos flexores
  d Karpalbeugesehnenscheide
  l vagina synovialis communis musculi flexorum

1015 **communicating branch** (*between nerves*)
  f rameau communicant
  e ramo comunicante
  d Verbindungsast
  l ramus communicans

1016 **compact substance of bone**
  f substance compacte de l'os
  e sustancia compacta del hueso
  d fester Knochenmantel
  l substantia compacta

1017 **complement**
  f complément
  e complemento
  d Komplement

1018 **complement fixation test; CF test**
  f épreuve de fixation du complément
  e prueba de fijación del complemento
  d Komplementbindungsprobe

1019 **compound fracture; open fracture**
  f fracture compliquée; fracture ouverte
  e fractura abierta
  d offene Fraktur

1020 **compound joint**
  f articulation composée
  e articulación compuesta
  d zusammengesetztes Gelenk
  l articulatio composita

1021 **conchal cartilage; auricular cartilage**
  f cartilage de la conque; cartilage auriculaire
  e cartílago de la oreja
  d Ohrmuschelknorpel
  l cartilago auriculae

1022 conchal crest
 f crête conchale
 e cresta conchal
 d Crista conchalis
 l crista conchalis

 * **concha of auricle** → 335

 * **conduction anaesthesia** → 2981

1023 condylar articulation; condyloid joint
 f articulation condylaire; charnière
    imparfaite
 e articulación condilar
 d Kondylengelenk
 l articulatio condylaris

1024 condylar fossa
 f fosse condylaire
 e fosa condilar
 d Fossa condylaris
 l fossa condylaris

1025 condylar process of mandible
 f processus condylaire de la mandibule;
    condyle mandibulaire
 e apófisis condilar
 d Gelenkfortsatz des Unterkiefers
 l processus condylaris mandibulae

 * **condylarthrosis** → 1442

1026 condyle
 f condyle
 e cóndilo
 d Kondylus; Gelenkfortsatz
 l condylus

1027 condyle of humerus
 f condyle de l'humérus
 e cóndilo del húmero
 d Gelenkknorren des Humerus
 l condylus humeri

 * **condyloid joint** → 1023

1028 congenital abnormality; malformation
 f malformation congénitale
 e anormalidad congénita; malformación
    congénita
 d Entwicklungsstörung; Mißbildung

 * **congenital absence of heart** → 24

1029 congenital tremor
 f tremblement congénital
 e temblor congénito
 d Zitterkrankheit

1030 conical papillae of tongue
 f papilles coniques
 e papilas cónicas
 d konische Zungenpapillen
 l papillae conicae

 * **conjugal ligament** → 2319

1031 conjunctiva; conjunctival tunic
 f conjonctive; tunique conjonctive
 e túnica conjuntiva
 d Bindehaut
 l tunica conjunctiva

1032 conjunctival arteries
 f artères conjonctivales
 e arterias conjuntivales
 d Arteriae conjunctivales; Bindehautarterien
 l arteriae conjunctivales

1033 conjunctival fornix
 f fornix de la conjonctive
 e fórnix de la conjuntiva
 d Fornix conjunctivae; Bindehautgewölbe
 l fornix conjunctivae

1034 conjunctival glands
 f glandes conjonctivales
 e glándulas conjuntivales
 d Lidbindehautdrüsen
 l glandulae conjunctivales

 * **conjunctival tunic** → 1031

1035 conjunctivitis
 f conjonctivite
 e conjuntivitis
 d Konjunktivitis; Bindehautentzündung

1036 connective tissue
 f tissu conjonctif
 e tejido conjuntivo
 d Bindegewebe

1037 constipation
 f constipation
 e estreñimiento
 d Verstopfung; Obstipation

1038 constrictor muscle
 f muscle constricteur
 e músculo constrictor
 d Schließmuskel
 l musculus constrictor

1039 constrictor muscle of neck
 f muscle constricteur du cou
 e músculo constrictor del cuello
 d Musculus constrictor colli
 l musculus constrictor colli

* contagious abortion → 527

1040 **contagious agalactia; Mycoplasma agalactiae infection**
 f agalaxie contagieuse; infection à Mycoplasma agalactiae
 e agalaxia contagiosa de ovejas y cabras; infección por Mycoplasma agalactiae
 d ansteckende Agalaktie; Mycoplasma-agalactiae-Infektion

1041 **contagious bovine pleuropneumonia; CBPP; Mycoplasma mycoides infection**
 f péripneumonie contagieuse bovine; infection à Mycoplasma mycoides
 e pleuroneumonía contagiosa bovina; infección por Mycoplasma mycoides
 d Lungenseuche des Rindes; Mycoplasma-mycoides-Infektion
 l pleuropneumonia contagiosa bovum

1042 **contagious caprine pleuropneumonia; CCPP; Mycoplasma mycoides var. capri infection**
 f pleuropneumonie contagieuse caprine; infection à Mycoplasma mycoides var. capri
 e pleuroneumonía contagiosa caprina; infección por Mycoplasma mycoides var. capri
 d Lungenseuche der Ziege; infektiöse Pleuropneumonie der Ziege
 l pleuropneumonia contagiosa caprarum

1043 **contagious disease; transmissible disease**
 f maladie contagieuse; maladie transmissible
 e enfermedad transmisible
 d kontagiöse Krankheit; ansteckende Krankheit

1044 **contagious ecthyma; contagious pustular dermatitis; orf**
 f ecthyma contagieux
 e ectima contagioso
 d Ecthyma contagiosum
 l ecthyma contagiosum

1045 **contagious ecthyma parapoxvirus; contagious pustular dermatitis virus**
 f parapoxvirus de l'ecthyma contagieux
 e virus del ectima contagioso
 d Dermatitis-pustulosa-Parapoxvirus

1046 **contagious equine metritis; CEM; Taylorella equigenitalis infection**
 f métrite contagieuse équine; infection à Taylorella equigenitalis

 e metritis contagiosa equina; infección por Taylorella equigenitalis
 d kontagiöse equine Metritis; Taylorella-equigenitalis-Infektion

1047 **contagiousness**
 f contagiosité
 e carácter contagioso
 d Kontagiosität

* contagious pustular dermatitis → 1044

* contagious pustular dermatitis virus → 1045

1048 **contamination**
 f contamination
 e contaminación
 d Kontamination; Verunreinigung

1049 **contour feathers**
 f pennes de contour
 e plumas de contorno
 d Konturfedern
 l pennae contourae

1050 **contraction**
 f contraction; resserrement
 e contracción
 d Kontraktion; Verkürzung

1051 **controlled experiment**
 f essai contrôlé; étude contrôlée
 e ensayo controlado
 d kontrollierter Versuch

1052 **convoluted part of renal tubules**
 f tubes contournés du rein
 e túbulos contorneados del riñón
 d gewundene Abschnitte der Nierenkanälchen
 l tubuli renales contorti

1053 **convoluted seminiferous tubules**
 f tubes séminifères contournés
 e túbulos seminíferos contorneados
 d gewundene Hodenkanälchen
 l tubuli seminiferi contorti

1054 **convulsion**
 f convulsion
 e convulsión
 d Krampf

* Cooperia infestation → 1055

1055 **cooperiosis; Cooperia infestation**
 f coopériose; infestation par Cooperia spp.
 e cooperiosis
 d Cooperiose; Befall mit Cooperia

1056 copper deficiency
 f carence en cuivre
 e carencia de cobre
 d Kupfermangel

1057 copper poisoning
 f intoxication au cuivre; cuprisme
 e intoxicación por cobre
 d Kupfervergiftung

1058 coprodeum
 f coprodéum
 e coprodeo
 d Coprodeum; Kotraum
 l coprodeum

1059 coracobrachial muscle
 f muscle coraco-brachial
 e músculo coracobraquial
 d Rabenschnabel-Armmuskel
 l musculus coracobrachialis

 * **coracoclavicular membrane** → 4210

1060 coracohumeral ligament
 f ligament coraco-huméral
 e ligamento coracohumeral
 d Ligamentum coracohumerale
 l ligamentum coracohumerale

1061 coracoid
 f coracoïde
 e coracoides
 d Rabenschnabelbein
 l coracoideum

1062 coracoid process of scapula
 f processus coracoïde de l'omoplate
 e apófisis coracoidea de la escápula
 d Rabenschnabelfortsatz
 l processus coracoideus

1063 coracoscapular interosseous ligament
 f ligament interosseux coraco-scapulaire
 e ligamento interóseo coracoescapular
 d Ligamentum interosseum coracoscapulare
 l ligamentum interosseum coracoscapulare

1064 coracoscapular symphysis
 f symphyse coraco-scapulaire
 e sínfisis coracoescapular
 d Symphysis coracoscapularis
 l symphysis coracoscapularis

1065 corium; dermis
 f chorion; derme
 e corión; dermis
 d Corium; Lederhaut
 l corium; dermis

1066 corium of frog; cuneal corium
 f chorion de la fourchette; derme de la fourchette
 e corión de la cuña; dermis de la cuña
 d Strahllederhaut
 l corium cunei; dermis cunei

 * **corium of hoof wall** → 2416

 * **corium of limbus** → 3404

1067 corium of sole
 f chorion de la sole
 e corión de la suela
 d Sohlenlederhaut
 l corium soleae

1068 cornea
 f cornée
 e córnea
 d Kornea; Hornhaut des Auges
 l cornea

1069 corneal opacity
 f opacité de la cornée
 e opacidad de la córnea
 d Hornhauttrübung

1070 corneal ulcer
 f ulcère dc la cornée
 e úlcera de la córnea
 d Hornhautgeschwür
 l ulcus corneae

1071 corniculate tubercle (*of larynx*)
 f tubercule corniculé
 e tubérculo corniculado
 d Tuberculum corniculatum
 l tuberculum corniculatum

1072 cornual artery; horn artery
 f artère cornuale; artère de la corne
 e arteria cornual
 d Arteria cornualis
 l arteria cornualis

 * **cornual process** → 2034

1073 coronal artery; coronet artery
 f artère coronale
 e arteria coronal
 d Arteria coronalis
 l arteria coronalis

1074 coronal border of hoof wall
 f bord coronal de la paroi du sabot
 e borde coronal
 d Kronrand
 l margo coronalis

1075 **coronal groove of hoof**
 f sillon coronal
 e surco coronal
 d Hufsaum; Kronsaum
 l sulcus coronalis

1076 **corona radiata of ovum**
 f couronne rayonnante
 e corona radiata
 d Corona radiata
 l corona radiata

1077 **coronary arteries**
 f artères coronaires
 e arterias coronarias
 d Arteriae coronariae; Herzkranzarterien
 l arteriae coronariae

* **coronary band → 1085**

1078 **coronary corium**
 f chorion coronaire; derme du bourrelet
 e corión de la corona; dermis de la corona
 d Kronlederhaut
 l corium coronae; dermis coronae

1079 **coronary groove of heart**
 f sillon coronarien
 e surco coronario
 d Herzkranzfurche
 l sulcus coronarius

1080 **coronary groove of rumen**
 f sillon coronaire du rumen
 e surco coronario del rumen
 d Kranzfurche des Rumens
 l sulcus coronarius (ruminis)

1081 **coronary ligament of liver**
 f ligament coronaire du foie
 e ligamento coronario del hígado
 d Ligamentum coronarium hepatis
 l ligamentum coronarium hepatis

1082 **coronary pillar of rumen**
 f pilier coronaire du rumen
 e pilar coronario del rumen
 d Kranzpfeiler des Rumens
 l pila coronaria ruminis

1083 **coronary sinus**
 f sinus coronarien
 e seno coronario
 d Koronarsinus
 l sinus coronarius

1084 **coronavirus**
 f coronavirus
 e coronavirus
 d Coronavirus

1085 **coronet; coronary band**
 f bourrelet
 e corona
 d Krone; Kronsegment
 l corona

* **coronet artery → 1073**

1086 **coronoid fossa of humerus**
 f fosse coronoïdienne
 e fosa coronoidea
 d Fossa coronoidea
 l fossa coronoidea humeri

1087 **coronoid process**
 f processus coronoïde
 e apófisis coronoidea
 d Processus coronoideus
 l processus coronoideus

* **corpora nigra → 1886**

1088 **corpus albicans**
 f corps blanc
 e cuerpo blanquecino
 d Corpus albicans
 l corpus albicans

1089 **corpus callosum**
 f corps calleux
 e cuerpo calloso
 d Corpus callosum; Gehirnbalken
 l corpus callosum

1090 **corpus luteum**
 f corps jaune
 e cuerpo lúteo
 d Gelbkörper
 l corpus luteum

1091 **corridor disease; Theileria lawrencei infection**
 f infection à Theileria lawrencei
 e infección por Theileria lawrencei
 d Theileria-lawrencei-Infektion

1092 **cortex**
 f cortex
 e corteza
 d Rinde; Kortex
 l cortex

* **cortex of cerebellum → 808**

* **cortex of kidney → 3811**

1093 **cortical substance of bone**
 f substance corticale de l'os

e sustancia cortical
d Knochenrinde
l substantia corticalis

1094 corticoreticular fibres
f fibres cortico-réticulaires
e fibras corticorreticulares
d Fibrae corticoreticulares
l fibrae corticoreticulares

* **corticospinal tract** → 3718

1095 corticosteroid
f corticostéroïde
e corticosteroide
d Kortikosteroid

1096 corticotrophin; corticotropin; ACTH
f corticotrophine; ACTH
e corticotropina; ACTH
d Corticotrophin; Kortikotropin; ACTH
l corticotrophinum

* **corticotropin** → 1096

* **Corti's ganglion** → 4154

* **cortisol** → 2060

1097 cortisone
f cortisone
e cortisona
d Cortison
l cortisonum

1098 corynebacterium
f corynébactérie
e corinebacteria
d Korynbakterie
l Corynebacterium

* **Corynebacterium ovis infection** → 700

* **Corynebacterium pseudotuberculosis infection** → 4726

1099 coryza
f coryza
e coriza
d Coryza; Schnupfen

1100 costal arch
f arc costal
e arco costal
d Rippenbogen
l arcus costalis

1101 costal cartilage
f cartilage costal
e cartílago costal
d Rippenknorpel
l cartilago costalis

1102 costal groove (*on avian lung*)
f sillon costal
e surco costal
d Rippenfurche
l sulcus costalis

1103 costal notches of sternum
f incisures costales du sternum
e escotaduras costales
d Incisurae costales sterni
l incisurae costales sterni

1104 costal pleura
f plèvre costale
e pleura costal
d Rippenfell
l pleura costalis

* **costal region** → 3872

1105 costo-abdominal nerve
f nerf costo-abdominal
e nervio costoabdominal
d Nervus costoabdominalis
l nervus costoabdominalis

1106 costocervical trunk
f tronc costo-cervical
e tronco costocervical
d Truncus costocervicalis
l truncus costocervicalis

1107 costocervical vein
f veine costo-cervicale
e vena costocervical
d Vena costocervicalis
l vena costocervicalis

1108 costochondral articulation
f articulation costo-chondrale
e articulación costocondral
d Rippen-Rippenknorpelgelenk
l articulatio costochondralis

1109 costomediastinal recess of pleura
f cul-de-sac pleural costo-médiastinal
e receso costomediastínico
d Recessus costomediastinalis pleurae
l recessus costomediastinalis pleurae

1110 costoseptal muscle
f muscle costo-septal

e músculo costoseptal
d Musculus costoseptalis
l musculus costoseptalis

1111 costosternal muscle
f muscle costo-sternal
e músculo costoesternal
d Musculus costosternalis
l musculus costosternalis

1112 costotransverse articulation
f articulation costo-transversaire
e articulación costotransversa
d Rippen-Brustwirbelquerfortsatz-Gelenk
l articulatio costotransversaria

1113 costotransverse ligament
f ligament costo-transversaire
e ligamento costotransverso
d Ligamentum costotransversarium
l ligamentum costotransversarium

1114 costovertebral articulation
f articulation costo-vertébrale
e articulación costovertebral
d Articulatio costovertebralis
l articulatio costovertebralis

1115 costoxiphoid ligaments
f ligaments costo-xiphoïdiens
e ligamentos costoxifoideos
d Ligamenta costoxiphoidea
l ligamenta costoxiphoidea

1116 cotyledon
f cotylédon
e cotiledón
d Kotyledone
l cotyledon

1117 cough
f toux
e tos
d Husten

1118 coumafos; coumaphos
f coumafos
e cumafos
d Coumafos; Coumaphos
l coumafosum

* **coumaphos** → 1118

1119 cover hair; guard hair
f jarre
e cabello
d Fellhaar; Deckhaar
l capillus

* **covert feather** → 499

* **Cowdria ruminantium infection** → 1975

* **Cowper's gland** → 606

1120 cowpox
f variole bovine
e viruela bovina
d Kuhpocken; Rinderpocken

1121 coxal tuberosity; point of hip
f tubérosité coxale; pointe de la hanche
e tuberosidad coxal; punta del anca
d Hüfthöcker
l tuber coxae

* **coxal tuberosity region** → 3529

* **Coxiella burnetii infection** → 3728

* **coxofemoral joint** → 2012

* **CPE** → 1213

1122 cranial abdominal artery
f artère abdominale crâniale
e arteria abdominal craneal
d Arteria abdominalis cranialis
l arteria abdominalis cranialis

1123 cranial abdominal region
f région abdominale crâniale
e región craneal del abdomen
d kraniale Bauchregion; Oberbauchgegend
l regio abdominis cranialis

1124 cranial abdominal vein
f veine abdominale crâniale
e vena abdominal craneal
d Vena abdominalis cranialis
l vena abdominalis cranialis

1125 cranial abductor muscle of lower leg
f muscle abducteur crânial de la jambe
e músculo abductor craneal de la pierna
d kranialer Auswärtsführer des Unterschenkels
l musculus abductor cruris cranialis

* **cranial aperture of pelvis** → 3364

* **cranial arachnoid** → 257

1126 cranial bones; bones of skull
f os du crâne
e huesos del cráneo
d Schädelknochen
l ossa cranii

1127 cranial cavity
 f cavité crânienne
 e cavidad del cráneo
 d Schädelhöhle
 l cavum cranii

1128 cranial cervical ganglion
 f ganglion cervical crânial
 e ganglio cervical craneal
 d kranialer sympathischer Halsnervenknoten
 l ganglion cervicale craniale

1129 cranial clunial nerves
 f nerfs cluniaux crâniaux
 e nervios craneales de los nalgas
 d Nervi clunium craniales
 l nervi clunium craniales

* **cranial edge of pubis → 3352**

1130 cranial epigastric vein
 f veine épigastrique crâniale
 e vena epigástrica craneal
 d Vena epigastrica cranialis
 l vena epigastrica cranialis

1131 cranial extremity of vertebra; vertebral head
 f extrémité crâniale de la vertèbre; tête vertébrale
 e extremidad craneal de la vértebra; cabeza de la vértebra
 d Wirbelkopf
 l extremitas cranialis vertebrae

1132 cranial fossa
 f fosse crânienne
 e fosa del cráneo
 d Schädelgrube
 l fossa cranii

1133 cranial gluteal artery
 f artère glutéale crâniale
 e arteria glútea craneal
 d Arteria glutea cranialis
 l arteria glutea cranialis

1134 cranial gluteal nerve
 f nerf glutéal crânial
 e nervio glúteo craneal
 d kranialer Gesäßnerv
 l nervus gluteus cranialis

1135 cranial gluteal vein
 f veine glutéale crâniale
 e vena glútea craneal
 d Vena glutea cranialis
 l vena glutea cranialis

1136 cranial groove of rumen
 f sillon crânial du rumen
 e surco craneal del rumen
 d kraniale Pansenfurche
 l sulcus cranialis (ruminis)

1137 cranial laryngeal artery
 f artère laryngée crâniale
 e arteria laríngea craneal
 d Arteria laryngea cranialis; vordere Kehlkopfarterie
 l arteria laryngea cranialis

1138 cranial laryngeal nerve
 f nerf laryngé crânial
 e nervio laríngeo craneal
 d Nervus laryngeus cranialis
 l nervus laryngeus cranialis

1139 cranial laryngeal vein
 f veine laryngée crâniale
 e vena laríngea cranial
 d vordere Kehlkopfvene
 l vena laryngea cranialis

* **cranial lobe → 247**

* **cranial margin of tibia → 4571**

1140 cranial mesenteric artery
 f artère mésentérique crâniale
 e arteria mesentérica craneal
 d Arteria mesenterica cranialis
 l arteria mesenterica cranialis

1141 cranial mesenteric ganglion
 f ganglion mésentérique crânial
 e ganglio mesentérico craneal
 d Ganglion mesentericum craniale
 l ganglion mesentericum craniale

1142 cranial mesenteric lymph nodes
 f nœuds lymphatiques mésentériques crâniaux
 e nódulos linfáticos mesentéricos craneales
 d vordere Mesenteriallymphknoten
 l lymphonodi mesenterici craniales

1143 cranial mesenteric lymphocentre
 f lymphocentre mésentérique crânial
 e linfocentro mesentérico craneal
 d Lymphocentrum mesentericum craniale
 l lymphocentrum mesentericum craniale

1144 cranial mesenteric vein
 f veine mésentérique crâniale
 e vena mesentérica craneal
 d Vena mesenterica cranialis
 l vena mesenterica cranialis

1145 cranial nerves
f nerfs crâniens
e nervios craneales
d Gehirnnerven
l nervi craniales

1146 cranial oblique muscle of head
f muscle oblique crânial de la tête
e músculo oblicuo craneal de la cabeza
d vorderer schiefer Kopfmuskel
l musculus obliquus capitis cranialis

1147 cranial pancreaticoduodenal artery
f artère pancréatico-duodénale crâniale
e arteria pancreaticoduodenal craneal
d Arteria pancreaticoduodenalis cranialis
l arteria pancreaticoduodenalis cranialis

1148 cranial phrenic artery
f artère phrénique crâniale
e arteria frénica craneal
d kraniale Zwerchfellarterie
l arteria phrenica cranialis

1149 cranial phrenic veins
f veines phréniques crâniales
e venas frénicas craneales
d Venae phrenicae craniales
l venae phrenicae craniales

1150 cranial pillar of rumen
f pilier crânial du rumen
e pilar craneal del rumen
d kranialer Pansenpfeiler
l pila cranialis ruminis

1151 cranial roots of accessory nerve
f racines crâniales du nerf accessoire
e raíces craneales del nervio accesorio
d kraniale Wurzeln des Nervus accessorius
l radices craniales nervi accessorii

1152 cranial scapulohumeral muscle
f muscle scapulo-huméral crânial
e músculo escapulohumeral craneal
d Musculus scapulohumeralis cranialis
l musculus scapulohumeralis cranialis

1153 cranial sinuses; venous sinuses of dura mater
f sinus crâniens; sinus de la dure-mère
e senos de la duramadre
d Sinus durae matris
l sinus durae matris

1154 cranial sutures; sutures of skull
f sutures crâniennes; sutures de la tête
e suturas del cráneo
d Schädelnähte
l suturae capitis; suturae cranii

1155 cranial thoracic air sac
f sac aérien thoracique crânial
e saco aéreo torácico craneal
d kranialer Brustluftsack
l saccus thoracicus cranialis

1156 cranial thyroid artery
f artère thyroïdienne crâniale
e arteria tiroidea craneal
d kraniale Schilddrüsenarterie
l arteria thyroidea cranialis

1157 cranial tibial artery
. *f* artère tibiale crâniale
e arteria tibial craneal
d kraniale Schienbeinarterie
l arteria tibialis cranialis

1158 cranial tibial muscle
f muscle tibial crânial
e músculo tibial craneal
d vorderer Schienbeinmuskel
l musculus tibialis cranialis

1159 cranial tibial vein
f veine tibiale crâniale
e vena tibial craneal
d vordere Tibialvene
l vena tibialis cranialis

1160 cranial vena cava
f veine cave crâniale
e vena cava craneal
d Vena cava cranialis; vordere Körperhohlvene
l vena cava cranialis

* cranial wing web → 3632

1161 cremasteric artery
f artère crémastérique
e arteria cremastérica
d Arteria cremasterica
l arteria cremasterica

1162 cremasteric vein
f veine crémastérique
e vena cremastérica
d Vena cremasterica
l vena cremasterica

1163 cremaster muscle
f muscle crémaster
e músculo cremáster
d Heber des Scheidenhautfortsatzes
l musculus cremaster

1164 crest
  f crête
  e cresta
  d Crista; Kamm
  l crista

1165 crest of humerus
  f crête de l'humérus
  e cresta del húmero
  d Crista humeri
  l crista humeri

1166 crest of ilium; iliac crest
  f crête iliaque
  e cresta ilíaca
  d Darmbeinkamm
  l crista iliaca

1167 cribriform plate
  f lame criblée
  e lámina cribosa
  d Siebplatte
  l lamina cribrosa

1168 cricoarytenoid articulation
  f articulation crico-aryténoïdienne
  e articulación cricoaritenoidea
  d Articulatio cricoarytaenoidea
  l articulatio cricoarytenoidea

1169 cricoarytenoid muscle
  f muscle crico-aryténoïdien
  e músculo cricoaritenoideo
  d Musculus cricoarytaenoideus
  l musculus cricoarytenoideus

1170 cricoid cartilage
  f cartilage cricoïde
  e cartílago cricoides
  d Schildknorpel
  l cartilago cricoidea

1171 cricopharyngeal muscle
  f muscle crico-pharyngien
  e músculo cricofaríngeo
  d Musculus cricopharyngeus
  l musculus cricopharyngeus

1172 cricothyroid articulation
  f articulation crico-thyroïdienne
  e articulación cricotiroidea
  d Articulatio cricothyreoidea
  l articulatio cricothyroidea

1173 cricothyroid ligament
  f ligament crico-thyroïdien
  e ligamento cricotiroideo
  d Ligamentum cricothyreoideum
  l ligamentum cricothyroideum

1174 cricothyroid muscle
  f muscle crico-thyroïdien
  e músculo cricotiroideo
  d Musculus cricothyreoideus
  l musculus cricothyroideus

1175 cricotracheal ligament
  f ligament crico-tracheal
  e ligamento cricotraqueal
  d Ligamentum cricotracheale
  l ligamentum cricotracheale

1176 crista galli of ethmoid bone
  f crista galli de l'os ethmoïde
  e cresta de gallo del hueso etmoideo
  d Crista galli; Hahnenkamm des Siebbeins
  l crista galli

1177 crop
  f jabot
  e buche
  d Kropf
  l ingluvies

* **crop bound** → 1178

1178 crop impaction; crop bound
  f obstruction du jabot; jabot dur; indigestion
      ingluviale
  e obstrucción del buche
  d Kropfverstopfung; harter Kropf

1179 crop region
  f région du jabot
  e región del buche; región ingluvial
  d Kropfregion
  l regio ingluvialis

1180 cross immunity
  f immunité croisée
  e inmunidad cruzada
  d Kreuzimmunität

1181 cross infection
  f infection croisée
  e infección cruzada
  d gekreuzte Infektion

* **croup** → 3966

1182 crown of head
  f vertex; sommet de la tête
  e vértice del cráneo
  d Scheitel
  l vertex

1183 crown of tooth
  f couronne de la dent

e corona del diente
d Zahnkrone
l corona dentis

1184 cruciate ligaments
f ligaments croisés
e ligamentos cruzados
d Kreuzbänder
l ligamenta cruciata

1185 crufomate
f crufomate
e crufomato
d Crufomat
l crufomatum

* **crural region** → 2582

1186 cryptococcosis; Cryptococcus neoformans infection
f cryptococcose; infection à Cryptococcus neoformans
e criptococosis; infección por Cryptococcus neoformans
d Kryptokokkose; Cryptococcose; Cryptococcus-neoformans-Infektion

* **Cryptococcus neoformans infection** → 1186

1187 cryptorchidism
f cryptorchidie
e criptorquidia
d Kryptorchismus; Hodenretention

1188 cryptosporidiosis; Cryptosporidium infection
f cryptosporidiose
e criptosporidiosis
d Kryptosporidiose

* **Cryptosporidium infection** → 1188

* **crypts of Lieberkühn** → 2309

* **cubital articulation** → 1440

* **cubital region** → 1441

1189 cucullary muscle
f muscle cucullaire
e músculo cucular
d Musculus cucullaris
l musculus cucullaris

1190 culmen
f culmen
e culmen
d Culmen
l culmen

1191 culture medium
f milieu de culture
e medio de cultura
d Nährboden; Kulturmedium

* **cuneal corium** → 1066

1192 cuneate nucleus
f noyau cunéiforme
e núcleo cuneado
d Nucleus cuneatus
l nucleus cuneatus

1193 cuneiform tubercle (*of larynx*); **Wrisberg's tubercle**
f tubercule cunéiforme
e tubérculo cuneiforme
d Tuberculum cuneiforme
l tuberculum cuneiforme

* **cuneonavicular joint** → 800

* **cupula of cochlea** → 242

1194 curvature of stomach
f courbure de l'estomac
e curvatura del estómago
d Magenkrümmung
l curvatura ventriculi

1195 cutaneous application; application to the skin
f application cutanée
e aplicación cutánea
d äußerliche Anwendung

1196 cutaneous caruncle
f caroncule cutanée
e carúncula cutánea
d Hautkarunkel
l caruncula cutanea

* **cutaneous glands** → 4101

* **cutaneous muscle of neck and face** → 3507

1197 cutaneous muscles
f muscles cutanés/peauciers
e músculos cutáneos
d Hautmuskeln
l musculi cutanei

* **cutaneous streptothricosis** → 1269

1198 cutaneous vein
f veine cutanée
e vena cutánea
d Hautvene
l vena cutanea

1199 **cyacetazide**
 f cyacétazide
 e ciacetacida
 d Cyacetazid
 l cyacetazidum

 * **cyanobacteria** → 495

1200 **cyanocobalamin; vitamin B12**
 f cyanocobalamine; vitamine B12
 e cianocobalamina; vitamina B12
 d Cyanocobalamin; Vitamin B12;
    Zyanokobalamin
 l cyanocobalaminum

1201 **cyanosis**
 f cyanose
 e cianosis
 d Zyanose

1202 **cyclophosphamide**
 f cyclophosphamide
 e ciclofosfamida
 d Cyclophosphamid; Zyklophosphamid
 l cyclophosphamidum

1203 **cyclopia**
 f cyclopie
 e ciclopía
 d Zyklopie

1204 **cyclopropane**
 f cyclopropane
 e ciclopropano
 d Cyclopropan
 l cyclopropanum

1205 **cyst**
 f kyste
 e quiste
 d Zyste

1206 **cystic artery; gallbladder artery**
 f artère cystique
 e arteria cística
 d Arteria cystica; Gallenblasenarterie
 l arteria cystica

1207 **cystic duct**
 f conduit cystique
 e conducto cístico
 d Gallenblasengang
 l ductus cysticus

1208 **cysticercosis**
 f cysticercose
 e cisticercosis
 d Zystizerkose

1209 **cysticercus; bladder worm**
 f cysticerque
 e cisticerca
 d Zystizerkus; Blasenwurm
 l cysticercus

1210 **cystitis**
 f cystite
 e cistitis
 d Zystitis; Harnblasenentzündung

1211 **cytauxzoonosis**
 f infection à Cytauxzoon
 e infección por Cytauxzoon
 d Cytauxzoon-Infektion

 * **Cytoecetes ondiri infection** → 3121

1212 **cytomegalovirus**
 f cytomégalovirus
 e citomegalovirus
 d Cytomegalovirus

1213 **cytopathic effect; CPE**
 f effet cytopathogène
 e efecto citopatogéno
 d zytopathogenischer Effekt

 * **Cytophaga columnaris infection** → 999

 * **Cytophaga psychrophila infection** → 1702

 * **cytostatic agent** → 225

1214 **cytotoxicity**
 f cytotoxicité
 e citotoxicidad
 d Zytotoxizität

D

1215 dartos
f tunique dartos
e túnica dartos
d Muskelhaut der Hodenhüllen; Fleischhaut des Hodensacks
l tunica dartos

1216 DDT; clofenotane; dicophane
f DDT; clofénotane
e DDT; clofenotano
d DDT; Clofenotan
l clofenotanum

* DDVP → 1293

1217 deafness
f surdité
e sordera
d Taubheit

1218 death
f mort
e muerte
d Tod

1219 debility
f débilité; faiblesse
e debilidad
d Schwäche
l debilitas

1220 deciduous teeth; temporary teeth; milk teeth
f dents déciduales; dents de lait
e dientes caducos
d Milchzähne
l dentes decidui

1221 decoquinate
f décoquinate
e decoquinato
d Decoquinat
l decoquinatum

1222 decussation of the pyramids
f décussation des pyramides
e decusación de las pirámides
d Pyramidenkreuzung
l decussatio pyramidum

1223 deep artery
f artère profonde
e arteria profunda
d tiefe Arterie
l arteria profunda

1224 deep cervical lymph nodes
f nœuds lymphatiques cervicaux profonds
e nódulos linfáticos cervicales profundos
d tiefe Halslymphknoten
l lymphonodi cervicales profundi

1225 deep cervical lymphocentre
f lymphocentre cervical profond
e linfocentro cervical profundo
d tiefes Lymphzentrum des Halses
l lymphocentrum cervicale profundum

1226 deep cervical vein
f veine cervicale profonde
e vena cervical profunda
d Vena cervicalis profunda
l vena cervicalis profunda

1227 deep digital flexor muscle
f muscle fléchisseur profond des doigts
e músculo flexor digital profundo
d tiefer Zehenbeuger
l musculus flexor digitorum profundus

* deep gland of third eyelid → 1959

1228 deep gluteal muscle
f fessier profond
e músculo glúteo profundo
d tiefer Kruppenmuskel
l musculus gluteus profundus

* deep inguinal lymph nodes → 2121

1229 deep parotid lymph nodes
f nœuds lymphatiques parotidiens profonds
e nódulos linfáticos parotídeos profundos
d tiefe Parotislymphknoten
l lymphonodi parotidei profundi

1230 deep pectoral muscle
f muscle pectoral profond/ascendant
e músculo pectoral profundo/ascendente
d tiefer Brustmuskel
l musculus pectoralis profundus; musculus pectoralis ascendens

1231 deep peroneal nerve
f nerf péronier profond
e nervio peroneo profundo
d Nervus peroneus profundus
l nervus peroneus profundus

1232 **deep petrosal nerve**
 f nerf pétreux profond
 e nervio petroso profundo
 d Nervus petrosus profundus
 l nervus petrosus profundus

1233 **deep pseudotemporal muscle**
 f muscle pseudotemporal profond
 e músculo pseudotemporal profundo
 d Musculus pseudotemporalis profundus
 l musculus pseudotemporalis profundus

1234 **deep temporal nerves**
 f nerfs temporaux profonds
 e nervios temporales profundos
 d tiefe Schläfennerven
 l nervi temporales profundi

1235 **deep vein**
 f veine profonde
 e vena profunda
 d tiefe Vene
 l vena profunda

1236 **defaecation**
 f défécation
 e defecación
 d Defäkation; Kotabsatz

1237 **defence mechanism**
 f mécanisme de défense
 e mecanismo de defensa
 d Abwehrmechanismus

1238 **deferent duct; vas deferens**
 f conduit déférent; canal déférent
 e conducto deferente
 d Samenleiter
 l ductus deferens

1239 **deferential plexus**
 f plexus déférentiel
 e plexo deferencial
 d Plexus deferentialis
 l plexus deferentialis

1240 **deficiency**
 f carence
 e carencia
 d Mangel

1241 **deficiency disease**
 f maladie carentielle
 e enfermedad carencial
 d Mangelkrankheit

 * **definitive host** → 1701

1242 **deformity**
 f malformation; difformité
 e deformidad
 d Mißbildung; Formveränderung

1243 **degeneration**
 f dégénérescence
 e degeneración
 d Degeneration; Entartung

1244 **dehorning**
 f écornage; décornage
 e descornación
 d Enthornung

1245 **dehydration**
 f déshydratation
 e deshidratación
 d Exsikkose; Dehydration

 * **Deiters' tract** → 4907

1246 **delmadinone**
 f delmadinone
 e delmadinona
 d Delmadinon
 l delmadinonum

1247 **deltoid muscle**
 f muscle deltoïde
 e músculo deltoideo
 d Deltamuskel
 l musculus deltoideus

1248 **deltoid tuberosity**
 f tubérosité deltoïdienne
 e tuberosidad deltoidea
 d Tuberositas deltoidea
 l tuberositas deltoidea

1249 **demodectic mange; demodicosis**
 f démodexose
 e sarna demodéctica
 d Demodikose

 * **demodicosis** → 1249

1250 **demyelination**
 f démyélinisation
 e desmielinización
 d Entmarkung; Demyelinisation

 * **dental arcade** → 1251

1251 **dental arch; dental arcade**
 f arcade dentaire
 e arco dentario
 d Zahnbogen
 l arcus dentalis

1252 **dental bone**
 f os dentaire
 e hueso dentario
 d Os dentale
 l os dentale

1253 **dental caries**
 f carie dentaire
 e caries dentaria
 d Zahnkaries; Zahnfäule

1254 **dental cement**
 f cément de la dent
 e cemento
 d Zement; Zahnzement
 l cementum

1255 **dental formula**
 f formule dentaire
 e fórmula dentaria
 d Zahnformel

1256 **dental plate** (*of ruminants*)
 f coussinet dentaire
 e almohadilla dentaria
 d Kauplatte
 l pulvinus dentalis

1257 **dental plexus**
 f plexus dentaire
 e plexo dentario
 d Plexus dentalis
 l plexus dentalis

1258 **dentate gyrus**
 f gyrus dentatus
 e giro dentado
 d Gyrus dentatus
 l gyrus dentatus

 * **dentate nucleus** → 2441

1259 **denticulate ligament**
 f ligament dentelé
 e ligamento denticulado
 d Ligamentum denticulatum
 l ligamentum denticulatum

1260 **dentine**
 f dentine
 e dentina
 d Zahnbein
 l dentinum

1261 **dentition**
 f dentition
 e dentición
 d Dentition

 * **deoxycortone** → 1274

 * **deoxynivalenol** → 4952

1262 **depigmentation**
 f dépigmentation
 e depigmentación
 d Depigmentierung; Pigmentschwund

1263 **depluming itch**
 f gale déplumante
 e sarna desplumante
 d Knemidokoptes-Körperräude

1264 **depluming mite**
 f acarien responsable de la gale déplumante
 e ácaro de las plumas
 d Federmilbe der Kammhühner
 l Knemidocoptes laevis gallinae

 * **depraved appetite** → 3466

1265 **depressor muscle of lower lip**
 f abaisseur de la lèvre inférieure
 e músculo depresor del labio mandibular
 d Niederzieher der Unterlippe
 l musculus depressor labii inferioris

1266 **depressor nerve**
 f nerf dépresseur
 e nervio depresor
 d Nervus depressor
 l nervus depressor

 * **dermal ridges** → 3878

1267 **dermatitis**
 f dermatite
 e dermatitis
 d Dermatitis; Hautentzündung

1268 **dermatomycosis; ringworm**
 f dermatomycose; dermatophytie
 e dermatomicosis; tricofitosis
 d Dermatomykose; Hautpilzerkrankung

1269 **dermatophilosis; cutaneous streptothricosis; mycotic dermatitis; Dermatophilus congolensis infection**
 f dermatophilose; streptothricose; infection à Dermatophilus congolensis
 e dermatofilosis; estreptotricosis; infección por Dermatophilus congolensis
 d Hautstreptotrichose; Dermatophilus-congolensis-Infektion

 * **Dermatophilus congolensis infection** → 1269

* **dermatosis** → 4099

* **dermis** → 1065

* **Descemet's membrane** → 3565

1270 descending aorta
f aorte descendante
e aorta descendente
d absteigende Aorta
l aorta descendens

1271 descending colon
f côlon descendant
e colon descendente
d absteigendes Kolon
l colon descendens

1272 descending pectoral muscle
f muscle pectoral descendant
e músculo pectoral descendente
d absteigender Brustmuskel
l musculus pectoralis descendens

1273 descent of testes
f descente des testicules
e descenso de los testículos
d Hodenabstieg
l descensus testis

1274 desoxycortone; deoxycortone
f désoxycortone
e desoxicortona
d Desoxycorton
l desoxycortonum

1275 detoxication
f détoxication
e desintoxicación
d Entgiftung

1276 dewclaw; accessory digit
f ergot; doigt accessoire
e paradígito; paraúngula
d Afterklaue; Afterkralle
l paradigitus; paraungula

1277 dewlap
f fanon
e papada
d Triel; Wamme
l palear

1278 dexamethasone
f dexaméthasone
e dexametasona
d Dexamethason
l dexamethasonum

1279 diagnosis
f diagnostic
e diagnóstico
d Diagnose

1280 diagnostic test
f épreuve diagnostique
e prueba diagnóstica
d diagnostische Probe

1281 diamfenetide; diamphenethide
f diamfénétide
e diamfenetida
d Diamfenetid
l diamfenetidum

* **diamphenethide** → 1281

1282 diaphragm
f diaphragme
e diafragma
d Diaphragma; Zwerchfell
l diaphragma

1283 diaphragmatic hernia
f hernie diaphragmatique
e hernia diafragmática
d Zwerchfellbruch; Diaphragmatozele
l hernia diaphragmatica

1284 diaphragmatic pleura
f plèvre diaphragmatique
e pleura diafragmática
d Pleura diaphragmatica
l pleura diaphragmatica

1285 diaphysis
f diaphyse
e diáfisis
d Diaphyse
l diaphysis

* **diarrhea** → 1286

1286 diarrhoea; diarrhea
f diarrhée
e diarrea
d Diarrhöe; Durchfall

1287 diarthrosis
f diarthrose
e diartrosis
d Diarthrose; echtes Gelenk
l diarthrosis

* **diastema** → 2236

1288 **diastole**
 f diastole
 e diástole
 d Diastole

1289 **diaveridine**
 f diavéridine
 e diaveridina
 d Diaveridin
 l diaveridinum

1290 **diazepam**
 f diazépam
 e diazepam
 d Diazepam
 l diazepamum

1291 **diazinon; dimpylate**
 f diazinon; dimpylate
 e diazinón; dimpilato
 d Diazinon; Dimpylat
 l dimpylatum

1292 **dichlorophen**
 f dichlorphène
 e diclorofeno
 d Dichlorphen
 l dichlorphenum

1293 **dichlorvos; DDVP**
 f dichlorvos; DDVP
 e diclorvos
 d Dichlorvos; DDVP
 l dichlorvosum

1294 **diclazuril**
 f diclazuril
 e diclazurilo
 d Diclazuril
 l diclazurilum

1295 **dicloxacillin**
 f dicloxacilline
 e dicloxacilina
 d Dicloxacilin
 l dicloxacillinum

* **dicophane** → 1216

1296 **dicoumarol**
 f dicoumarol
 e dicumarol
 d Dicoumarol
 l dicoumarolum

1297 **dicrocoeliosis; Dicrocoelium infestation**
 f dicrocœliose
 e dicroceliosis
 d Dikrozöliose

* **Dicrocoelium infestation** → 1297

1298 **dictyocaulosis; Dictyocaulus infestation**
 f dictyocaulose
 e dicticaulosis
 d Diktyokaulose; Befall mit Lungenwürmer

* **Dictyocaulus infestation** → 1298

1299 **dieldrin**
 f dieldrine
 e dieldrina
 d Dieldrin
 l dieldrinum

1300 **diencephalon**
 f diencéphale
 e diencéfalo
 d Dienzephalon; Zwischenhirn
 l diencephalon

1301 **dienestrol; dienoestrol**
 f diénestrol
 e dienestrol
 d Dienestrol
 l dienestrolum

* **dienoestrol** → 1301

1302 **diethylcarbamazine**
 f diéthylcarbamazine
 e dietilcarbamazina
 d Diethylcarbamazin
 l diethylcarbamazina

1303 **diethylstilbestrol; stilboestrol**
 f diéthylstilbestrol; diéthylstilbœstrol
 e dietilestilbestrol
 d Diethylstilbestrol
 l diethylstilbestrolum

1304 **differential diagnosis**
 f diagnostic différentiel
 e diagnóstico diferencial
 d Differentialdiagnose

1305 **diffuse placenta**
 f placenta diffus
 e placenta difusa
 d Placenta diffusa
 l placenta diffusa

1306 **digastric muscle**
 f muscle digastrique
 e músculo digástrico
 d zweibäuchiger Muskel
 l musculus digastricus

1307 digestive disorder
  f trouble digestif
  e trastorno digestivo
  d Verdauungsstörung

1308 digestive system
  f appareil digestif
  e aparato digestivo; sistema digestivo
  d Verdauungsorgane
  l apparatus digestorius; systema digestorium

1309 digestive system disease
  f maladie de l'appareil digestif
  e enfermedad del aparato digestivo
  d Krankheit des Verdauungssystems;
     Gastroenteropathie

1310 digital arteries
  f artères digitales
  e arterias digitales
  d Arteriae digitales; Zehenarterien
  l arteriae digitales

1311 digital bones of fore limb
  f os des doigts
  e huesos de los dedos de la mano
  d Vorderzehenknochen
  l ossa digitorum manus

1312 digital bones of hind limb
  f os des orteils
  e huesos de los dedos del pie
  d Hinterzehenknochen
  l ossa digitorum pedis

1313 digital cushion; toe pad
  f coussinet digital
  e almohadilla digital
  d Zehenballen
  l torus digitalis

1314 digital cushion of hoof
  f coussinet digital de l'onglon
  e almohadilla ungular
  d Hufballen
  l torus ungulae

1315 digital synovial sheaths of forelimb
  f gaines synoviales des tendons de la main
  e vainas sinoviales de los tendinos de los
     dedos de la mano
  d Fesselbeugesehnenscheiden; distale
     Beugesehnenscheiden
  l vaginae synoviales tendinum digitorum
     manus

1316 digital veins
  f veines digitales
  e venas digitales
  d Venae digitales
  l venae digitales

* digits of fore limb → 1703

* digits of hind limb → 4588

1317 dihydrostreptomycin
  f dihydrostreptomycine
  e dihidroestreptomicina
  d Dihydrostreptomycin
  l dihydrostreptomycinum

1318 dilator muscle
  f muscle dilatateur
  e músculo dilatador
  d erweiternder Muskel
  l musculus dilatator

1319 dilator muscle of pupil
  f muscle dilatateur de la pupille
  e músculo dilatador de la pupila
  d radiärer Erweiterer der Pupille
  l musculus dilatator pupillae

1320 dilepididosis; infestation with Dilepididae
  f dilépididose
  e dilepididosis
  d Dilepididose

1321 dimercaprol
  f dimercaprol
  e dimercaprol
  d Dimercaprol
  l dimercaprolum

1322 dimetridazole
  f dimétridazole
  e dimetridazol
  d Dimetridazol
  l dimetridazolum

1323 diminazene
  f diminazène
  e diminazeno
  d Diminazen; Berenil
  l diminazenum

* dimpylate → 1291

1324 dinitolmide
  f dinitolmide
  e dinitolmida
  d Dinitolmid
  l dinitolmidum

1325 dinoprost
  f dinoprost
  e dinoprost
  d Dinoprost
  l dinoprostum

* Dioctophyme renale infestation → 1326

1326 dioctophymosis; Dioctophyme renale
infestation
f dioctophymose
e dioctofimosis
d Befall mit Dioctophyme renale

* Dipetalonema infestation → 1327

1327 dipetalonemosis; Dipetalonema infestation
f dipétalonémose
e dipetalonemosis
d Dipetalonemose

1328 diphyllobothriosis; Diphyllobothrium
infestation
f diphyllobothriose
e difilobotriosis
d Diphyllobothriose

* Diphyllobothrium infestation → 1328

1329 diploic veins
f veines diploïques
e venas diploicas
d Venae diploicae
l venae diploicae

1330 dirofilariosis; heartworm disease
f dirofilariose cardio-vasculaire
e dirofilariosis
d Dirofilariose; Herzwurmbefall

1331 disease; illness
f maladie
e enfermedad
d Krankheit; Erkrankung

1332 disease control
f contrôle des maladies
e control de las enfermedades
d Tierseuchenbekämpfung

1333 disease prevalence
f prévalence d'une maladie
e prevalencia de una enfermedad
d Prävalenz einer Krankheit;
Durchseuchung

1334 disease resistance; resistance to disease
f résistance aux maladies
e resistencia a la enfermedad
d Infektionsresistenz; Krankheitsabwehr

1335 disease transmission
f transmission d'une maladie
e transmisión de una enfermedad
d Krankheitsübertragung

* disease vector → 4831

1336 disinfectant
f désinfectant
e desinfectante
d Desinfektionsmittel

1337 disinfection
f désinfection
e desinfección
d Desinfektion

* disk of snout → 3921

1338 disorder
f désordre; trouble
e trastorno
d Störung

* displacement of the heart → 1428

1339 distal interphalangeal articulations; coffin
joint
f articulations interphalangiennes distales;
articulation du pied
e articulaciones interfalangianas distales
d dritte Zehengelenke; Hufgelenk
l articulationes interphalangeae distales

1340 distal phalanx; third phalanx; coffin bone
f phalange distale
e falange distal; hueso ungular; hueso
unguicular
d Hufbein; Klauenbein; Krallenbein
l phalanx distalis; os ungulare

1341 distal sesamoid bone; navicular bone
f os sésamoïde distal; os petit sésamoïde
e hueso sesamoideo distal
d Strahlbein
l os sesamoideum distale

* distemper → 639

1342 distemper morbillivirus; canine distemper
virus
f virus de la maladie de Carré
e virus del moquillo; virus de la enfermedad
de Carré
d Staupe-Virus

1343 dithiazanine iodide
f iodure de dithiazanine
e ioduro de ditiazanina
d Dithiazanin-Jodid
l dithiazanini iodidum

1344 diverticulum
f diverticule
e divertículo
d Divertikel
l diverticulum

* **diverticulum of auditory tube** → 1914

1345 docking; tail amputation
f écourtage; amputation de la queue
e truncamiento de la cola
d Kaudektomie; Schwanzkupieren

1346 dog disease
f maladie des chiens
e enfermedad de los perros
d Hundekrankheit

1347 dog flea
f puce du chien
e pulga del perro
d Hundefloh
l Ctenocephalides canis

1348 dog hydatid tapeworm
f Taenia hydatigena
e tenia hidatigena; tenia marginada
d Taenia hydatigena
l Taenia hydatigena

1349 dog louse
f pou du chien
e piojo del perro
d Hundelaus
l Linognathus setosus

1350 doping (*of racing animals*)
f dopage
e narcotización
d Dopen; Doping

1351 dorsal arteries of nose
f artères dorsales du nez
e arterias dorsales de la nariz
d Arteriae dorsales nasi;
  Nasenrückenarterien
l arteriae dorsales nasi

* **dorsal bone** → 3029

1352 dorsal branch
f rameau dorsal
e ramo dorsal
d dorsaler Ast
l ramus dorsalis

1353 dorsal cerebellar veins
f veines cérébelleuses dorsales
e venas dorsales del cerebelo

d Venae cerebelli dorsales; dorsale
  Kleinhirnvenen
l venae cerebelli dorsales

1354 dorsal colon
f côlon dorsal
e colon dorsal
d dorsales Kolon
l colon dorsale

1355 dorsal column of spinal cord
f cordon dorsal de la moelle épinière
e cordón dorsal del médula espinal
d Dorsalstrang des Rückenmarks
l funiculus dorsalis

1356 dorsal costo-abdominal artery
f artère costo-abdominale dorsale
e arteria costoabdominal dorsal
d Arteria costoabdominalis dorsalis
l arteria costoabdominalis dorsalis

1357 dorsal costo-abdominal vein
f veine costo-abdominale dorsale
e vena costoabdominal dorsal
d Vena costoabdominalis dorsalis
l vena costoabdominalis dorsalis

1358 dorsal digital nerves
f nerfs digitaux dorsaux
e nervios digitales dorsales
d Nervi digitales dorsales
l nervi digitales dorsales

1359 dorsal external ophthalmic vein
f veine ophtalmique externe dorsale
e vena oftálmica externa dorsal
d Vena ophthalmica externa dorsalis
l vena ophthalmica externa dorsalis

1360 dorsal horn of spinal cord
f corne dorsale de la moelle épinière
e asta dorsal del médula espinal
d Dorsalhorn der grauen Substanz
l cornu dorsale

1361 dorsal ligaments
f ligaments dorsaux
e ligamentos dorsales
d Ligamenta dorsalia
l ligamenta dorsalia

1362 dorsal metatarsal nerves
f nerfs métatarsiens dorsaux
e nervios metatarsianos dorsales
d Nervi metatarsei dorsales
l nervi metatarsei dorsales

1363 dorsal nasal concha; dorsal turbinate
 f cornet nasal dorsal
 e concha nasal dorsal
 d dorsale Nasenmuschel
 l concha nasalis dorsalis

1364 dorsal nerve of penis
 f nerf dorsal du pénis
 e nervio dorsal del pene
 d Nervus dorsalis penis
 l nervus dorsalis penis

1365 dorsal nerve of scapula
 f nerf dorsal de la scapula
 e nervio dorsal de la espalda
 d Nervus dorsalis scapulae
 l nervus dorsalis scapulae

1366 dorsal oblique muscle (*of eyeball*)
 f muscle oblique dorsal (*du bulbe de l'œil*)
 e músculo oblicuo dorsal (*del ojo*)
 d oberer schräger Augenmuskel
 l musculus obliquus dorsalis

1367 dorsal omental recess
 f récessus omental dorsal
 e receso dorsal del omento
 d dorsale Netzbeutelhöhle
 l recessus dorsalis omentalis

1368 dorsal perineal artery
 f artère périnéale dorsale
 e arteria perineal dorsal
 d Arteria perinealis dorsalis
 l arteria perinealis dorsalis

1369 dorsal petrosal sinus
 f sinus pétreux dorsal
 e seno petroso dorsal
 d Sinus petrosus dorsalis
 l sinus petrosus dorsalis

1370 dorsal prominence of tongue (*ruminants*)
 f torus lingual
 e almohadilla lingual
 d Zungenrückenwulst
 l torus linguae

1371 dorsal rectus muscle (*of eyeball*)
 f muscle droit dorsal (*du bulbe de l'oeil*)
 e músculo recto dorsal (*del ojo*)
 d oberer gerader Augenmuskel
 l musculus rectus dorsalis

1372 dorsal region of neck
 f région dorsale du cou
 e región dorsal del cuello
 d dorsale Halsregion
 l regio colli dorsalis

1373 dorsal sac of rumen
 f sac dorsal du rumen
 e saco dorsal del rumen
 d dorsaler Pansensack
 l saccus dorsalis ruminis

1374 dorsal sagittal sinus
 f sinus sagittal dorsal
 e seno sagital dorsal
 d Sinus sagittalis dorsalis
 l sinus sagittalis dorsalis

1375 dorsal scapular artery
 f artère scapulaire dorsale
 e arteria escapular dorsal
 d Arteria scapularis dorsalis
 l arteria scapularis dorsalis

1376 dorsal scapular vein
 f veine scapulaire dorsale
 e vena escapular dorsal
 d Vena scapularis dorsalis
 l vena scapularis dorsalis

1377 dorsal serrated muscle
 f muscle dentelé dorsal
 e músculo serrato dorsal
 d Musculus serratus dorsalis
 l musculus serratus dorsalis

1378 dorsal spinal artery
 f artère spinale dorsale
 e arteria espinal dorsal
 d Arteria spinalis dorsalis
 l arteria spinalis dorsalis

1379 dorsal spinocerebellar tract; Gowers' tract
 f tractus spino-cérébelleux dorsal; faisceau de Gowers
 e tracto espinocerebeloso dorsal
 d Tractus spinocerebellaris dorsalis; Gowers'scher Strang
 l tractus spinocerebellaris dorsalis

1380 dorsal surface of tongue
 f dos de la langue
 e dorso de la lengua
 d Zungenrücken
 l dorsum linguae

1381 dorsal thoracic lymphocentre
 f lymphocentre thoracique dorsal
 e linfocentro torácico dorsal
 d Lymphocentrum thoracicum dorsale
 l lymphocentrum thoracicum dorsale

* **dorsal turbinate** → 1363

1382 dorsal vagal trunk
f tronc vagal dorsal
e tronco vago dorsal
d dorsaler Vagusstamm
l truncus vagalis dorsalis

1383 dorsal veins
f veines dorsales
e venas dorsales
d Dorsalvenen
l venae dorsales

1384 dorsolateral tract
f tractus dorso-latéral
e tracto dorsolateral
d Tractus dorsolateralis
l tractus dorsolateralis

1385 dorsomedial nucleus of hypothalamus
f noyau dorso-médial de l'hypothalamus
e núcleo hipotalámico dorsomedial
d Nucleus hypothalamicus dorsomedialis
l nucleus hypothalamicus dorsomedialis

1386 dosage; dose rate
f posologie
e dosificación
d Dosierung

1387 dose
f dose
e dosis
d Dosis
l dosis

* dose rate → 1386

1388 double-pored dog tapeworm
f dipylidium du chien
e dipilidio canino
d Dipylidium; gurkenkernförmiger Bandwurm
l Dipylidium caninum

1389 dourine; Trypanosoma equiperdum infection
f dourine; infection par Trypanosoma equiperdum
e durina; mal de coito
d Beschälseuche; Dourine
l exanthema coitale paralyticum

* down feathers → 3519

1390 doxycycline
f doxycycline
e doxiciclina
d Doxycyclin
l doxycyclinum

1391 drug residues
f résidus de médicament
e residuos de fármacos
d Arzneimittelrückstand

1392 drug therapy; medicinal treatment
f pharmacothérapie; traitement médical
e terapia por fármacos; tratamiento medicinal
d Pharmakotherapie

1393 drug toxicity
f toxicité des médicaments
e toxicidad de los fármacos
d Arzneimitteltoxizität

1394 duck hepatitis enterovirus
f entérovirus de l'hépatite du canard
e virus de la hepatitis del pato
d Entenhepatitis-Enterovirus

1395 duck plague; duck viral enteritis
f peste du canard; entérite virale du canard
e enteritis vírica del pato
d Virusenteritis der Enten; Entenpest

1396 duck plague herpesvirus
f herpèsvirus de la peste du canard
e virus de la enteritis vírica del pato
d Entenpest-Herpesvirus

* duck viral enteritis → 1395

1397 duck viral hepatitis
f hépatite virale du canard
e hepatitis vírica del pato
d Virushepatitis der Enten; Entenhepatitis

1398 duct
f conduit; canal
e conducto
d Gang
l ductus

1399 ductless glands; endocrine glands
f glandes endocrines
e glándulas sin conductos; glándulas endocrinas
d endokrine Drüsen
l glandulae sine ductibus

1400 duct of epididymis
f conduit épididymaire
e conducto del epidídimo
d Nebenhodengang
l ductus epididymidis

* duct of epoophoron → 1804

* **duct of mandibular salivary gland** → 2667

1401 ductule
f canalicule; ductule
e conductillo
d kleiner Gang
l ductulus

1402 ductules of prostate gland
f canalicules prostatiques
e conductillos prostáticos
d Ductuli prostatici
l ductuli prostatici

1403 duodenal glands
f glandes duodénales (de Brunner)
e glándulas duodenales (de Brunner)
d Duodenaldrüsen; Brunnersche Drüsen
l glandulae duodenales

1404 duodenal papilla
f papille duodénale
e papila del duodeno
d Papilla duodeni
l papilla duodeni

1405 duodenum
f duodénum
e duodeno
d Duodenum; Zwölffingerdarm
l duodenum

1406 dwarfism
f nanisme
e enanismo
d Zwergwuchs

1407 dyschondroplasia
f dyschondroplasie
e discondroplasia
d Dyschondroplasie

1408 dysentery
f dysenterie
e disentería
d Dysenterie; Ruhr

* **dysfunction** → 1779

1409 dysplasia
f dysplasie
e displasia
d Dysplasie

* **dyspnea** → 1410

1410 dyspnoea; dyspnea
f dyspnée
e dispnea
d Dyspnöe

1411 dystocia
f dystocie
e distocia
d Dystokie; Schwergeburt

1412 dystrophy
f dystrophie
e distrofia
d Dystrophie
l dystrophia

E

1413 ear
f oreille
e oído
d Ohr
l auris

1414 ear disease
f maladie de l'appareil auditif
e enfermedad de la oreja
d Ohrenkrankheit

* **ear flap** → 3478

1415 ear lobe
f lobe auriculaire; oreillon
e lóbulo de la oreja; orejilla
d Wangenlappen; Ohrläppchen; Ohrscheibe
l lobus auricularis

* **ear mange** → 3187

1416 ear mite of dogs
f acarien auriculaire du chien
e ácaro auricular del perro
d Ohrräudemilbe
l Otodectes cynotis

1417 ear muscles; auricular muscles
f muscles auriculaires
e músculos auriculares
d Ohrmuskeln
l musculi auriculares

1418 ear region
f région auriculaire
e región auricular
d Ohrgegend
l regio auricularis

1419 ear tick
f tique des oreilles
e garrapata auricular
d Otobius; Ohrenzecke
l Otobius megnini

1420 ear tip
f apex de l'auricule; pointe de l'oreille
e vértice de la oreja; punta de la oreja
d Ohrspitze
l apex auriculae

1421 ear veins; auricular veins
f veines auriculaires
e venas auriculares
d Ohrvenen
l venae auriculares

**1422 East Coast fever; bovine theileriosis;
     Theileria parva infection**
f theilériose bovine à Theileria parva
e teileriosis bovina por Theileria parva
d Ostküstenfieber

* **EBL** → 534

* **EBL virus** → 535

* **Ebner's glands** → 1911

1423 echinococcosis; hydatid disease
f échinococcose; hydatidose
e equinococosis; hidatidosis
d Echinokokkose; Hydatidose

1424 echinococcus; hydatid tapeworm
f ténia échinocoque
e equinococo
d Echinokokk; dreigliedriger
     Hundebandwurm
l Echinococcus granulosus

* **ecternal parasite** → 1426

1425 ectethmoidal bone; lateroethmoidal bone
f os ectethmoïde
e hueso ectetmoides
d Os ectethmoidale
l os ectethmoidale

1426 ectoparasite; ecternal parasite
f ectoparasite; parasite externe
e ectoparásito; parásito externo
d Ektoparasit; Außenparasit

1427 ectoparasitosis
f ectoparasitose
e ectoparasitosis
d Ektoparasitose

1428 ectopia cordis; displacement of the heart
f ectocardie
e ectopia del corazón
d Ektopie des Herzens
l ectopia cordis

1429 ectromelia; mouse pox
f ectromélie
e ectromelia de los ratones
d Ectromelia; Mäusepocken

1430 ectromelia orthopoxvirus
f orthopoxvirus de l'ectromélie
e virus de la ectromelia
d Ectromelia-Virus; Mäusepockenvirus

1431 eczema
f eczéma
e eccema
d Ekzem

* edema → 3074

* edematous → 3075

* EDS → 1433

1432 efferent ducts of testis
f canalicules efférents du testicule
e conductillos eferentes del testículo
d Ausführungsgänge des Hodens
l ductuli efferentes testis

* egg → 3209

1433 egg drop syndrome; EDS
f syndrome chute de ponte
e síndrome de baja de la puesta
d Legeabfall-Syndrom

* Ehrlichia infection → 1434

* Ehrlichia phagocytophilia infection → 4580

1434 ehrlichiosis; Ehrlichia infection
f ehrlichiose; infection à Ehrlichia
e erliquiosis; fiebre por garrapatas; infección por Ehrlichia spp.
d Ehrlichiose; Ehrlichia-Infektion

* EHV-1 → 1522

* EHV-4 → 1523

* Eimeria infection → 1435

1435 eimeriosis; Eimeria infection
f eimériose
e eimeriosis
d Eimeriose

1436 ejaculatory duct
f conduit éjaculateur; canal éjaculateur
e conducto eyaculador
d Ductus ejaculatorius
l ductus ejaculatorius

1437 elastic cartilage
f cartilage élastique
e cartílago elástico
d elastische Knorpel
l cartilago elastica

1438 elastic tissue
f tissu élastique
e tejido elástico
d elastisches Gewebe

1439 elbow
f coude
e cúbito
d Ellbogen
l cubitus

* elbow hygroma → 651

1440 elbow joint; cubital articulation
f articulation du coude
e articulación del codo; articulación cubital
d Ellbogengelenk
l articulatio cubiti

1441 elbow region; cubital region
f région du coude
e región del codo
d Ellbogenregion
l regio cubiti

* ELISA → 1485

1442 ellipsoidal articulation; condylarthrosis
f condylarthrose
e articulación elipsoidal
d Ellipsoidgelenk
l articulatio ellipsoidea

1443 elliptical recess of vestibule
f récessus elliptique
e receso elíptico del vestíbulo
d elliptische Mulde des Vorhofs
l recessus ellipticus vestibuli

1444 emaciation; cachexia; wasting
f émaciation; amaigrissement
e demacración
d Abmagerung; Kachexie

1445 emasculator
f pince à castrer
e emasculador
d Kastrierzange

1446 embryo
f embryon
e embrión
d Embryo

1447 embryology
  f embryologie
  e embriología
  d Embryologie

1448 embryonic
  f embryonnaire
  e embrionario
  d embryonal

1449 embryonic mortality
  f mortalité embryonnaire
  e mortalidad embrionaria
  d Embryonenverlust; embryonaler Fruchttod

1450 embryo transfer
  f transfert d'embryons
  e transferencia de embriones
  d Embryo-Transfer

  * emergency slaughter → 705

1451 emissary veins
  f veines émissaires
  e venas emisarias
  d Venae emissariae
  l venae emissariae

1452 enamel of tooth
  f émail de la dent
  e esmalte
  d Zahnschmelz
  l enamalum

  * enarthrosis → 403

  * encapsulated nerve endings → 4493

1453 encephalic dura mater
  f dure-mère encéphalique
  e duramadre del encéfalo
  d harte Hirnhaut
  l dura mater encephali

1454 encephalitis
  f encéphalite
  e encefalitis
  d Enzephalitis; Gehirnentzündung

1455 encephalitozoonosis
  f encéphalitozoonose
  e infección por Encephalitozoon
  d Encephalitozoon-Infektion

1456 encephalomalacia
  f encéphalomalacie
  e encefalomalacia
  d Enzephalomalazie

1457 encephalomyelitis
  f encéphalomyélite
  e encefalomielitis
  d Enzephalomyelitis; Gehirn-Rückenmarkentzündung

  * encephalomyocarditis virus → 678

1458 encephalon; brain
  f encéphale
  e encéfalo
  d Enzephalon; Gehirn
  l encephalon

  * encephalopathy → 566

  * endbrain → 4465

1459 endemic
  f endémique
  e endémico
  d endemisch

1460 endemic
  f endémie
  e endemia
  d Endemie

1461 endocarditis
  f endocardite
  e endocarditis
  d Endokarditis

1462 endocardium
  f endocarde
  e endocardio
  d Endokard
  l endocardium

  * endocrine glands → 1399

1463 endocrine system
  f système endocrinien
  e sistema endocrino
  d endokrines System

1464 endolymph
  f endolymphe
  e endolinfa
  d Endolymphe
  l endolympha

1465 endolymphatic duct
  f conduit endolymphatique
  e conducto endolinfático
  d Ductus endolymphaticus
  l ductus endolymphaticus

1466 endolymphatic sac
 f sac endolymphatique
 e saco endolinfático
 d Saccus endolymphaticus
 l saccus endolymphaticus

1467 endometritis
 f endométrite
 e endometritis
 d Endometritis

1468 endometrium; uterine mucosa
 f endomètre
 e endometrio
 d Endometrium; Schleimhaut der
   Gebärmutter
 l endometrium

1469 endoparasite; internal parasite
 f endoparasite; parasite interne
 e endoparásito; parásito interno
 d Endoparasit; Innenparasit

1470 endoscopy
 f endoscopie
 e endoscopía
 d Endoskopie

1471 endothelium
 f endothélium
 e endotelio
 d Endothel
 l endothelium

1472 endotoxin
 f endotoxine
 e endotoxina
 d Endotoxin

 * **enlargement of heart → 675**

 * **enlargement of spleen → 4173**

1473 enrofloxacin
 f enrofloxacine
 e enrofloxacino
 d Enrofloxacin; Baytril
 l enrofloxacinum

1474 enteric plexus
 f plexus entérique
 e plexo entérico
 d Eingeweidegeflecht
 l plexus entericus

1475 enteric redmouth disease; Yersinia ruckeri infection
 f infection à Yersinia ruckeri; yersiniose des salmonidés

 e infección por Yersinia ruckeri
 d Rotmaulseuche der Forelle; Yersinia-ruckeri-Infektion

1476 enteritis
 f entérite
 e enteritis
 d Enteritis; Darmentzündung

1477 enterotoxaemia
 f entérotoxémie
 e enterotoxemia
 d Enterotoxämie

1478 enterotoxin
 f entérotoxine
 e enterotoxina
 d Enterotoxin

1479 enterovirus
 f entérovirus
 e enterovirus
 d Enterovirus

1480 entoglossal bone; paraglossal bone
 f os entoglosse; os paraglosse
 e hueso endogloso
 d Os entoglossum
 l os entoglossum; os paraglossum

1481 entropion
 f entropion
 e entropio
 d Entropium; Lideinstülpung

1482 enzootic
 f enzootique
 e enzoótico
 d enzootisch

1483 enzootic
 f enzootie
 e enzootia
 d Enzootie

1484 enzootic abortion of sheep; chlamydial abortion of ewes
 f avortement enzootique du mouton; chlamydiose ovine
 e aborto enzoótico de la oveja; aborto clamidial
 d Chlamydienabort des Schafes

 * **enzootic ataxia of lambs → 4380**

 * **enzootic bovine leukosis → 534**

1485 enzyme immunoassay; enzyme-linked
     immunosorbent assay; ELISA
  f dosage immuno-enzymatique
  e dosificación inmuno-enzimática
  d Enzymimmunoassay; ELISA

  * enzyme-linked immunosorbent assay →
    1485

1486 eosinophil granulocyte
  f granulocyte éosinophile
  e granulocito eosinófilo
  d eosinophiler Granulozyt

1487 eosinophilia
  f éosinophilie
  e eosinofilia
  d Eosinophilie

1488 ependyma
  f épendyme
  e epéndimo
  d Ependym
  l ependyma

  * Eperythrozoon infection → 1489

1489 eperythrozoonosis; Eperythrozoon
     infection
  f épérythrozoonose; infection à
    Eperythrozoon
  e eperitrozoonosis; infección por
    Eperythrozoon spp.
  d Eperythrozoonose; Eperythrozoon-
    Infektion

1490 ephemeral fever of cattle
  f fièvre éphémère
  e fiebre efémera
  d Ephemeralfieber

  * epicardium → 4932

1491 epicondyle
  f épicondyle
  e epicóndilo
  d Epikondylus
  l epicondylus

1492 epidemic
  f épidémique
  e epidémico
  d epidemisch

1493 epidemic
  f épidémie
  e epidemia
  d Epidemie

  * epidemic tremor → 348

1494 epidemiology
  f épidémiologie
  e epidemiología
  d Epidemiologie

1495 epidermis
  f épiderme
  e epidermis
  d Epidermis; Oberhaut
  l epidermis

  * epidermis of limbus → 3403

1496 epididymis
  f épididyme
  e epidídimo
  d Nebenhoden
  l epididymis

1497 epididymitis
  f épididymite
  e epididimitis
  d Epididymitis; Nebenhodenentzündung

1498 epidural cavity; epidural space
  f cavum épidural; espace épidural
  e cavidad epidural
  d Epiduralraum
  l cavum epidurale

  * epidural space → 1498

1499 epigastric arteries
  f artères épigastriques
  e arterias epigástricas
  d Oberbaucharterien
  l arteriae epigastricae

  * epigastrium → 2826

  * epiglottal depression → 1500

1500 epiglottal furrow; epiglottal depression
  f vallécule épiglottique
  e valécula epiglótica
  d Vallecula epiglottica
  l vallecula epiglottica

1501 epiglottis
  f épiglotte
  e epiglotis
  d Kehldeckel
  l epiglottis

1502 epinephrine; adrenaline
  f épinéphrine; adrénaline
  e epinefrina; adrenalina
  d Epinephrin; Adrenalin
  l epinephrinum

1503 **epiphyseal cartilage**
  f cartilage épiphysaire
  e cartílago epifisario
  d Epiphysen(fugen)knorpel
  l cartilago epiphysialis

1504 **epiphyseal line**
  f ligne épiphysaire
  e línea epifisaria
  d Epiphysenlinie
  l linea epiphysialis

1505 **epiphysis**
  f épiphyse
  e epífisis
  d Epiphyse
  l epiphysis

1506 **epiploic foramen; foramen of Winslow**
  f foramen épiploïque; hiatus de Winslow
  e agujero epiploico; hiato de Winslow
  d Netzbeutelloch
  l foramen epiploicum

1507 **epiploon; greater omentum**
  f épiploon; grand omentum
  e epiplón; omento mayor
  d Epiploon; großes Netz
  l epiploon; omentum majus

1508 **epistaxis; nosebleed**
  f épistaxis; saignement du nez
  e epistaxis
  d Epistaxis; Nasenbluten

1509 **epithalamus**
  f épithalamus
  e epitálamo
  d Epithalamus
  l epithalamus

1510 **epithelium**
  f épithélium
  e epitelio
  d Epithel

1511 **epitympanic recess**
  f récessus épitympanique
  e receso epitimpánico
  d Recessus epitympanicus
  l recessus epitympanicus

1512 **epizootic**
  f épizootique
  e epizoótico
  d epizootisch

1513 **epizootic**
  f épizootie
  e epizootia
  d Epizootie

1514 **epizootic lymphangitis; Histoplasma farciminosum infection**
  f lymphangite épizootique; infection à Histoplasma farciminosum
  e linfangitis epizoótica; infección por Histoplasma farciminosum
  d seuchenhafte Lymphangitis der Einhufer; Pseudorotz
  l lymphangitis epizootica

1515 **epoophoron**
  f époophoron; organe de Rosenmüller
  e epoóforo; cuerpo de Rosenmüller
  d Epoophoron
  l epoophoron

* **equine abortion virus** → 1522

1516 **equine arteritis virus**
  f virus de l'artérite équine
  e virus de la arteritis vírica equina
  d equines Arteriitis-Virus

* **equine azoturia** → 3285

1517 **equine babesiosis; piroplasmosis of horses**
  f babésiose équine
  e babesiosis equina
  d Babesiose des Pferdes

1518 **equine coital exanthema**
  f exanthème coïtal équin
  e exantema genital equino
  d equines Koitalexanthem
  l exanthema coitale equorum

1519 **equine encephalomyelitis**
  f encéphalomyélite équine
  e encefalomielitis equina
  d Pferdeenzephalomyelitis

1520 **equine encephalomyelitis virus**
  f virus de l'encéphalomyélite équine
  e virus de la encefalomielitis equina
  d equines Enzephalomyelitis-Virus

1521 **equine herpesvirus**
  f herpèsvirus équin
  e herpesvirus equino
  d equines Herpesvirus

1522 **equine herpesvirus type 1; EHV-1; equine abortion virus**
  f herpèsvirus equin type 1

e herpesvirus equino tipo 1
d equines Herpesvirus Typ 1; EHV-1;
  equines Abortvirus

1523 equine herpesvirus type 4; EHV-4; equine
     rhinopneumonitis virus
f herpèsvirus equin type 4
e herpesvirus equino tipo 4; virus de la
  rinoneumonitis equina
d equines Herpesvirus Typ 4; equines
  Rhinopneumonitis-Virus; EHV-4

1524 equine infectious anaemia
f anémie infectieuse équine
e anemia infecciosa equina
d infektiöse Anämie der Einhufer
l anaemia infectiosa equorum

1525 equine infectious anaemia virus
f virus de l'anémie infectieuse équine
e virus de la anemia infecciosa equina
d equines infektiöses Anämie-Virus

1526 equine influenza
f influenza équin; grippe équine
e influenza equina
d Pferdeinfluenza

1527 equine influenzavirus
f influenzavirus équin
e virus de la influenza equina
d equines Influenzavirus

* equine rhinopneumonitis virus → 1523

1528 equine viral abortion
f avortement viral de la jument
e aborto vírico de la yegua
d Virusabort der Stuten

1529 equine viral arteritis
f artérite virale du cheval
e arteritis vírica del caballo
d virale Arteriitis des Pferdes

1530 equine viral rhinopneumonitis
f rhinopneumonie virale du cheval
e rinoneumonia vírica del caballo
d infektiöse Rhinopneumonitis des Pferdes

1531 eradication
f éradication
e erradicación
d Eradikation; Tilgung; Ausrottung

1532 erector muscle of spine
f muscle erector spinae
e músculo erector de la espina
d Musculus erector spinae
l musculus erector spinae

* erector muscles of hairs → 268

1533 ergocalciferol; vitamin D2
f ergocalciférol; vitamine D2
e ergocalciferol; vitamina D2
d Ergocalciferol; Vitamin D2
l ergocalciferolum

1534 ergometrine
f ergométrine
e ergometrina
d Ergometrin
l ergometrinum

1535 ergot
f ergot; éperon
e espolón del metacarpo
d Sporn
l calcar metacarpeum; calcar metatarseum

1536 ergot alkaloid; Claviceps purpurea toxin
f alcaloïde de l'ergot de seigle
e alcaloide ergotamina; toxina de Claviceps
  purpurea
d Mutterkornalkaloid

1537 ergotism; Claviceps purpurea intoxication
f ergotisme; intoxication à Claviceps
  purpurea
e ergotismo
d Ergotismus; Mutterkornvergiftung

1538 erysipelas bacterium
f bactérie du rouget du porc
e bacteria del mal rojo porcino
d Rotlaufbakterium
l Erysipelothrix rhusiopathiae

* Erysipelothrix rhusiopathiae infection →
  4384

1539 erythroblastosis
f érythroblastose
e eritroblastosis
d Erythroblastose

1540 erythrocyte; red blood cell
f érythrocyte; globule rouge; hématie
e eritrocito; glóbulo rojo
d Erythrozyt; rotes Blutkörperchen

1541 erythrocyte count
f nombre d'hématies; numération globulaire
e recuento de los glóbulos rojos
d Erythrozytenzahl

1542 erythrocytosis
f érythrocytose
e eritrocitosis
d Erythrozytose

1543 **erythromycin**
  f érythromycine
  e eritromicina
  d Erythromycin
  l erythromycinum

1544 **erythropoiesis**
  f érythropoïèse
  e eritropoyesis
  d Erythropoiese; Erythrozytenbildung

* **Escherichia coli infection** → 977

* **esophageal glands** → 3076

* **esophageal groove** → 1808

* **esophageal hiatus** → 3077

* **esophageal impression** → 3078

* **esophageal obstruction** → 3079

* **esophageal veins** → 3080

* **esophagus** → 3082

* **estradiol** → 3083

* **estrogen** → 3085

* **estrone** → 3086

* **estrus synchronization** → 3087

1545 **ethmoidal arteries**
  f artères ethmoïdales
  e arterias etmoidales
  d Arteriae ethmoidales; Siebbeinarterien
  l arteriae ethmoidales

1546 **ethmoidal bone**
  f os ethmoïde
  e hueso etmoides
  d Siebbein
  l os ethmoidale

1547 **ethmoidal cells**
  f cellules ethmoïdales osseuses
  e celdillas etmoidales
  d Siebbeinzellen
  l cellulae ethmoidales osseae

1548 **ethmoidal crest**
  f crête ethmoïdale
  e cresta etmoidal
  d Crista ethmoidalis
  l crista ethmoidalis

1549 **ethmoidal foramen**
  f foramen ethmoïdal
  e agujero etmoidal
  d Foramen ethmoidale; Siebbeinloch
  l foramen ethmoidale

1550 **ethmoidal fossa**
  f fosse ethmoïdale
  e fosa etmoidal
  d Fossa ethmoidalis
  l fossa ethmoidalis

1551 **ethmoidal labyrinth**
  f labyrinthe ethmoïdal
  e laberinto etmoidal
  d Siebbeinlabyrinth
  l labyrinthus ethmoidalis

1552 **ethmoidal nerve**
  f nerf ethmoïdal
  e nervio etmoidal
  d Nervus ethmoidalis
  l nervus ethmoidalis

1553 **ethmomandibular muscle**
  f muscle ethmo-mandibulaire
  e músculo etmomandibular
  d Musculus ethmomandibularis
  l musculus ethmomandibularis

1554 **ethmoturbinals; turbinate bones**
  f ethmoturbinaux; volutés ethmoïdales
  e etmoturbinados
  d Muschelbeine; Siebbeinmuscheln
  l ethmoturbinalia

* **etiology** → 87

1555 **etisazole**
  f étisazole
  e etisazol
  d Etisazol
  l etisazolum

1556 **etomidate**
  f étomidate
  e etomidato
  d Etomidat
  l etimodatum

1557 **European foulbrood; Melissococcus pluton infection**
  f loque européenne; loque bénigne; infection à Melissococcus pluton
  e loque europea; infección por Melissococcus pluton
  d gutartige Faulbrut; Melissococcus-pluton-Infektion

* eustachian tube → 330

1558 **euthanasia**
  f euthanasie
  e eutanasia
  d Euthanasie

1559 **exanthema**
  f exanthème
  e exantema
  d Exanthem

* excessive growth → 1835

1560 **excreta**
  f excreta
  e excreta
  d Exkret

1561 **excretory duct**
  f conduit excréteur
  e conducto excretor
  d Ausführungsgang
  l ductus excretorius

1562 **excretory ductules of lacrimal glands**
  f canalicules excréteurs de la glande
    lacrymale
  e conductillos excretores de las glándulas
    lagrimales
  d Ausführungsgänge der Tränendrüsen
  l ductuli excretorii glandulae lacrimales

1563 **exertional rhabdomyolysis; tying-up
  syndrome of racehorses**
  f rhabdomyolyse d'effort
  e rabdomiolisis de ejercicio
  d Steifmuskelkrankheit der Rennpferde

1564 **exoccipital bone**
  f os exoccipital
  e hueso exoccipital
  d Os exoccipitale
  l os exoccipitale

1565 **exocrine gland**
  f glande exocrine; glande à sécretion externe
  e glándula exocrina
  d exokrine Drüse

1566 **exophthalmos**
  f exophtalmie
  e exoftalmía
  d Exophthalmus

1567 **exostosis**
  f exostose
  e exostosis
  d Exostose; Knochenauswuchs

1568 **experimental infection**
  f infection expérimentale
  e infección experimental
  d experimentelle Infektion;
    Infektionsversuch

1569 **extensor muscle of first digit**
  f muscle extenseur du doigt I
  e músculo extensor del dedo 1°
  d Musculus extensor digiti I
  l musculus extensor digiti I

1570 **extensor muscle of second digit**
  f muscle extenseur du doigt II
  e músculo extensor del dedo 2°
  d Musculus extensor digiti II
  l musculus extensor digiti II

1571 **external acoustic meatus**
  f méat acoustique externe; conduit auditif
    externe
  e meato acústico externo
  d äußerer Gehörgang
  l meatus acusticus externus

1572 **external anal sphincter**
  f sphincter externe de l'anus
  e músculo esfínter externo del ano
  d äußerer Afterschließmuskel
  l musculus sphincter ani externus

1573 **external carotid artery**
  f artère carotide externe
  e arteria carótida externa
  d Arteria carotis externa
  l arteria carotis externa

1574 **external carotid nerves**
  f nerfs carotidiens externes
  e nervios carotídeos externos
  d Nervi carotici externi
  l nervi carotici externi

1575 **external coat of vessels**
  f tunique externe des vaisseaux
  e túnica externa
  d Außenschicht der Blutgefäßwand
  l tunica externa vasorum

1576 **external ear**
  f oreille externe
  e oído externo
  d äußeres Ohr
  l auris externa

1577 **external ethmoidal vein**
  f veine ethmoïdale externe
  e vena etmoidal externa
  d Vena ethmoidalis externa
  l vena ethmoidalis externa

1578 **external genitalia**
 f organes génitaux externes
 e partes genitales externas
 d äußere Geschlechtsorgane
 l partes genitales externae

1579 **external iliac artery**
 f artère iliaque externe
 e arteria ilíaca externa
 d Arteria iliaca externa
 l arteria iliaca externa

1580 **external iliac vein**
 f veine iliaque externe
 e vena ilíaca externa
 d Vena iliaca externa
 l vena iliaca externa

1581 **external iliofemoral muscle**
 f muscle ilio-fémoral externe
 e músculo iliofemoral externo
 d Musculus iliofemoralis externus
 l musculus iliofemoralis externus

1582 **external jugular vein**
 f veine jugulaire externe
 e vena yugular externa
 d Vena jugularis externa
 l vena jugularis externa

1583 **external layer of hoof wall**
 f couche externe de la paroi cornée
 e estrato externo de la pared
 d Glasurschicht des Hufes
 l stratum externum parietis cornei

1584 **external mandibular adductor muscle**
 f muscle adducteur externe de la mandibule
 e músculo aductor externo de la mandíbula
 d Musculus adductor mandibulae externus
 l musculus adductor mandibulae externus

1585 **external nose**
 f partie extérieure du nez
 e nariz externa
 d äußere Nase
 l nasus externus

1586 **external oblique muscle of abdomen**
 f muscle oblique externe de l'abdomen
 e músculo oblicuo externo del abdomen
 d äußerer schiefer Bauchmuskel
 l musculus obliquus externus abdominis

1587 **external obturator muscle**
 f muscle obturateur externe
 e músculo obturador externo
 d äußerer Verstopfungsmuskel
 l musculus obturatorius externus

1588 **external occipital protuberance**
 f protubérance occipitale externe
 e protuberancia occipital externa
 d Protuberantia occipitalis externa
 l protuberantia occipitalis externa

* **external opening of cloaca** → 4842

1589 **external opening of uterus**
 f orifice externe de l'utérus
 e orificio externo del útero
 d äußerer Muttermund
 l ostium uteri externum

1590 **external ophthalmic artery**
 f artère ophtalmique externe
 e arteria oftálmica externa
 d Arteria ophthalmica externa
 l arteria ophthalmica externa

1591 **external orifice of urethra**
 f ostium externe de l'urètre
 e orificio externo de la uretra
 d äußere Harnröhrenmündung
 l ostium urethrae externum

1592 **external pudendal artery**
 f artère honteuse externe
 e arteria pudenda externa
 d Arteria pudenda externa
 l arteria pudenda externa

1593 **external sagittal crest**
 f crête sagittale externe
 e cresta sagital externa
 d Scheitelkamm
 l crista sagittalis externa

1594 **external spiral sulcus**
 f sillon spiral externe
 e surco espiral externo
 d Sulcus spiralis externus
 l sulcus spiralis externus

1595 **external thoracic artery**
 f artère thoracique externe
 e arteria torácica externa
 d Arteria thoracica externa
 l arteria thoracica externa

1596 **external thoracic vein**
 f veine thoracique externe
 e vena torácica externa
 d äußere Brustkorbvene
 l vena thoracica externa

1597 **external vane**
 f vexille externe
 e estandarte externo
 d Außenfahne; äußere Federfahne
 l vexillum externum

1598 extracellular fluid
  f liquide extracellulaire
  e fluido extracelular
  d extrazelluläre Flüssigkeit

1599 extraocular muscles
  f muscles du bulbe de l'œil
  e músculos del globo
  d Augenmuskeln
  l musculi bulbi

1600 extremity
  f extrémité
  e extremidad
  d Extremität
  l extremitas

1601 exudative diathesis
  f diathèse exsudative
  e diatesis exudativa
  d exsudative Diathese

1602 eye
  f œil
  e ojo
  d Auge
  l oculus

1603 eyeball; bulb of eye
  f bulbe oculaire; globe oculaire

  e globo del ojo
  d Augapfel
  l bulbus oculi

1604 eye disease
  f maladie de l'œil
  e enfermedad del ojo
  d Augenkrankheit

1605 eyelash
  f cil palpébral
  e cilio; pestaña
  d Augenwimper
  l cilium

1606 eyelid
  f paupière
  e párpado
  d Augenlid
  l palpebra

  * **eyelid muscle → 4447**

1607 eyeworm
  f filaire lacrymale
  e filaria ocular
  d Augenwurm
  l Thelazia

  * **eyeworm infestation → 4513**

F

1608 face
f face
e cara
d Gesicht
l facies

1609 face fly
f mouche d'automne
e mosca autumnal
d Augenfliege
l Musca autumnalis

1610 facial artery
f artère faciale
e arteria facial
d Arteria facialis; Gesichtsarterie
l arteria facialis

1611 facial bones
f os de la face
e huesos de la cara
d Gesichtsknochen
l ossa faciei

1612 facial crest
f crête faciale
e cresta facial
d Gesichtsleiste
l crista facialis

1613 facial eczema; sporidesmin intoxication
f eczéma facial du mouton; intoxication par la sporidesmine
e eccema facial de la oveja; intoxicación por esporidesmina
d Fazialekzem des Schafes; Sporidesmin-Vergiftung

1614 facial nerve
f nerf facial
e nervio facial
d Nervus facialis; Gesichtsnerv
l nervus facialis

1615 facial tuber
f tuber facial; tubercule facial
e tuberosidad facial
d Tuber faciale
l tuber faciale

1616 facial vein
f veine faciale
e vena facial
d Vena facialis; Gesichtsvene
l vena facialis

1617 faeces; feces
f fèces
e heces
d Fäzes; Kot

1618 falciform ligament; broad ligament of liver
f ligament falciforme du foie
e ligamento falciforme del hígado
d sichelförmiges Leberband
l ligamentum falciforme hepatis

* **fallopian tube → 3204**

* **false nostril → 2937**

* **false pregnancy → 3661**

* **false rib → 309**

1619 falx of cerebrum
f faux du cerveau
e hoz cerebral
d Großhirnsichel
l falx cerebri

1620 fascia
f fascia
e fascia
d Faszie; Muskelbinde
l fascia

* **fascia of forearm → 197**

* **fascia of neck → 832**

1621 fasciculi proprii
f faisceaux propres
e fascículos propios
d Fasciculi proprii
l fasciculi proprii

1622 fasciculus cuneatus; Burdach's column
f faisceau cunéiforme (de Burdach)
e fascículo cuneado; columna de Burdach
d Fasciculus cuneatus; Burdachscher Strang
l fasciculus cuneatus

1623 fasciculus gracilis; Goll's column
f faisceau gracile (de Goll)
e fascículo gracil; cordón de Goll
d Fasciculus gracilis; Gollscher Strang
l fasciculus gracilis

* **Fasciola infestation → 1625**

* fascioliasis → 1625

* Fascioloides infestation → 1624

1624 fascioloidosis; Fascioloides infestation
f fascioloïdose
e fascioloidosis
d Faszioloidose

1625 fasciolosis; fascioliasis; liver fluke
infestation; Fasciola infestation
f fasciolose; distomatose
e fasciolosis; distomatosis hepática
d Fasziolose; Leberegelkrankheit
l hepatitis distomatosa

1626 fastigial nucleus
f noyau fastigial; noyau du toit
e núcleo fastigial; núcleo de la cima
d Dachkern
l nucleus fastigii

1627 fatal outcome
f issue fatale
e resultado fatal
d tödlicher Ausgang
l exitus letalis

1628 fatty degeneration; steatosis
f dégénérescence graisseuse
e esteatosis
d fettige Degeneration; Steatose

1629 fatty liver; hepatic steatosis
f stéatose hépatique
e hígado graso; esteatosis hepática
d Fettleber; Lebersteatose

1630 fauces; throat
f gosier
e fauces
d Fauces
l fauces

1631 favus; Trichophyton gallinae infection
f favus; infection à Trichophyton gallinae
e tiña; infección por Trichophyton gallinae
d Favus; Kammgrind; Trichophyton-
gallinae-Infektion

1632 feather
f plume
e pluma
d Feder
l penna

1633 feather follicle
f follicule plumaire

e folículo de pluma
d Federfollikel; Federbalg
l folliculus pennae

1634 feather ligaments
f ligaments des pennes
e ligamentos de las plumas
d Federbänder
l ligamenta pennarum

1635 feather muscles
f muscles des plumes
e músculos de las plumas
d Federmuskeln
l musculi pennales

* feather of horse → 1674

1636 feather shaft
f tige de la plume
e escapo de la pluma; tallo de la pluma
d Federkiel
l scapus plumae

1637 feather tracts
f ptérylies
e pterilias
d Federfluren
l pterylae

1638 febantel
f fébantel
e febantel
d Febantel
l febantelum

1639 febrile; feverish
f fébrile
e febril
d febril; fieberhaft

* feces → 1617

1640 feline calicivirus
f calicivirus félin
e calicivirus felino
d felines Calicivirus

1641 feline dysautonomia; Key-Gaskell
syndrome
f dysautonomie féline; syndrome de Key-
Gaskell
e disautonomía felina; síndrome de Key-
Gaskell
d feline Dysautonomie; Key-Gaskell-
Syndrom

1642 feline herpesvirus; feline rhinotracheitis virus
f herpèsvirus félin; virus de la rhinotrachéite féline
e virus de la rinotraqueitis del gato; herpesvirus felino
d felines Herpesvirus; felines Rhinotracheitis-Virus

1643 feline immunodeficiency virus; FIV
f virus de l'immunodéficience féline
e virus de la inmunodepresión felina
d felines Immundefizienz-Virus

1644 feline infectious anaemia; Haemobartonella felis infection
f anémie infectieuse du chat; infection à Haemobartonella felis
e anemia infecciosa felina; infección por Haemobartonella felis
d feline infektiöse Anämie; Haemobartonella-felis-Infektion

1645 feline infectious peritonitis; FIP
f péritonite virale du chat
e peritonitis vírica del gato
d feline infektiöse Peritonitis; FIP

1646 feline leukaemia
f leucémie féline
e leucemia felina
d Katzenleukämie

1647 feline leukaemia virus; FeLV
f virus de la leucémie féline
e virus de leucose felina
d Katzenleukämie-Virus

1648 feline panleukopenia
f panleucopénie féline
e panleucopenia felina
d Katzenpanleukopenie; Katzenseuche

1649 feline panleukopenia parvovirus
f parvovirus de la panleucopénie féline
e virus de la panleucopenia felina
d felines Panleukopenie-Virus

1650 feline rhinotracheitis
f rhinotrachéite féline
e rinotraqueitis felina
d Katzenrhinotracheitis; Katzenschnupfen

* **feline rhinotracheitis virus** → 1642

1651 feline sarcoma virus
f virus du sarcome félin
e virus del sarcoma felino
d felines Sarkom-Virus

1652 feline urological syndrome; FUS
f syndrome urologique félin
e síndrome urólogico del gato
d Urolithiasis-Syndrom der Katze

* **FeLV** → 1647

1653 female genital disease
f maladie de l'appareil génital femelle
e enfermedad de los órganos genitales femeninos
d Gynäkopathie

1654 female genital organs
f organes génitaux femelles
e órganos genitales femininos
d weibliche Geschlechtsorgane
l organa genitalia feminina

1655 femoral artery
f artère fémorale
e arteria femoral
d Arteria femoralis; Oberschenkelarterie
l arteria femoralis

1656 femoral canal
f canal fémoral
e canal femoral
d Schenkelkanal
l canalis femoralis

1657 femoral nerve
f nerf fémoral
e nervio femoral
d Oberschenkelnerv
l nervus femoralis

1658 femoral plexus
f plexus fémoral
e plexo femoral
d Nervengeflecht des Oberschenkels
l plexus femoralis

1659 femoral ring
f anneau fémoral
e anillo femoral
d Schenkelring
l anulus femoralis

1660 femoral vein
f veine fémorale
e vena femoral
d Oberschenkelvene
l vena femoralis

1661 femoropatellar articulation
f articulation fémoro-patellaire
e articulación femororrotuliana
d Kniescheibengelenk
l articulatio femoropatellaris

1662 femoropatellar ligaments
 f ligaments fémoro-patellaires
 e ligamentos femororrotulianos
 d Ligamenta femoropatellare
 l ligamenta femoropatellare

1663 femorotibial articulation
 f articulation fémoro-tibiale
 e articulación femorotibial
 d Kniekehlgelenk
 l articulatio femorotibialis

1664 femorotibial muscles
 f muscles fémoro-tibiaux
 e músculos femorotibiales
 d Musculi femorotibiales
 l musculi femorotibiales

1665 femur; thighbone
 f fémur; os femoris
 e hueso femur
 d Oberschenkelbein
 l os femoris

1666 fenbendazole
 f fenbendazole
 e fenbendazol
 d Fenbendazol; Panacur
 l fenbendazolum

 * **fenchlorphos** → 1667

1667 fenclofos; fenchlorphos; ronnel
 f fenclophos
 e fenclofos
 d Fenclofos
 l fenclofosum

 * **fenestra of cochlea** → 3930

1668 fenprostalene
 f fenprostalène
 e fenprostaleno
 d Fenprostalen
 l fenprostalenum

1669 fentanyl
 f fentanyl
 e fentanilo
 d Fentanyl
 l fentanylum

1670 fetal death
 f mort du fœtus
 e muerte fetal
 d Fruchttod

1671 fetal membranes
 f membranes fœtales
 e membranas fetales
 d Eihäute; Fruchthüllen
 l membranae fetales

 * **fetal mummification** → 2878

1672 fetal placenta
 f placenta fœtal
 e porción fetal; placenta fetal
 d fetaler Anteil der Plazenta
 l pars fetalis placentae

 * **fetlock** → 2805, 3657

1673 fetlock joint; metacarpophalangeal articulations
 f articulation du boulet; articulations métacarpo-phalangiennes
 e articulaciones metacarpo-falangianas
 d Articulationes metacarpophalangeae; Fesselgelenk
 l articulationes metacarpophalangeae

 * **fetlock joint** → 2804

 * **fetlock of ruminants** → 2791

1674 fetlock tuft; feather of horse
 f fanon métacarpien; fanon métatarsien
 e cerda metacarpiana; cerda metatarsiana
 d Kötenschopf
 l cirrus metacarpeus; cirrus metatarseus

1675 fetus
 f fœtus
 e feto
 d Fötus; Frucht

1676 fever
 f fièvre; état fébril
 e fiebre
 d Fieber
 l febris

 * **feverish** → 1639

 * **FGA** → 1725

 * **fiber** → 1677

1677 fibre; fiber
 f fibre
 e fibra
 d Faser
 l fibra

1678 fibril
f fibrille
e fibrilla
d Fibrille
l fibrilla

1679 fibrin
f fibrine
e fibrina
d Fibrin

1680 fibrinogen
f fibrinogène
e fibrinogeno
d Fibrinogen

1681 fibroblast
f fibroblaste
e fibroblasto
d Fibroblast

1682 fibrocartilage
f fibrocartilage; cartilage fibreux
e fibrocartílago
d Faserknorpel
l fibrocartilago

* **fibrocartilaginous junction → 4399**

1683 fibrocartilaginous ring of tympanic membrane
f anneau fibro-cartilagineux
e anillo fibrocartilaginoso de la membrana del tímpano
d Faserring um das Trommelfell
l anulus fibrocartilagineus membranae tympani

1684 fibroma
f fibrome
e fibroma
d Fibrom

1685 fibrous capsule
f capsule fibreuse
e cápsula fibrosa
d fibröse Kapsel
l capsula fibrosa

* **fibrous joint → 4401**

1686 fibrous pericardium
f péricarde fibreux
e pericardio fibroso
d Fibrosa des Perikards
l pericardium fibrosum

1687 fibrous rings of heart
f anneaux fibreux du cœur
e anillos fibrosos del corazón
d Atrioventrikular-Faserringe
l anuli fibrosi cordis

1688 fibrous tunic
f tunique fibreuse
e túnica fibrosa
d Faserhaut
l tunica fibrosa

1689 fibrous tunic of eyeball
f tunique fibreuse du bulbe de l'œil
e túnica fibrosa del globo
d äußere Augenhaut
l tunica fibrosa bulbi

1690 fibula
f fibula; péroné
e peroné
d Fibula; Wadenbein
l fibula

1691 fibular incisure of tibia
f incisure fibulaire
e escotadura peronea
d Wadenbeinausschnitt
l incisura fibularis tibiae

1692 fibular tarsal bone; heel bone; calcaneum
f calcaneum
e calcáneo; hueso del talón
d Fersenbein
l calcaneus; os calcis

1693 filament
f filament
e filamento
d Filamentum
l filamentum

1694 filaria
f filaire
e filaria
d Filarie
l Filaroidea

* **filariasis → 1695**

1695 filariidosis; filariasis
f filariose
e filariosis
d Filariose

1696 filiform papillae of tongue
f papilles filiformes
e papilas filiformes
d fadenförmige Zungenpapillen
l papillae filiformes

1697 filoplume
 f filoplume
 e filopluma
 d Fadenfeder
 l filopluma

1698 fimbria
 f frange
 e fimbria; franje
 d Fimbrie; Franse
 l fimbria

1699 fimbriae of uterine tube
 f franges de la trompe utérine
 e fimbrias de la trompa uterina
 d Fimbriae tubae uterinae
 l fimbriae tubae uterinae

1700 fimbriated fold of tongue
 f pli frangé
 e pliegue frangeado
 d Plica fimbriata
 l plica fimbriata

1701 final host; definitive host
 f hôte définitif
 e huésped definitivo
 d Endwirt

1702 fin and tail rot; Cytophaga psychrophila infection
 f infection à Cytophaga psychrophila
 e infección por Cytophaga psychrophila
 d Cytophaga-psychrophila-Infektion

* **finger bone → 3431**

1703 fingers; digits of fore limb
 f doigts (de la main)
 e dedos de la mano
 d Finger; Vorderzehen
 l digiti manus

* **FIP → 1645**

1704 first carpal bone
 f os carpal I; os trapèze
 e hueso carpiano 1°; hueso trapecio
 d Os carpale I
 l os carpale I; os trapezium

* **first cervical vertebra → 317**

* **first digit of fore limb → 4551**

1705 first phalanx; proximal phalanx; large pastern bone
 f phalange proximale

 e primera falange; hueso compedal
 d Fesselbein
 l phalanx proximalis; os compedale

* **first premolar → 4979**

1706 first tarsal bone; tarsal bone I
 f os tarsal I; os cunéiforme médial
 e hueso tarsiano 1°; hueso cuneiforme medial
 d erster Tarsalknochen
 l os tarsale I; os cuneiforme mediale

1707 fish disease
 f maladie des poissons
 e enfermedad de los peces
 d Fischkrankheit

1708 fish rhabdovirus
 f rhabdovirus du poisson
 e rhabdovirus de peces
 d Rhabdovirus de Fische

1709 fissure
 f fissure; scissure
 e cisura
 d Fissur; Spalte
 l fissura

1710 fistula
 f fistule
 e fístula
 d Fistel
 l fistula

* **fistula of lateral cartilage of hoof → 3737**

* **FIV → 1643**

1711 flank
 f flanc
 e lado; flanco
 d Flanke; Weiche
 l latus

1712 flank fold; fold of flank
 f corde du flanc
 e pliegue del flanco
 d Flankenfalte; Kniefalte
 l plica lateris

* **flank region → 2438**

1713 flat bone
 f os plat
 e hueso plano
 d platter Knochen
 l os planum

1714 flavivirus
 f flavivirus
 e flavivirus
 d Flavivirus

1715 flea
 f puce; siphonaptère
 e pulga
 d Floh
 l Siphonaptera

1716 flea infestation; siphonapterosis
 f pulicose
 e pulicosis
 d Flohbefall

 * **Flechsig's tract → 4862**

1717 flesh fly
 f sarcophagide
 e moscarda
 d Fleischfliege
 l Sarcophaga

1718 fleshy crest; comb (*of fowl*)
 f crête charnue
 e cresta carnosa
 d Hahnenkamm; Kamm der Hühner
 l crista carnosa

1719 flexor muscles of lower leg (*of birds*)
 f muscles fléchisseurs de la jámbe
 e músculos flexores lateral y medial de la pierna
 d Musculus flexor cruris lateralis/medialis
 l musculus flexor cruris lateralis/medialis

1720 flexure
 f inflexion
 e flexura
 d Flexur; Biegung
 l flexura

1721 flight feather of tail; tail feather
 f rectrice
 e rectríz
 d Steuerfeder
 l rectrix

1722 flight feather of wing; wing feather
 f rémige
 e remera; pluma de vuelo
 d Schwungfeder
 l remex

1723 flocculonodular lobe
 f lobe flocculo-nodulaire
 e lóbulo floculonodular
 d Lobus flocculonodularis
 l lobus flocculonodularis

1724 fludrocortisone
 f fludrocortisone
 e fludrocortisona
 d Fludrocortison
 l fludrocortisonum

1725 flugestone; fluorogestone acetate; FGA
 f flugestone
 e flugestona
 d Flugeston
 l flugestonum

 * **fluke → 4651**

1726 flumetasone; flumethasone
 f flumétasone
 e flumetasona
 d Flumetason
 l flumetasonum

 * **flumethasone → 1726**

 * **fluorescent antibody test → 2144**

 * **fluorine poisoning → 1727**

 * **fluorogestone acetate → 1725**

1727 fluorosis; fluorine poisoning
 f fluorose
 e fluorosis; intoxicación por fluor
 d Fluorose; Fluorvergiftung

1728 fly
 f mouche
 e mosca
 d Fliege
 l Diptera

 * **fly strike → 2908**

 * **FMD → 1740**

 * **foamy virus → 4186**

1729 focus of infection
 f foyer d'infection
 e foco de infección
 d Infektionsherd

1730 fold
 f pli; repli
 e pliegue
 d Falte
 l plica

 * **fold of flank → 1712**

1731 folds of reticulum
f crêtes réticulaires
e crestas del retículo
d Cristae reticuli; Netzmagenleisten
l cristae reticuli

1732 foliate papillae of tongue
f papilles filiformes
e papilas foliadas
d blattförmige Papillen
l papillae foliatae

1733 follicle
f follicule
e folículo
d Follikel
l folliculus

1734 follicle-stimulating hormone; follitropin; FSH
f hormone folliculo-stimulante
e hormona estimulante folicular
d Follikelhormon; follikelstimulierendes Hormon; FSH

1735 follicular cyst
f kyste folliculaire
e quiste folicular
d Follikelzyste

* **follitropin** → 1734

1736 food hygiene
f hygiène alimentaire
e higiene de los alimentos
d Lebensmittelhygiene

1737 food inspection
f inspection des denrées alimentaires
e inspección de los alimentos
d Lebensmittelüberwachung

1738 food poisoning
f intoxication alimentaire
e intoxicación alimentaria
d Futtervergiftung; Lebensmittelvergiftung

1739 foot; hind foot
f pied
e pie
d Fuß; Hinterfuß
l pes

1740 foot and mouth disease; FMD
f fièvre aphteuse
e fiebre aftosa
d Maul- und Klauenseuche; MKS
l aphthae epizooticae

* **foot covering** → 3527

1741 foot louse of sheep
f pou du pied du mouton
e piojo del pie de la oveja
d Linognathus pedalis
l Linognathus pedalis

* **foot rot** → 2177

1742 foramen
f foramen; trou
e agujero
d Foramen; Loch

* **foramen of transverse process** → 4638

* **foramen of Winslow** → 1506

1743 forearm
f avant-bras
e antebrazo
d Unterarm
l antebrachium

1744 forearm region
f région de l'avant-bras
e región del antebrazo
d Unterarmgegend
l regio antebrachii

* **forebrain** → 3637

1745 forehead
f front
e frente
d Stirn
l frons

1746 foreign body
f corps étranger
e cuerpo extraño
d Fremdkörper
l corpus alienum

* **foreign substance** → 4984

* **foreleg** → 4538

* **forelock** → 1747

* **forestomach** → 3654

1747 foretop; forelock
f toupet
e tupé; copete; cerda de la cabeza
d Stirnschopf
l cirrus capitis

1748 formaldehyde; formalin; formol
f formaldéhyde; formol
e formaldehído
d Formaldehyd; Formalin

* **formalin** → 1748

* **formol** → 1748

1749 fornix
f fornix; trigone
e fórnix
d Fornix; Gewölbe
l fornix

1750 fossa
f fosse; fossette
e fosa
d Fossa; Vertiefung
l fossa

* **fossa of atlas wing** → 314

1751 fourth carpal bone; hamate bone
f os carpal IV; os hamatum
e hueso carpiano 4°; hueso unciforme
d Os carpale IV
l os carpale IV; os hamatum

1752 fourth tarsal bone; tarsal bone IV
f os tarsal IV; os cuboïde
e hueso tarsiano 4°; hueso cuboideo
d vierter Tarsalknochen; Würfelbein
l os tarsale IV; os cuboideum

1753 fourth ventricle
f quatrième ventricule
e cuarto ventrículo
d vierte Hirnkammer
l ventriculus quartus

1754 fowl cholera; avian pasteurellosis
f choléra aviaire; pasteurellose aviaire
e colera aviar; pasteurelosis aviar
d Hühnercholera; Geflügelpasteurellose

1755 fowl plague virus
f virus de la peste aviaire
e virus de la peste aviar
d klassisches Geflügelpest-Virus

1756 fowl pox
f variole aviaire
e viruela aviar
d Geflügelpocken
l variola avium

1757 fowl tick; poultry tick
f tique de la poule
e garrapata de las gallinas
d Geflügelzecke; Taubenzecke
l Argas persicus

1758 fowl typhoid; Salmonella gallinarum infection
f salmonellose des volailles; infection à Salmonella gallinarum
e tifosis aviar; salmonelosis de las gallinas
d Salmonellose des Geflügels; Salmonella-gallinarum-Infektion

1759 fracture
f fracture
e fractura
d Fraktur; Knochenbruch

1760 fracture fixation; osteosynthesis
f ostéosynthèse
e osteosíntesis
d Frakturbehandlung; Osteosynthese

1761 framycetin
f framycétine
e framicetina
d Framycetin
l framycetinum

* **Francisella tularensis infection** → 4703

1762 free bodies; joint mice
f corps libres; arthrolithes
e cuerpos libros
d freie Körper; Gelenkmäuse
l corpora libera

1763 freemartinism
f free-martinisme
e free-martinismo
d Zwickenbildung

1764 frenulum
f frein
e frenillo
d Frenulum
l frenulum

1765 fringed tapeworm
f ténia des ovins
e tenia frangeada
d Thysanosoma actinioides
l Thysanosoma actinioides

1766 frog of hoof
f fourchette
e cuña de la úngula; ranilla
d Hufstrahl
l cuneus ungulae

* frog stay → 4146

1767 frontal bone
f os frontal
e hueso frontal
d Stirnbein
l os frontale

1768 frontal lobe
f lobe frontal
e lóbulo frontal
d Stirnlappen
l lobus frontalis

1769 frontal muscle
f muscle frontal
e músculo frontal
d Frontalmuskel
l musculus frontalis

1770 frontal nerve
f nerf frontal
e nervio frontal
d Nervus frontalis
l nervus frontalis

1771 frontal process (*of turkey*)
f processus frontal (*du dindon*)
e proceso frontal
d Stirnzapfen (*des Truthuhns*)
l processus frontalis

1772 frontal process of maxilla
f processus frontal du maxillaire
e apófisis frontal
d Stirnfortsatz des Oberkiefers
l processus frontalis maxillae

1773 frontal region
f région frontale
e región frontal
d Stirnregion
l regio frontalis

1774 frontal sinus
f sinus frontal
e seno frontal
d Stirnhöhle
l sinus frontalis

1775 frontal squama
f écaille frontale
e escama frontal
d Stirnbeinschuppe
l squama frontalis

1776 frontal tuber
f tubérosité frontale
e tuberosidad frontal
d Stirnhöcker
l tuber frontale

1777 frontomaxillary aperture
f aperture fronto-maxillaire
e abertura frontomaxilar
d Stirnbein-Oberkiefer-Öffnung
l apertura frontomaxillaris

1778 frontoscutular muscle
f muscle fronto-scutulaire
e músculo frontoescutular
d Musculus frontoscutularis
l musculus frontoscutularis

* frusemide → 1790

* FSH → 1734

1779 functional disorder; dysfunction
f trouble fonctionnel; dysfonctionnement
e trastorno funcional; disfunción
d Funktionsstörung; Dysfunktion

1780 fundus
f fond
e fondo
d Fundus
l fundus

1781 fundus glands; gastric glands
f glandes fundiques; glandes gastriques du fundus
e glándulas gástricas
d Fundusdrüsen
l glandulae gastricae propriae

1782 fundus of abomasum
f fond de l'abomasum
e fondo del abomoso
d Fundus abomasi
l fundus abomasi

1783 fundus of reticulum
f fond du réseau
e fondo del retículo
d Fundus reticuli
l fundus reticuli

1784 fundus of stomach
f fundus; fond de l'estomac
e fondo del estómago
d Magengrund; Magenfundus
l fundus ventriculi

1785 fundus of uterus
f fond de l'utérus
e fondo del útero
d Fundus uteri
l fundus uteri

* fungal disease → 2900

1786 **fungiform papillae of tongue**
 f papilles fongiformes
 e papilas fungiformes
 d pilzförmige Zungenpapillen
 l papillae fungiformes

1787 **fungus**
 f champignon
 e hongo
 d Pilz

1788 **furaltadone**
 f furaltadone
 e furaltadona
 d Furaltadon
 l furaltadonum

1789 **furazolidone**
 f furazolidone
 e furazolidona
 d Furazolidon
 l furazolidonum

1790 **furosemide; frusemide**
 f furosémide
 e furosemida
 d Furosemid
 l furosemidum

1791 **furunculosis**
 f furonculose
 e furunculosis
 d Furunkulose

 * **furunculosis of Salmonidae** → 85

 * **FUS** → 1652

1792 **fusariotoxicosis; Fusarium mycotoxicosis**
 f mycotoxicose à Fusarium
 e fusariotoxicosis; micotoxicosis por Fusarium spp.
 d Fusarium-Mykotoxikose

 * **Fusarium mycotoxicosis** → 1792

 * **Fusarium roseum toxin** → 4952, 4995

 * **Fusarium T2 toxin** → 4425

1793 **fusiform muscle; spindle-shaped muscle**
 f muscle fusiforme
 e músculo fusiforme
 d spindelförmiger Muskel
 l musculus fusiformis

 * **Fusobacterium necrophorum infection** → 2965

G

1794 gallbladder
 f vésicule biliaire
 e vesícula biliar
 d Gallenblase
 l vesica fellea

* **gallbladder artery** → 1206

* **gallid herpesvirus type 1** → 2684

1795 gallstone; biliary calculus; cholelith
 f calcul biliaire
 e cálculo biliar; colelito
 d Gallenstein; Cholelith
 l calculus biliare

* **gamma benzene hexachloride** → 2520

1796 ganglia of sympathetic trunk
 f ganglions du tronc sympathique
 e ganglios del tronco simpático
 d Ganglia trunci sympathici
 l ganglia trunci sympathici

1797 ganglion
 f ganglion nerveux
 e ganglio
 d Ganglion; Nervenknoten
 l ganglion

1798 ganglionic layer of cerebellum
 f stratum ganglionnaire cérébelleux
 e estrato ganglionar del cerebelo
 d Purkinje-Zellschicht
 l stratum gangliosum cerebelli

1799 ganglion of trigeminal nerve; semilunar ganglion; Gasser's ganglion
 f ganglion trigéminal (de Gasser)
 e ganglio trigeminal (de Gasser)
 d Ganglion trigeminale; halbmondförmiges Ganglion
 l ganglion trigeminale

1800 gangliosidosis
 f gangliosidose
 e gangliosidosis
 d Gangliosidose

1801 gangrene
 f gangrène
 e gangrena
 d Gangrän; Gewebsnekrose

1802 gangrenous mastitis
 f mammite gangréneuse
 e mastitis gangrenosa
 d Mastitis gangraenosa

1803 gapeworm of poultry
 f syngame des volailles
 e singamo traqueal; gusano traqueal
 d Syngamus; Luftröhrenwurm
 l Syngamus trachea

1804 Gärtner's duct; duct of epoophoron
 f canal de Gärtner
 e conducto epoóforo longitudinal
 d Gärtnerscher Gang
 l ductus epoophori longitudinalis

1805 gas gangrene; Clostridium perfringens type A infection
 f gangrène gazeuse; infection à Clostridium perfringens type A
 e gangrena gascosa; infccción por Clostridium perfringens tipo A
 d Gasbrand; Gasgangrän; Clostridium-perfringens-Typ-A-Infektion

* **Gasser's ganglion** → 1799

1806 gastric
 f gastrique
 e gástrico
 d Magen-

1807 gastric diverticulum (*of pig*)
 f diverticule gastrique (*du porc*)
 e divertículo del estómago (*del cerdo*)
 d Diverticulum ventriculi
 l diverticulum ventriculi

* **gastric glands** → 1781

1808 gastric groove; oesophageal groove; esophageal groove
 f sillon gastrique; gouttière œsophagienne
 e surco del estómago
 d Magenrinne; Schlundrinne
 l sulcus ventriculi

1809 gastric impression on liver
 f empreinte gastrique du foie
 e impresión gástrica
 d Mageneindruck
 l impressio gastrica

1810 **gastric pits**
 f cryptes muqueuses de l'estomac
 e foveolas gástrica
 d Magengrübchen
 l foveolae gastricae

1811 **gastric plexuses**
 f plexus gastriques
 e plexos gástricos
 d Plexus gastrici
 l plexus gastrici

 * **gastric ulcer** → 4227

1812 **gastric veins**
 f veines gastriques
 e venas gástricas
 d Magenvenen
 l venae gastricae

1813 **gastritis**
 f gastrite
 e gastritis
 d Gastritis; Magenentzündung

1814 **gastrocnemius muscle**
 f muscle gastrocnémien
 e músculo gastrocnemio
 d Gastroknemius; Wadenmuskel
 l musculus gastrocnemius

1815 **gastroduodenal artery**
 f artère gastro-duodénale
 e arteria gastroduodenal
 d Arteria gastroduodenalis
 l arteria gastroduodenalis

1816 **gastroduodenal vein**
 f veine gastro-duodénale
 e vena gastroduodenal
 d Vena gastroduodenalis
 l vena gastroduodenalis

1817 **gastroenteritis**
 f gastro-entérite
 e gastroenteritis
 d Gastroenteritis; Magendarmentzündung

1818 **gastro-intestinal**
 f gastro-intestinal
 e gastrointestinal
 d Magen-Darm-

1819 **gemellus muscles**
 f muscle jumeaux
 e músculos gemelos
 d Zwillingsmuskeln
 l musculi gemelli

1820 **genicular arteries**
 f artères du genou
 e arterias de la rodilla
 d Kniearterien
 l arteriae genus

1821 **genicular veins**
 f veines du genou
 e venas de la rodilla
 d Knievenen
 l venae genus

1822 **geniculate body**
 f corps géniculé
 e cuerpo geniculado
 d Corpus geniculatum
 l corpus geniculatum

1823 **geniculate ganglion**
 f ganglion géniculé
 e ganglio geniculado
 d Ganglion geniculi
 l ganglion geniculi

1824 **genioglossal muscle**
 f muscle génio-glosse
 e músculo geniogloso
 d Kinn-Zungen-Muskel
 l musculus genioglossus

1825 **geniohyoid muscle**
 f muscle génio-hyoïdien
 e músculo genihioideo
 d Musculus geniohydoideus
 l musculus geniohydoideus

1826 **genital system; genital tract**
 f appareil génital
 e tracto genital
 d Genitaltrakt; Geschlechtsapparat

1827 **genital system disease**
 f maladie de l'appareil génital
 e enfermedad de los órganos genitales
 d Genitalkrankheit

 * **genital tract** → 1826

1828 **genitofemoral nerve**
 f nerf génito-fémoral
 e nervio genitofemoral
 d Scham-Schenkel-Nerv
 l nervus genitofemoralis

1829 **genome**
 f génome
 e genoma
 d Genom

1830 **gentamicin; gentamycin**
f gentamicine
e gentamicina
d Gentamicin
l gentamicinum

* **gentamycin** → 1830

1831 **germ-free; gnotobiotic**
f axénique; gnotobiotique
e libre de gérmenes
d keimfrei; gnotobiotisch

1832 **germinal disk**
f disque germinatif
e disco germinal
d Keimscheibe
l discus germinalis

\ 1833 **germinal epithelium**
f épithélium germinal
e epitelio germinal
d Keimepithel
l epithelium germinativum

* **germinative layer** → 412

* **gestation** → 3596

* **Giardia infection** → 1834

1834 **giardiosis; Giardia infection**
f giardiose
e giardiosis
d Giardiose; Giardia-Infektion

* **gid of sheep** → 975

* **gid worm** → 976

1835 **gigantism; excessive growth**
f gigantisme
e gigantismo
d Riesenwuchs

* **Gigantobilharzia infestation** → 1836

1836 **gigantobilharziosis; Gigantobilharzia infestation**
f gigantobilharziose
e gigantobilharziosis
d Gigantobilharziose

1837 **gingivitis**
f gingivite
e gingivitis
d Gingivitis; Zahnfleischentzündung

* **ginglymus** → 2008

1838 **gizzard; muscular part of stomach**
f gésier; partie musculaire de l'estomac
e molleja
d Muskelmagen
l ventriculus; pars muscularis ventriculi

1839 **gizzard erosion**
f érosion du gésier
e erosión de la molleja
d Muskelmagenerosion

1840 **gland**
f glande
e glándula
d Glandula; Drüse
l glandula

1841 **glanders; Pseudomonas mallei infection**
f morve; infection à Pseudomonas mallei
e muermo; infección por Pseudomonas mallei
d Rotz; Pseudomonas-mallei-Infektion
l malleus

1842 **glands of renal pelvis**
f glandes du bassinet
e glándulas de la pelvis renal
d Nierenbeckendrüsen
l glandulae pelvis renalis

1843 **glands of the cardia** (*of stomach*)
f glandes cardiales
e glándulas cardiales
d Kardiadrüsen
l glandulae cardiacae

* **glands of the mouth** → 3139

1844 **glands of urinary bladder**
f glandes de la vessie urinaire
e glándulas vesicales
d Harnblasedrüsen
l glandulae vesicales

1845 **glandular stomach; proventriculus**
f proventricule; partie glandulaire de l'estomac
e proventrículo
d Drüsenmagen
l pars glandularis ventriculi

1846 **glans penis**
f gland du pénis
e glande del pene
d Eichel
l glans penis

* Glässer's disease → 3560

1847 **glaucoma**
f glaucome
e glaucoma
d Glaukom

1848 **glenohumeral ligaments**
f ligaments gléno-huméraux
e ligamentos glenohumerales
d Ligamenta glenohumeralia
l ligamenta glenohumeralia

1849 **glenoid cavity**
f cavité glénoïde
e cavidad glenoidal
d Gelenkpfanne
l cavitas glenoidalis

* glenoid fossa → 2669

1850 **glenoid lip of shoulder joint**
f bourrelet glénoïdal
e labro glenoidal
d Labrum glenoidale
l labrum glenoidale

1851 **glenoid notch of scapula**
f incisure glénoïdale
e escotadura glenoidal
d Incisura glenoidalis
l incisura glenoidalis

* gliding joint → 3494

1852 **globulin**
f globuline
e globulina
d Globulin

1853 **glomerular capsule; Bowman's capsule**
f capsule du glomérule
e cápsula del glomérulo
d Bowmansche Kapsel
l capsula glomeruli

1854 **glomerulonephritis**
f glomérulonéphrite
e glomerulonefritis
d Glomerulonephritis

1855 **glomerulus**
f glomérule
e glomérulo
d Gefäßknäuel
l glomerulus

1856 **glossitis**
f glossite
e glositis
d Glossitis; Zungenentzündung

1857 **glosso-epiglottal folds**
f plis glosso-épiglottiques
e pliegues glosoepiglóticos
d Plicae glossoepiglotticae
l plicae glossoepiglotticae

1858 **glossopharyngeal nerve**
f nerf glosso-pharyngien
e nervio glosofaríngeo
d Nervus glossopharyngeus
l nervus glossopharyngeus

1859 **glottal cleft**
f fente de la glotte
e hendidura de la glotis
d Stimmritze
l rima glottidis

1860 **glottis**
f glotte
e glotis
d Glottis
l glottis

1861 **gloxazone**
f gloxazone
e gloxazona
d Gloxazon
l gloxazonum

1862 **glucagon**
f glucagon
e glucagón
d Glucagon
l glucagonum

1863 **glucocorticoid**
f glucocorticoïde
e glucocorticoide
d Glukokortikoid

1864 **gluteal region**
f région glutéale; région fessière
e región glútea
d Hinterbackengegend
l regio glut(a)ea

1865 **gluteobiceps muscle**
f muscle glutéobiceps
e músculo gluteobíceps
d Musculus glutaeobiceps
l musculus glut(a)eobiceps

1866 glycoprotein
  f glycoprotéine
  e glicoproteína
  d Glykoprotein

1867 glycosuria
  f glycosurie
  e glucosuria
  d Glukosurie

1868 gnat; midge
  f moucheron
  e cínife
  d Gnitze
  l Culicoides

  * **Gnathostoma infestation** → 1869

1869 gnathostomosis; Gnathostoma infestation
  f gnathostomose
  e gnatostomosis
  d Gnathostomose

  * **gnotobiotic** → 1831

  * **GnRH** → 1877

1870 goat disease
  f maladie des caprins
  e enfermedad de las cabras
  d Ziegenkrankheit

1871 goat louse
  f pou de la chèvre
  e piojo de la cabra
  d Ziegenlaus
  l Linognathus stenopsis

1872 goat pox
  f variole caprine
  e viruela caprina
  d Ziegenpocken

  * **goiter** → 1873

1873 goitre; goiter
  f goitre
  e bocio
  d Kropf; Struma
  l struma

1874 goitrogenic
  f goitrogène
  e bociogénico
  d kropferzeugend

  * **Goll's column** → 1623

1875 gomphosis
  f gomphose
  e gonfosis
  d Gomphose
  l gomphosis

1876 gonad
  f gonade
  e gonada
  d Gonade; Keimdrüse

1877 gonadorelin; GnRH
  f gonadoréline
  e gonadorelina
  d Gonadorelin; GnRH
  l gonadorelinum

1878 gonadotrophin; gonadotropin
  f gonadotrophine; gonadotropine
  e gonadotrofina; gonadotropina
  d Gonadotrophin; Gonadotropin

1879 gonadotropic hormone
  f hormone gonadotrope
  e hormona gonadotrópica
  d gonadotropes Hormon

  * **gonadotropin** → 1878

  * **Gongylonema infestation** → 1880

1880 gongylonemosis; Gongylonema infestation
  f gongylonémose
  e gongilonemosis
  d Gongylonemose; Befall mit Gongylonema

  * **Gowers' tract** → 1379

1881 graafian follicles; vesicular ovarian follicles
  f follicules ovariques vésiculaires; vésicules de De Graaf
  e folículos ováricos vesiculosos; folículos de De Graaf
  d Tertiärfollikeln; Graaf-Follikeln
  l folliculi ovarici vesiculosi

1882 gracilis muscle
  f muscle gracile
  e músculo grácilis
  d schlanker Schenkelmuskel
  l musculus gracilis

1883 Grahamella infection
  f grahamellose; infection à Grahamella
  e grahamelosis; infección por Grahamella spp.
  d Grahamellose; Grahamella-Infektion

1884 Gram-negative bacterium
f bactérie Gram-négative
e bacteria Gram-negativa
d gramnegative Bakterie

1885 Gram-positive bacterium
f bactérie Gram-positive
e bacteria Gram-positiva
d grampositive Bakterie

* **Grandry's corpuscle → 440**

1886 granula iridica; corpora nigra
f granulations iriennes; grains de suie
e gránulos del iris
d Traubenkörner
l granula iridica; corpora nigra

1887 granular layer of cerebellum
f stratum granulaire cérébelleux
e estrato granuloso del cerebelo
d Körnerschicht der Kleinhirnrinde
l stratum granulosum cerebelli

1888 granular layer of epidermis
f stratum granulaire de l'épiderme
e estrato granuloso
d Stratum granulosum; Körnerschicht
l stratum granulosum epidermidis

1889 granulocyte
f granulocyte
e granulocito
d Granulozyt

1890 granuloma
f granulome
e granuloma
d Granulom

1891 grass sickness of horses
f maladie de l'herbe du cheval
e enfermedad de la hierba de los caballos
d Graskrankheit des Pferdes

1892 grass tetany
f tétanie d'herbage
e tetania de la hierba
d Weidetetanie

1893 great adductor muscle
f muscle grand adducteur
e músculo aductor magno
d großer Einwärtszieher
l musculus adductor magnus

1894 great cardiac vein
f grande veine coronaire

e vena magna del corazón
d Vena cordis magna
l vena cordis magna

1895 great cerebral vein
f grande veine cérébrale (de Galien)
e vena magna del cerebro
d Vena cerebri magna
l vena cerebri magna

* **great cistern → 810**

1896 greater occipital nerve
f nerf grand occipital
e nervio occipital mayor
d Nervus occipitalis major
l nervus occipitalis major

* **greater omentum → 1507**

1897 greater psoas muscle
f muscle grand psoas
e músculo psóas mayor
d großer Lendenmuskel
l musculus psoas major

1898 greater ring of iris; outer ring of iris
f grande circonférence de l'iris
e anillo mayor del iris
d äußerer Irisring
l anulus iridis major

1899 green blowfly; green bottle-fly
f mouche verte; lucilie verte
e moscarda verde
d Goldfliege
l Lucilia

* **green bottle-fly → 1899**

1900 grey commissure
f commissure grise
e comisura gris
d graue Kommissur
l commissura grisea

1901 grey matter
f substance grise
e sustancia gris
d graue Substanz
l substantia grisea

1902 griseofulvin
f griséofulvine
e griseofulvina
d Griseofulvin
l griseofulvinum

1903 groin
 f aine
 e ingle
 d Leiste
 l inguen

 * **groin region** → 2214

1904 groove; sulcus
 f sillon
 e surco
 d Furche; Rinne
 l sulcus

1905 ground border of hoof wall
 f bord soléaire du sabot
 e borde solear del casco
 d Sohlenrand; Tragrand
 l margo solearis

1906 growth
 f croissance
 e crecimiento
 d Wachstum

1907 growth disorder
 f trouble de la croissance
 e trastorno del crecimiento
 d Wachstumstörung

 * **growth hormone** → 4120

 * **guaiacol glyceryl ether** → 1908

1908 guaifenesin; guaiphenesin; guaiacol glyceryl ether
 f guaifénésine; guaiphénésine; gaïacol éther glycérique
 e guaifenesina
 d Guaifenesin; Gujakolglyzerinäther
 l guaifenesinum

 * **guaiphenesin** → 1908

 * **guard hair** → 1119

1909 gubernaculum of testis
 f gubernaculum testis
 e gubernáculo del testículo
 d Gubernaculum testis; Hodenleitband
 l gubernaculum testis

 * **Gumboro disease** → 2174

1910 gums
 f gencives
 e encías
 d Zahnfleisch
 l gingivae

1911 gustatory glands; Ebner's glands
 f glandes gustatives
 e glándulas gustativas
 d Ebnersche Drüsen
 l glandulae gustatoriae

1912 gustatory organ; organ of taste
 f organe du goût
 e órgano del gusto
 d Geschmacksorgan
 l organum gustus

 * **gustatory papillae** → 2526

1913 gustatory pore
 f pore gustatif
 e poro gustativo
 d Geschmackspore
 l porus gustatorius

 * **gut flora** → 2311

1914 guttural pouch; diverticulum of auditory tube
 f poche gutturale; diverticule de la trompe auditive
 e divertículo del la trompa auditiva
 d Luftsack (des Pferdes)
 l diverticulum tubae auditivae

H

1915 habenula
f habénula
e habénula
d Habenula
l habenula

1916 habenular nuclei
f noyaux de l'habénula
e núcleos habenularos
d Habenulakerne
l nuclei habenulares

* **Habronema infestation** → 1917

1917 habronemosis; Habronema infestation
f habronémose
e habronemosis
d Habronemose; Befall mit Habronema

1918 haemagglutination; hemagglutination
f hémagglutination
e hemoaglutinación
d Hämagglutination

1919 haemagglutination-inhibition test; HI test
f épreuve d'inhibition de l'hémagglutination
e prueba de inhibición de la
    hemoaglutinación
d Hämagglutinations-Hemmungsreaktion;
    HAH-Reaktion

1920 haemagglutination test; hemagglutination test
f épreuve d'hémagglutination
e prueba de hemoaglutinación
d Hämagglutinationstest

1921 haemal arch; hemal arch
f arc hémal
e arco hemal
d Hämalbogen
l arcus hemalis

1922 haemal node; hemal node; haemolymph node
f ganglion hématique; ganglion hémo-
    lymphatique
e nódulos linfáticos hemales
d Hämalknoten; Hämolymphknoten
l lymphonodus hemalis

1923 haemal process; hemal process
f processus hémal
e apófisis hemal
d Hämalfortsatz
l processus hemalis

1924 haematocrit; hematocrit
f hématocrite
e hematocrito
d Hämatokrit

1925 haematology; hematology
f hématologie
e hematología
d Hämatologie

1926 haematoma; hematoma
f hématome
e hematoma
d Hämatom; Blutbeule

* **haematopoiesis** → 476

1927 haematuria; hematuria
f hématurie
e hematuria
d Hämaturie; Blutharnen

* **Haemobartonella felis infection** → 1644

* **Haemobartonella infection** → 1928

1928 haemobartonellosis; Haemobartonella infection
f hémobartonellose
e hemobartonelosis; infección por
    Haemobartonella
d Hämobartonellose

1929 haemoglobin; hemoglobin
f hémoglobine
e hemoglobina
d Hämoglobin

1930 haemoglobin content
f teneur en hémoglobine
e concentración de hemoglobina
d Hämoglobingehalt

1931 haemoglobinuria; hemoglobinuria
f hémoglobinurie
e hemoglobinuria
d Hämoglobinurie

* **haemolymph node** → 1922

1932 haemolysis; hemolysis
f hémolyse
e hemolisis
d Hämolyse

1933 haemolytic; hemolytic
f hémolytique
e hemolítico
d hämolytisch

1934 haemonchosis; Haemonchus infestation
f hémonchose
e hemoncosis
d Hämonchose; Befall mit Haemonchus

* **Haemonchus infestation** → **1934**

1935 haemophilia; hemophilia
f hémophilie
e hemofilia
d Hämophilie

* **Haemophilus paragallinarum infection** → **2175**

* **Haemophilus parasuis infection** → **3560**

1936 haemorrhage; hemorrhage; bleeding
f hémorragie
e hemorragia
d Hämorrhagie; Blutung

1937 haemorrhagic diathesis; hemorrhagic diathesis
f diathèse hémorragique
e diatesis hemorrágica
d hämorrhagische Diathese

1938 haemorrhagic enteritis; hemorrhagic enteritis
f entérite hémorragique
e enteritis hemorrágica
d hämorrhagische Enteritis

1939 haemorrhagic septicaemia of cattle; hemorrhagic septicemia; Pasteurella multocida infection
f septicémie hémorragique à Pasteurella multocida; pasteurellose bovine
e septicemia hemorrágica de los bovinos; infección por Pasteurella multocida
d hämorrhagische Septikämie der Rinder; Pasteurella-multocida-Infektion

1940 haemorrhagic septicaemia rhabdovirus; VHS virus
f rhabdovirus de la septicémie hémorragique
e rhabdovirus de la septicemia hemorrágica; virus VHS
d Rhabdovirus der hämorrhagischen Septikämie

1941 haemothorax; hemothorax
f hémothorax
e hemotórax
d Hämatothorax

1942 hair
f poil
e pelo
d Haar
l pilus

1943 hair bulb
f bulbe pileux
e bulbo del pelo
d Haarzwiebel
l bulbus pili

1944 hair cross
f croix de poils
e cruz de los pelos
d Haarkreuz
l crux pilorum

1945 hair follicle
f follicule pileux
e folículo del pelo
d Haarfollikel; Haarbalg
l folliculus pili

* **hairlessness** → **321**

* **hair loss** → **127**

1946 hair papilla
f papille du poil
e papila del pelo
d Haarpapille
l papilla pili

1947 hair parting
f épi divergent
e línea divergente de los pelos
d Haarscheitel
l linea pilorum divergens

1948 hair ridge
f épi convergent
e línea convergente de los pelos
d Haarkamm; Haarleiste
l linea pilorum convergens

1949 hair root
f racine du poil
e raíz del pelo
d Haarwurzel
l radix pili

1950 hair shaft
 f tige du poil
 e tallo de pelo
 d Haarschaft
 l scapus pili

1951 hair stream; hair tract
 f fleuve pileux
 e trayecto de los pelos
 d Haarstrich
 l flumina pilorum

 * hair tract → 1951

1952 hair vortex
 f vortex des poils; tourbillon des poils
 e remolino de los pelos
 d Haarwirbel
 l vortex pilorum

 * hairworm infestation → 648

1953 hairworms
 f Capillaria
 e vermes capilares
 d Kapillarien; Haarwürmer
 l Capillaria

1954 halofuginone
 f halofuginone
 e halofuginona
 d Halofuginon
 l halofuginonum

1955 halothane
 f halothane
 e halotano
 d Halothan
 l halothanum

1956 haloxon
 f haloxone
 e haloxon
 d Haloxon
 l haloxonum

 * hamate bone → 1751

 * hammer → 2654

 * hamstring tendon → 1004

 * hamulus → 2028

1957 hand
 f main
 e mano
 d Hand
 l manus

1958 haplomycosis; Chrysosporium infection
 f haplomycose; infection à Chrysosporium
 e infección por Chrysosporium
 d Haplomykose; Chrysosporium-Infektion

1959 harderian gland; deep gland of third eyelid
 f glande profonde de la troisième paupière;
   glande de Harder
 e glándula profunda
 d tiefe Nickhautdrüse; Hardersche Drüse
 l glandula profunda palpebrae tertiae

1960 hard palate
 f palais dur
 e paladar duro
 d harter Gaumen
 l palatum durum

 * hard tick → 2351

 * haunch bone → 2133

 * haversian canal → 3046

 * HCG → 887

 * HCH → 2520

1961 head
 f tête
 e cabeza
 d Kopf
 l caput

1962 head cap of spermatozoon
 f capuchon céphalique du spermatozoaire
 e capuchón del espermatozoario
 d Kopfkappe des Spermatozoons
 l galea capitis

1963 head of epididymis
 f tête de l'épididyme
 e cabeza del epidídimo
 d Nebenhodenkopf
 l caput epididymidis

1964 head of femur
 f tête fémorale
 e cabeza del hueso fémur
 d Oberschenkelkopf
 l caput ossis femoris

1965 head of rib
 f tête de la côte
 e cabeza de la costilla
 d Rippenköpfchen
 l caput costae

1966 health
 f santé
 e salud; sanidad
 d Gesundheit

1967 health risk
 f risque sanitaire
 e riesgo sanitario
 d Gesundheitsrisiko

1968 healthy
 f sain
 e sano
 d gesund

1969 heart
 f cœur
 e corazón
 d Herz
 l cor

1970 heart disease
 f cardiopathie
 e cardiopatía
 d Kardiopathie; Herzkrankheit

1971 heart failure
 f défaillance cardiaque; mort cardiaque
 e colapso cardíaco
 d Herzschlag; Herztod

* heart muscle → 2912

1972 heart rate
 f fréquence cardiaque
 e frecuencia cardíaca
 d Herzfrequenz

1973 heart region
 f région du cœur
 e región cardíaca
 d Herzgegend
 l regio cardiaca

1974 heart valve
 f valve cardiaque; valvule cardiaque
 e valva del corazón; válvula cardíaca
 d Herzklappe
 l valva cordis; valvula cordis

1975 heartwater; Cowdria ruminantium infection
 f cowdriose; infection à Cowdria ruminantium
 e hidropericarditis infecciosa; infección por Cowdria ruminantium
 d Herzwasser; Cowdria-ruminantium-Infektion

1976 heartworm
 f dirofilaire
 e filaria hemática
 d Herzwurm
 l Dirofilaria

* heartworm disease → 1330

1977 heat exhaustion; heat stroke
 f coup de chaleur
 e golpe de calor
 d Hitzeschlag; Hitzekollaps

1978 heat resistance
 f résistance à la chaleur
 e resistencia al calor
 d Hitzeresistenz

1979 heat stress
 f choc thermique; stress thermique
 e stress por calor
 d Hitzebelastung

* heat stroke → 1977

1980 heel
 f talon
 e talón
 d Ferse
 l calx

* heel bone → 1692

1981 helicotrema
 f hélicotrème
 e helicotrema
 d Helicotrema
 l helicotrema

1982 helix
 f hélix
 e helix
 d Helix; Ohrkrempe
 l helix

1983 helminth
 f helminthe
 e helminto
 d Helminth

* helminthiasis → 1985

1984 helminthology
 f helminthologie
 e helmintología
 d Helminthologie

1985 **helminthosis; helminthiasis; worm infestation**
f helminthose
e helmintosis
d Helminthose; Wurmbefall

* **hemagglutination** → 1918

* **hemagglutination test** → 1920

* **hemal arch** → 1921

* **hemal node** → 1922

* **hemal process** → 1923

* **hematocrit** → 1924

* **hematology** → 1925

* **hematoma** → 1926

* **hematuria** → 1927

* **hemoglobin** → 1929

* **hemoglobinuria** → 1931

* **hemolysis** → 1932

* **hemolytic** → 1933

* **hemophilia** → 1935

* **hemorrhage** → 1936

* **hemorrhagic diathesis** → 1937

* **hemorrhagic enteritis** → 1938

* **hemorrhagic septicemia** → 1939

* **hemothorax** → 1941

1986 **hepatic artery**
f artère hépatique
e arteria hepática
d Arteria hepatica; Leberarterie
l arteria hepatica

1987 **hepatic plexus**
f plexus hépatique
e plexo hepático
d Nervengeflecht der Leber
l plexus hepaticus

1988 **hepatic portal; portal fissure**
f porte hépatique; hile du foie
e surco transverso del hígado; hilio del hígado
d Leberpforte
l porta hepatis

* **hepatic steatosis** → 1629

1989 **hepatic veins**
f veines hépatiques
e venas hepáticas
d Lebervenen
l venae hepaticae

1990 **hepatitis**
f hépatite
e hepatitis
d Hepatitis; Leberentzündung

1991 **hepatocyte; liver cell**
f hépatocyte; cellule hépatique
e hepatocito; célula hepática
d Hepatozyt; Leberzelle

1992 **hepatoduodenal ligament**
f ligament hépato-duodénal
e ligamento hepatoduodenal
d Ligamentum hepatoduodenale
l ligamentum hepatoduodenale

1993 **hepatogastric ligament**
f ligament hépato-gastrique
e ligamento hepatogástrico
d Ligamentum hepatogastricum
l ligamentum hepatogastricum

1994 **hepatomegaly; liver enlargement**
f hépatomégalie
e hepatomegalía
d Hepatomegalie; Lebervergrößerung

1995 **hepatorenal ligament**
f ligament hépato-rénal
e ligamento hepatorrenal
d Ligamentum hepatorenale
l ligamentum hepatorenale

* **Hepatozoon infection** → 1996

1996 **hepatozoonosis; Hepatozoon infection**
f hépatozoonose; infection par Hepatozoon
e hepatozoonosis; infección por Hepatozoon
d Hepatozoonose; Hepatozoon-Infektion

* **Herbst corpuscle** → 2410

1997 hereditary defect
 f défaut héréditaire
 e defecto hereditario; tara hereditaria
 d Erbfehler

1998 hermaphroditism
 f hermaphrodisme
 e hermafroditismo
 d Hermaphroditismus

1999 hernia
 f hernie
 e hernia
 d Hernie; Bruch

2000 herpesvirus
 f herpèsvirus
 e herpesvirus
 d Herpesvirus

2001 heterakiosis; Heterakis infestation
 f hétérakiose
 e heterakiosis
 d Heterakiose

 * Heterakis infestation → 2001

 * hexachlorophane → 2002

2002 hexachlorophene; hexachlorophane
 f hexachlorophène
 e hexaclorofeno
 d Hexachlorphen
 l hexachlorphenum

 * Hexamita meleagridis infection → 2003

2003 hexamitosis; Hexamita meleagridis infection
 f infection par Hexamita meleagridis
 e hexamitosis
 d Hexamitose

2004 hiatus
 f hiatus
 e hiato
 d Spalt; Schlitz
 l hiatus

 * hilus of kidney → 3813

2005 hilus of lung; pulmonary hilus
 f hile du poumon
 e hilio del pulmón
 d Lungenhilus
 l hilus pulmonis

2006 hilus of ovary
 f hile de l'ovaire
 e hilio del ovario
 d Eierstockhilus
 l hilus ovarii

2007 hilus of spleen
 f hile de la rate
 e hilio del bazo
 d Milzhilus
 l hilus lienis

 * hindbrain → 3864

 * hind foot → 1739

 * hindleg → 3365

2008 hinge joint; ginglymus
 f ginglyme; articulation trochléenne
 e gínglimo; articulación en charnela
 d Scharniergelenk; Wechselgelenk
 l ginglymus

2009 hip
 f hanche
 e cadera
 d Hüfte
 l coxa

2010 hip bone; pelvic bone
 f os coxal; os iliaque
 e hueso de la cadera
 d Hüftbein
 l os coxae

2011 hip dysplasia
 f dysplasie de la hanche
 e displasia de la cadera
 d Hüftgelenkdysplasie
 l dysplasia coxofemoralis

2012 hip joint; coxofemoral joint
 f articulation de la hanche; articulation coxo-fémorale
 e articulación de la cadera
 d Hüftgelenk
 l articulatio coxae

2013 hip joint region
 f région de l'articulation de la hanche
 e región de la articulación de la cadera
 d Hüftgelenksgegend
 l regio articulationis coxae

 * hippocampal gyrus → 3280

2014 **hippocampus; Ammon's horn**
 f hippocampe; corne d'Ammon
 e hipocampo; asta de Ammon
 d Hippocampus; Ammonshorn
 l hippocampus; cornu ammonis

 * **histamine antagonist** → 222

2015 **histocompatibility**
 f histocompatibilité
 e histocompatibilidad
 d Histokompatibilität

2016 **histological**
 f histologique
 e histológico
 d histologisch

2017 **histology**
 f histologie
 e histología
 d Histologie

2018 **histomonosis; blackhead of turkeys**
 f histomonose
 e histomonosis
 d Histomonose; enzootische
 Typhlohepatitis; Schwarzkopfkrankheit
 l enterohepatitis infectiosa avium

2019 **histopathology**
 f histopathologie
 e histopatología
 d Histopathologie

 * **Histoplasma capsulatum infection** → 2020

 * **Histoplasma farciminosum infection** →
 1514

2020 **histoplasmosis; Histoplasma capsulatum
 infection**
 f histoplasmose; infection à Histoplasma
 capsulatum
 e histoplasmosis; infección por Histoplasma
 capsulatum
 d Histoplasmose; Histoplasma-capsulatum-
 Infektion

 * **HI test** → 1919

2021 **hock; tarsus**
 f tarse; jarret
 e tarso
 d Fußwurzel
 l tarsus

2022 **hock joint; tarsal joint; ankle joint**
 f articulation du tarse; articulation du jarret
 e articulación del tarso
 d Sprunggelenk; Fußwurzelgelenk
 l articulatio tarsi

 * **hock region** → 4448

 * **hog cholera** → 4385

2023 **hollow of flank; paralumbar fossa**
 f fosse paralombaire; creux du flanc
 e fosa paralumbar
 d Hungergrube
 l fossa paralumbalis

2024 **homidium bromide**
 f bromure d'homidium
 e bromuro de homidio
 d Homidiumbromid
 l homidii bromidum

2025 **hoof**
 f sabot
 e úngula; casco
 d Huf
 l ungula

2026 **hoof capsule**
 f capsule de l'onglon
 e cápsula de la úngula
 d Hufkapsel
 l capsula ungulae

2027 **hoof region; claw region**
 f région de l'onglon
 e región ungular
 d Hufgegend; Klauengegend
 l regio ungulae

2028 **hook; hamulus**
 f hamulus; crochet
 e ganchillo
 d Häkchen
 l hamulus

2029 **hookworm; ancylostome**
 f ankylostome
 e anquilostoma; ancilostoma
 d Hakenwurm
 l Ancylostoma

2030 **horizontal plate of palatine bone**
 f lame horizontale de l'os palatin
 e lámina horizontal del hueso palatino
 d Horizontalteil des Gaumenbeins
 l lamina horizontalis ossis palatini

2031 hormonal
  f hormonal
  e hormonal
  d hormonal; hormonell

2032 hormone
  f hormone
  e hormona
  d Hormon

2033 horn
  f corne
  e cuerno
  d Horn
  l cornu

* horn artery → 1072

2034 horn core; cornual process
  f processus cornual
  e apófisis cornual
  d Hornzapfen
  l processus cornualis

2035 horn fly
  f mouche des cornes
  e mosca de los cuernos
  d kleine Weidestechfliege
  l Haematobia irritans

2036 horn gland
  f glande cornuale
  e glándula cornual
  d Horndrüse
  l glandula cornualis

* horn plate → 4956

2037 horn region
  f région cornuale
  e región cornual
  d Horngegend
  l regio cornualis

2038 horns of uterus; uterine horns
  f cornes utérines
  e cuernos del útero
  d Uterushörner
  l cornua uteri

2039 horn tip
  f apex de la corne; pointe de la corne
  e vértice del cuerno; punta del cuerno
  d Hornspitze
  l apex cornus

2040 horn tubules
  f tubules de la corne

  e túbulos epidérmicos
  d Hornröhrchen
  l tubuli epidermales

2041 horny covering of beak
  f rhamphothèque
  e ranfoteca
  d Hornscheiden des Schnabels
  l rhamphotheca

2042 horny layer of epidermis
  f stratum corneum de l'épiderme
  e estrato córneo
  d Stratum corneum; Hornschicht
  l stratum corneum epidermidis

2043 horse botfly; stomach botfly
  f gastérophilidés
  e gastrófilo
  d Magenbremse; Magendasselfliege
  l Gasterophilidae

2044 horse disease
  f maladie des équidés
  e enfermedad del caballo
  d Pferdekrankheit

2045 horse fly; cleg
  f taon; tabanide
  e tábano
  d Bremse
  l Tabanus

2046 horse louse-fly
  f hippobosque du cheval
  e mosca borriquera; hipobóscido del caballo
  d Pferdelausfliege
  l Hippobosca equina

2047 horse sucking louse
  f Haematopinus asini
  e piojo grande del caballo
  d Pferdelaus
  l Haematopinus asini

2048 host
  f hôte
  e huésped
  d Wirt

2049 host-parasite relationship
  f relation hôte-parasite
  e relación parásito/hospedador
  d Parasit-Wirt-Beziehung; Wirt-Parasit-Verhältnis

2050 house fly
  f mouche domestique

e mosca común
d Stubenfliege
l Musca domestica

2051 humerocarpal ligament
f ligament huméro-carpal
e ligamento humerocarpiano
d Ligamentum humerocarpale
l ligamentum humerocarpale

2052 humeroradial articulation
f articulation huméro-radiale
e articulación humerorradial
d Articulatio humeroradialis
l articulatio humeroradialis

2053 humeroulnar articulation
f articulation huméro-ulnaire
e articulación humerocubital
d Articulatio humeroulnaris
l articulatio humeroulnaris

2054 humerus
f humérus
e húmero
d Humerus; Oberarmknochen
l humerus

2055 humoral antibody
f anticorps humoral
e anticuerpo humoral
d humoraler Antikörper

2056 hyaline cartilage
f cartilage hyalin
e cartílago hialino
d hyaliner Knorpel
l cartilago hyalina

2057 hyaloid artery
f artère hyaloïdienne
e arteria hialoidea
d Glaskörperarterie
l arteria hyaloidea

* hyaloid membrane → 4942

2058 hyaluronidase
f hyaluronidase
e hialuronidasa
d Hyaluronidase
l hyaluronidasum

* hydatid disease → 1423

* hydatid tapeworm → 1424

2059 hydrocephalus
f hydrocéphalie
e hidrocefalia
d Hydrocephalus; Wasserkopf

2060 hydrocortisone; cortisol
f hydrocortisone; cortisol
e hidrocortisona
d Hydrocortison; Cortisol
l hydrocortisonum

2061 hydrothorax
f hydrothorax
e hidrotórax
d Hydrothorax

2062 hygiene
f hygiène
e higiene
d Hygiene

2063 hygroma
f hygroma
e higroma
d Hygrom

2064 hymenolepiosis; Hymenolepis infestation
f hyménolépiose
e himenolepiosis
d Hymenolepiose

* Hymenolepis infestation → 2064

2065 hyoepiglottal muscle
f muscle hyo-épiglottique
e músculo hioepiglótico
d Musculus hyoepiglotticus
l musculus hyoepiglotticus

2066 hyoglossal muscle
f muscle hyoglosse
e músculo hiogloso
d Musculus hyoglossus
l musculus hyoglossus

2067 hyoid bone
f os hyoïde
e hueso hioideo; aparato hioideo
d Zungenbein
l os hyoideum; apparatus hyoideus

2068 hyoid muscles
f muscles de l'os hyoïde
e músculos hioideos
d Zungenbeinmuskeln
l musculi hyoidei

2069 **hyostrongylosis; Hyostrongylus rubidus infection**
  f hyostrongylose
  e hiostrongilosis; verminosis gástrica del cerdo
  d Hyostrongylose; Befall mit Hyostrongylus rubidus

*   **Hyostrongylus rubidus infection → 2069**

2070 **hypercalcaemia**
  f hypercalcémie
  e hipercalcemia
  d Hyperkalzämie

2071 **hyperfunction**
  f hyperfonctionnement
  e hiperfunción
  d Hyperfunktion

2072 **hyperglycaemia**
  f hyperglycémie
  e hiperglicemia
  d Hyperglykämie

2073 **hyperimmune serum**
  f sérum hyperimmun
  e suero hiperinmune
  d Hyperimmunserum

2074 **hyperimmunization**
  f hyperimmunisation
  e hiperinmunización
  d Hyperimmunisierung

2075 **hyperlipaemia**
  f hyperlipémie
  e hiperlipemia
  d Hyperlipämie

2076 **hyperplasia**
  f hyperplasie
  e hiperplasia
  d Hyperplasie

2077 **hypersensitivity**
  f hypersensibilité
  e hipersensibilidad
  d Überempfindlichkeit

2078 **hyperthermia**
  f hyperthermie
  e hipertermia
  d Hyperthermie

2079 **hypertrophy**
  f hypertrophie
  e hipertrofia
  d Hypertrophie

*   **Hyphomyces destruens infection → 2080**

2080 **hyphomycosis; Hyphomyces destruens infection**
  f hyphomycose
  e hifomicosis
  d Hyphomykose

2081 **hypocalcaemia**
  f hypocalcémie
  e hipocalcemia
  d Hypokalzämie

2082 **hypochondriac region**
  f région hypocondriaque
  e región hipocondríaca
  d Unterrippenregion
  l regio hypochondriaca

*   **hypodermic needle → 2217**

*   **hypodermis → 4271**

2083 **hypodermosis; warble fly infestation**
  f hypodermose
  e hipodermosis
  d Hypodermose; Dasselbefall

2084 **hypogammaglobulinaemia**
  f hypogammaglobulinémie
  e hipogammaglobulinemia
  d Hypogammaglobulinämie

2085 **hypogastric nerve**
  f nerf hypogastrique
  e nervio hipogástrico
  d Nervus hypogastricus
  l nervus hypogastricus

2086 **hypoglossal canal** (*in occipital bone*)
  f canal du nerf hypoglosse
  e canal del nervio hipogloso
  d Canalis hypoglossi
  l canalis hypoglossi

2087 **hypoglossal nerve**
  f nerf (grand) hypoglosse
  e nervio hipogloso
  d Nervus hypoglossus; Unterzungennerv
  l nervus hypoglossus

2088 **hypoglycaemia**
  f hypoglycémie
  e hipoglicemia
  d Hypoglykämie

2089 **hypomagnesaemia**
  f hypomagnésémie
  e hipomagnesemia
  d Hypomagnesämie

* hypophyseal → 3483

2090 hypophyseal cavity
 f cavité de l'hypophyse
 e cavidad de la hipófisis
 d Hypophysenhöhle
 l cavum hypophysis

2091 hypophyseal fossa; pituitary fossa
 f fosse pituitaire
 e fosa hipofisaria
 d Fossa hypophysialis
 l fossa hypophysialis

2092 hypophysis; pituitary gland
 f hypophyse; glande pituitaire
 e hipófisis; glándula pituitaria
 d Hypophyse; Hirnanhangsdrüse
 l hypophysis (cerebri); glandula pituitaria

2093 hypoplasia
 f hypoplasie
 e hipoplasia
 d Hypoplasie

2094 hypoproteinaemia
 f hypoprotéinémie
 e hipoproteinemia
 d Hypoproteinämie

2095 hypostasis; hypostatic congestion
 f hypostase; congestion hypostatique
 e hipóstasis; congestión hipostática
 d Hypostase; Senkungshyperämie

* hypostatic congestion → 2095

2096 hypothalamus
 f hypothalamus
 e hipotálamo
 d Hypothalamus
 l hypothalamus

2097 hypothermia
 f hypothermie
 e hipotermia
 d Hypothermie

2098 hypotrichosis
 f hypotrichose
 e hipotricosis
 d Hypotrichose

2099 hypoxia
 f hypoxie
 e hipoxia
 d Hypoxia; Sauerstoffmangel

2100 hysterectomy
 f hystérectomie
 e histerectomía
 d Hysterektomie

I

* IBR → 2172

2101 IBR/IPV virus; bovine herpesvirus type 1
f herpèsvirus bovin type 1
e virus de la rinotraqueitis infecciosa bovina;
  virus IBR/IPV; herpesvirus bovino tipo 1
d IBR-Virus; bovines Herpesvirus Typ 1;
  BHV-1

2102 Ichthyophonus infection
f infection à Ichthyophonus
e ictiofonosis; infección por Ichthyophonus
d Ichthyophonuskrankheit;
  Ichthyosporidiasis

* icterus → 2353

2103 idoxuridine
f idoxuridine
e idoxuridina
d Idoxuridin
l idoxuridinum

* Ig → 2147

2104 ileal arteries
f artères iléales
e arterias iliales
d Arteriae ilei
l arteriae ilei

* ileal papilla → 2109

2105 ileal sphincter
f sphincter iléal
e músculo esfínter del íleon
d Hüftdarmsphinkter
l musculus sphincter ilei

2106 ileal veins
f veines iléales
e venas del íleon
d Venae ilei; Hüftdarmvenen
l venae ilei

2107 ileocaecal fold
f pli iléo-cæcal
e pliegue ileocecal
d Plica ileocaecalis
l plica ileoc(a)ecalis

2108 ileocaecal opening
f orifice iléo-cæcal
e orificio ileocecal
d Hüftdarm-Blinddarmöffnung
l ostium ileoc(a)ecale

2109 ileocaecal valve; ileal papilla
f papille iléale; valvule iléo-cæcale
e papila ileal; valva ileocecal (de Bauhin)
d Blinddarmklappe; Papilla ilealis
l papilla ilealis; valva ileoc(a)ecalis

2110 ileocolic artery
f artère iléo-colique
e arteria ileocólica
d Arteria ileocolica
l arteria ileocolica

2111 ileocolic vein
f veine iléo-colique
e vena ileocólica
d Vena ileocolica
l vena ileocolica

2112 ileum
f iléon
e íleon
d Ileum; Hüftdarm
l ileum

* iliac crest → 1166

2113 iliac fascia
f fascia iliaque
e fascia ilíaca
d Fascia iliaca
l fascia iliaca

2114 iliac muscle
f muscle iliaque
e músculo ilíaco
d Darmbeinmuskel
l musculus iliacus

2115 iliac plexus
f plexus iliaque
e plexo ilíaco
d Plexus iliacus
l plexus iliacus

2116 iliac spine
f épine iliaque
e espina ilíaca
d Darmbeinstachel
l spina iliaca

2117 iliac tuberosity
f tubérosité iliaque

e tuberosidad ilíaca
d Tuberositas iliaca
l tuberositas iliaca

2118 iliocostal muscle
f muscle ilio-costal
e músculo iliocostal
d Darmbein-Rippenmuskel;
gemeinschaftlicher Rippenmuskel
l musculus iliocostalis

2119 iliocostal muscle of loins
f muscle sacro-lombaire
e músculo iliocostal lumbar
d Musculus iliocostalis lumborum
l musculus iliocostalis lumborum

2120 iliofemoral ligament
f ligament ilio-fémoral
e ligamento iliofemoral
d Ligamentum iliofemorale
l ligamentum iliofemorale

**2121 iliofemoral lymph nodes; deep inguinal
lymph nodes**
f nœuds lymphatiques ilio-fémoraux; nœuds
lymphatiques inguinaux profonds
e nódulos linfáticos iliofemorales
d Lymphonodi iliofemorales
l lymphonodi iliofemorales

2122 iliofemoral lymphocentre
f lymphocentre ilio-fémoral; lymphocentre
inguinal profond
e linfocentro iliofemoral; linfocentro
inguinal profundo
d Lymphocentrum iliofemorale
l lymphocentrum iliofemorale

2123 iliofibular muscle
f muscle ilio-fibulaire
e músculo ilioperóneo
d Musculus iliofibularis
l musculus iliofibularis

2124 iliohypogastric nerve
f nerf ilio-hypogastrique
e nervio iliohipogástrico
d Nervus iliohypogastricus
l nervus iliohypogastricus

2125 ilioischiatic synchondrosis
f synchondrose ilio-ischiatique
e sincondrosis ilioisquiática
d Synchondrosis ilioischiadica
l synchondrosis ilioischiadica

2126 iliolumbar artery
f artère ilio-lombaire

e arteria iliolumbar
d Arteria iliolumbalis
l arteria iliolumbalis

2127 iliolumbar ligament
f ligament ilio-lombaire
e ligamento iliolumbar
d Ligamentum iliolumbale
l ligamentum iliolumbale

2128 iliolumbar vein
f veine ilio-lombaire
e vena iliolumbar
d Vena iliolumbalis
l vena iliolumbalis

2129 iliopsoas muscle
f muscle psoas iliaque
e músculo iliopsóas
d Darmbein-Lendenmuskel
l musculus iliopsoas

2130 iliosacral lymphocentre
f lymphocentre ilio-sacral
e linfocentro iliosacro
d Lymphocentrum iliosacrale
l lymphocentrum iliosacrale

2131 iliotibial muscles
f muscles ilio-tibiaux
e músculos iliotibiales
d Musculi iliotibiales
l musculi iliotibiales

2132 iliotrochanteric muscles
f muscles ilio-trochantériques
e músculos iliotrocantéricos
d Musculi iliotrochanterici
l musculi iliotrochanterici

2133 ilium; haunch bone
f ilion
e hueso íleon
d Darmbein
l os ilium

2134 ill
f malade
e enfermo
d krank

* **illness** → 1331

* **ILT** → 353

* **ILT virus** → 357

2135 **imidocarb**
 f imidocarbe
 e imidocarbo
 d Imidocarb
 l imidocarbum

2136 **immune**
 f immun
 e inmune
 d immun

2137 **immune complex**
 f complexe immun; immuncomplexe
 e complejo inmune
 d Immunkomplex

2138 **immune response**
 f réponse immunitaire
 e respuesta inmunitaria
 d Immunantwort

2139 **immune serum; antiserum**
 f immunsérum; antisérum
 e antisuero
 d Immunserum; Antiserum

2140 **immunity**
 f immunité
 e inmunidad
 d Immunität

2141 **immunization**
 f immunisation
 e inmunización
 d Immunisierung

2142 **immunodeficiency; immunological deficiency**
 f immunodéficience; déficit immunitaire
 e inmunodeficiencia; deficiencia inmunitaria
 d Immuninsuffizienz; Immunschwäche

2143 **immunodiffusion**
 f immunodiffusion
 e inmunodifusión
 d Immundiffusion

2144 **immunofluorescence; fluorescent antibody test**
 f immunofluorescence
 e inmunofluorescencia
 d Immunfluoreszenz

2145 **immunogenic**
 f immunogène
 e inmunogenico
 d immunogen

2146 **immunogenicity**
 f immunogénicité
 e inmunogenicidad
 d Immunogenität

2147 **immunoglobulin; Ig**
 f immunoglobuline; Ig
 e inmunoglobulina; Ig
 d Immunglobulin; Ig

2148 **immunological**
 f immunologique
 e inmunológico
 d immunologisch

* **immunological deficiency** → 2142

2149 **immunological disease**
 f maladie immunologique; immunopathologie
 e enfermedad inmunológica
 d Immunkrankheit

2150 **immunology**
 f immunologie
 e inmunología
 d Immunologie

2151 **immunostimulation**
 f immunostimulation
 e inmunoestimulación
 d Immunstimulierung

* **immunosuppressant** → 2153

2152 **immunosuppression**
 f immunosuppression
 e inmunosupresión
 d Immunsuppression

2153 **immunosuppressive agent; immunosuppressant**
 f immunosuppresseur
 e agente inmunosupresor
 d Immunosuppressivum

* **impaired growth** → 4753

2154 **inactivated vaccine**
 f vaccin inactivé
 e vacuna inactivada
 d inaktivierte Vakzine; Totvakzine

2155 **inappetence**
 f inappétence
 e inapetencia
 d Inappetenz

2156 **incisive bone; premaxilla**
 f os incisif; os prémaxillaire
 e hueso incisivo; premaxilar
 d Zwischenkieferbein
 l os incisivum

2157 **incisive duct**
 f conduit incisif
 e conducto incisivo
 d Nasen-Gaumenkanal
 l ductus incisivus

2158 **incisive muscles**
 f muscles incisifs
 e músculos incisivos
 d Musculus incisivus
   mandibularis/maxillaris
 l musculus incisivus
   mandibularis/maxillaris

2159 **incisive papilla**
 f papille incisive
 e papila incisiva
 d Papilla incisiva
 l papilla incisiva

2160 **incisor teeth**
 f dents incisives
 e dientes incisivos
 d Schneidezähne
 l dentes incisivi

2161 **incisure; notch**
 f incisure; échancrure
 e escotadura
 d Inzisur; Einschnitt
 l incisura

2162 **incisure of scapula; suprascapular notch**
 f incisure scapulaire; échancrure
   coracoïdienne
 e escotadura de la escápula
 d Incisura scapulae
 l incisura scapulae

2163 **inclination of pelvis; angle of pelvis**
 f inclinaison pelvienne
 e inclinación de la pelvis
 d Beckenneigung
 l inclinatio pelvis

2164 **inclusion body**
 f corps d'inclusion
 e cuerpo de inclusión
 d Einschlußkörperchen

2165 **incoordination**
 f incoordination
 e incoordinación
 d Inkoordination; Koordinationsstörung

2166 **incubation patch; brood patch**
 f plaque incubatrice
 e área de incubación
 d Brutfleck
 l area incubationis

2167 **incus; anvil**
 f enclume
 e yunque
 d Amboß
 l incus

2168 **infarct**
 f infarctus
 e infarto
 d Infarkt

2169 **infected**
 f infecté
 e infectado
 d infiziert

2170 **infection**
 f infection
 e infección
 d Infektion

2171 **infectious**
 f infectieux
 e infeccioso
 d infektiös

2172 **infectious bovine rhinotracheitis; IBR**
 f rhinotrachéite infectieuse bovine
 e rinotraqueitis infecciosa bovina
 d infektiöse bovine Rhinotracheitis; IBR

2173 **infectious bronchitis coronavirus**
 f coronavirus de la bronchite infectieuse
   aviaire
 e virus de la bronquitis infecciosa aviar
 d infektiöses Bronchitis-Coronavirus

2174 **infectious bursal disease; Gumboro disease**
 f bursite infectieuse aviaire; maladie de
   Gumboro
 e bursitis infecciosa; enfermedad de
   Gumboro
 d ansteckende (virale) Bursakrankheit des
   Huhnes; Gumboro-Krankheit

2175 **infectious coryza; avian infectious coryza;
   Haemophilus paragallinarum infection**
 f coryza infectieux aviaire; infection à
   Haemophilus paragallinarum
 e coriza aviar
 d Geflügelschnupfen; Haemophilus
   paragallinarum-Infektion
 l coryza contagiosa avium; rhinitis infectiosa
   avium

2176 infectious disease
f maladie infectieuse
e enfermedad infecciosa
d infektiöse Krankheit; Infektionskrankheit

**2177 infectious footrot of sheep; foot rot;
Bacteroides nodosus infection**
f piétin; infection à Bacteroides nodosus
e pedero; panadizo de los óvidos; infección
por Bacteroides nodosus
d Moderhinke; Fußfäule; Bacteroides-
nodosus-Infektion

**2178 infectious haematopoietic necrosis of
Salmonidae**
f nécrose hématopoïétique des salmonidés
e necrosis hematopoyética infecciosa de los
salmónidos
d infektiöse hämatopoietische Nekrose der
Salmoniden

**2179 infectious haematopoietic necrosis
rhabdovirus**
f rhabdovirus de la nécrose
hématopoïétique infectieuse
e rhabdovirus de la necrosis hematopoyética
infecciosa
d Rhabdovirus der infektiösen
hämatopoietischen Nekrose

2180 infectious pancreatic necrosis; IPN
f nécrose pancréatique infectieuse
e necrosis pancreática infecciosa; IPN
d infektiöse Pankreasnekrose

**2181 infectious pancreatic necrosis virus; IPN
virus**
f virus de la nécrose pancréatique
infectieuse
e virus de la necrosis pancreática infecciosa;
virus IPN
d Rhabdovirus der infektiösen
Pankreasnekrose

2182 infectious pustular vulvovaginitis; IPV
f vulvovaginite pustuleuse infectieuse; IPV
e vulvovaginitis pustulosa infecciosa; IPV
d infektiöse pustulöse Vulvovaginitis; IPV

2183 infectivity
f infectiosité; pouvoir infectant
e infecciosidad
d Infektiosität

2184 inferior vestibular area; acoustic area
f aire vestibulaire inférieure; aire acoustique
e área vestibular inferior
d untere Vestibularfläche; akustischer
Cortex
l area vestibularis inferior

2185 infertile
f infertile
e infecundo
d unfruchtbar

2186 infertility; sterility
f infertilité; stérilité
e infertilidad; esterilidad
d Unfruchtbarkeit; Infertilität; Sterilität

2187 infestation
f infestation
e infestación
d Befall

* infestation with Dilepididae → 1320

2188 inflammation
f inflammation
e inflamación
d Entzündung

* inflammation of abomasum → 16

* inflammation of the crop → 2209

* inflammation of the reticulum → 3838

2189 influenzavirus
f influenzavirus; virus de la grippe
e influenzavirus; virus de la gripe
d Influenzavirus

2190 inframamillary recess
f récessus inframamillaire
e receso inframamilar
d Recessus inframamillaris
l recessus inframamillaris

2191 infraorbital artery
f artère infra-orbitaire
e arteria infraorbitaria
d Arteria infraorbitalis
l arteria infraorbitalis

2192 infraorbital canal
f canal infra-orbitaire
e canal infraorbitario
d Oberkieferkanal
l canalis infraorbitalis

2193 infraorbital foramen
f foramen infra-orbitaire
e agujero infraorbitario
d Foramen infraorbitale
l foramen infraorbitale

2194 infraorbital nerve
f nerf sous-orbitaire
e nervio infraorbitario
d Nervus infraorbitalis
l nervus infraorbitalis

* infraorbital pouch → 2196

2195 infraorbital region
f région infra-orbitaire
e región infraorbitaria
d Unteraugenbogenregion
l regio infraorbitalis

2196 infraorbital sinus; infraorbital pouch
f sinus sous-orbitaire
e seno infraorbitario
d Tränengrube
l sinus infraorbitalis

2197 infraorbital tactile hairs
f poils tactiles infra-orbitaires
e pelos infraorbitarios
d Infraorbitalhaare
l pili infraorbitales

2198 infraorbital vein
f veine infra-orbitaire
e vena infraorbitaria
d Vena infraorbitalis
l vena infraorbitalis

2199 infrapalpebral sulcus
f sillon infrapalpébral
e surco infrapalpebral
d Sulcus infrapalpebralis
l sulcus infrapalpebralis

2200 infraspinous fossa
f fosse infra-épineuse
e fosa infraespinosa
d Fossa infraspinata; Untergrätengrube
l fossa infraspinata

2201 infraspinous muscle
f muscle sous-épineux/infra-épineux
e músculo infraespinoso
d unterer Grätenmuskel
l musculus infraspinatus

2202 infraspinous region
f région infraspinale
e región infraespinosa
d Infraspinalgegend
l regio infraspinata

2203 infratrochlear nerve
f nerf infratrochléaire; nerf nasal externe
e nervio infratroclear

d Nervus infratrochlearis
l nervus infratrochlearis

2204 infundibular nucleus
f noyau infundibulaire
e núcleo infundibular
d Nucleus infundibularis
l nucleus infundibularis

2205 infundibular part of adenohypophysis
f partie infundibulaire de l'adénohypophyse
e porción infundibular de la adenohipófisis
d Infundibulum der Hypophyse
l pars infundibularis adenohypophysis; pars tuberalis

* infundibular recess → 2995

2206 infundibulum
f infundibulum
e infundíbulo
d Infundibulum; Trichter
l infundibulum

* infundibulum of hypothalamus → 3484

2207 infundibulum of tooth
f cornet dentaire
e infundíbulo del diente
d Zahneinstülpung; Kunde
l infundibulum dentis

2208 infundibulum of uterine tube
f infundibulum de la trompe utérine
e infundíbulo de la trompa uterina
d Eileitertrichter
l infundibulum tubae uterinae

2209 ingluvitis; inflammation of the crop
f ingluvite; inflammation du jabot
e ingluvitis
d Ingluviitis; Kropfentzündung

2210 inguinal canal
f canal inguinal
e canal inguinal; espacio inguinal
d Leistenkanal
l canalis inguinalis; spatium inguinale

2211 inguinal hernia
f hernie inguinale
e hernia inguinal
d Inguinalhernie; Leistenbruch
l hernia inguinalis

2212 inguinal mammary gland
f mamelle inguinale
e mama inguinal
d inguinale Milchdrüse
l mamma inguinalis

2213 inguinal pouch of sheep
f sinus inguinal du mouton
e seno inguinal
d Inguinaltasche des Schafes
l sinus inguinalis

2214 inguinal region; groin region
f région inguinale
e región inguinal
d Leistengegend
l regio inguinalis

2215 inguinofemoral lymphocentre
f lymphocentre inguino-fémoral;
    lymphocentre inguinal superficiel
e linfocentro inguinofemoral; linfocentro
    inguinal superficial
d Lymphocentrum inguinofemorale
l lymphocentrum inguinofemorale

2216 injection
f injection
e inyección
d Injektion; Einspritzung

2217 injection needle; hypodermic needle
f aiguille hypodermique
e aguja para inyección hipodérmica
d Hohlnadel; Kanüle

2218 injury
f blessure
e herida
d Verletzung

* **inner coat of vessels** → 2316

* **inner ring of iris** → 2498

2219 inoculation
f inoculation
e inoculación
d Inokulation; Impfung

2220 insect
f insecte
e insecto
d Insekt
l Insecta

2221 insect bite
f piqûre d'insecte
e picadura de insecto
d Insektenstich

2222 insecticide
f insecticide
e insecticida
d Insektizid

2223 insufficiency
f insuffisance
e insuficiencia
d Insuffizienz

* **interarcuate ligaments** → 4991

2224 interatrial septum
f septum interatrial; cloison interauriculaire
e septo interatrial
d Vorkammerscheidewand
l septum interatriale

* **interbrain** → 4508

2225 intercarotid arteries
f artères intercarotidiennes
e arterias intercarotídeas
d Interkarotisarterien
l arteriae intercaroticae

2226 intercarpal joints
f articulations intercarpiennes
e articulaciones intercarpianas
d Zwischenkarpalgelenke
l articulationes intercarpeae

2227 intercarpal ligaments
f ligaments intercarpiens
e ligamentos intercarpianos
d karpale Querbänder
l ligamenta intercarpea

2228 intercornual ligament
f ligament intercornual
e ligamento intercornual
d Ligamentum intercornuale
l ligamentum intercornuale

2229 intercostal arteries
f artères intercostales
e arterias intercostales
d Zwischenrippenarterien
l arteriae intercostales

2230 intercostal lymph nodes
f nœuds lymphatiques intercostaux
e nódulos linfáticos intercostales
d Interkostallymphknoten
l lymphonodi intercostales

2231 intercostal muscles
f muscles intercostaux
e músculos intercostales
d Zwischenrippenmuskeln
l musculi intercostales

2232 intercostal nerves
 f nerfs intercostaux
 e nervios intercostales
 d Interkostalnerven
 l nervi intercostales

2233 intercostal space
 f espace intercostal
 e espacio intercostal
 d Zwischenrippenraum
 l spatium intercostale

2234 intercostal veins
 f veines intercostales
 e venas intercostales
 d Interkostalvenen
 l venae intercostales

2235 intercurrent disease
 f maladie intercurrente
 e enfermedad intercurrente
 d interkurrente Krankheit; Nebenkrankheit

2236 interdental space; diastema
 f diastème
 e diastema
 d Diastema; Lade
 l diastema

2237 interdigital artery
 f artère interdigitale
 e arteria interdigital
 d Zwischenzehenarterie
 l arteria interdigitalis

2238 interdigital ligaments
 f ligaments interdigitaux
 e ligamentos interdigitales
 d Ligamenta interdigitalia
 l ligamenta interdigitalia

* interdigital pouch of sheep → 2239

2239 interdigital sinus; interdigital pouch of sheep
 f sinus interdigital du mouton
 e seno interdigital
 d Zwischenklauensäckchen (des Schafes)
 l sinus interdigitalis

2240 interdigital space
 f espace interdigital
 e espacio interdigital
 d Spatium interdigitale;
   Zwischenklauenspalt
 l spatium interdigitale

2241 interferon
 f interféron
 e interferon
 d Interferon
 l interferonum

2242 interflexor muscle
 f muscles interfléchisseurs
 e músculos interflexores
 d Musculi interflexorii
 l musculi interflexorii

2243 interincisive fissure
 f fissure interincisive
 e cisura interincisiva
 d Fissura interincisiva
 l fissura interincisiva

2244 interlobar fissure of lung
 f scissure interlobaire
 e cisura interlobar
 d Fissura interlobaris
 l fissura interlobaris

2245 interlobular ductules (*of liver*)
 f canalicules interlobulaires (*du foie*)
 e conductillos interlobulares
 d interlobuläre Gallenkanäle
 l ductuli interlobulares

2246 intermammary groove
 f sillon intermammaire
 e surco intermamario
 d mediane Euterfurche
 l sulcus intermammarius

2247 intermandibular articulation
 f articulation intermandibulaire
 e articulación intermandibular
 d Articulatio intermandibularis
 l articulatio intermandibularis

2248 intermandibular region
 f région intermandibulaire
 e región intermandibular
 d Kehlganggegend
 l regio intermandibularis

2249 intermandibular suture; mandibular symphysis
 f suture intermandibulaire; symphyse mandibulaire
 e sutura intermandibular
 d Sutura intermandibularis
 l sutura intermandibularis

2250 intermandibular synchondrosis; mandibular symphysis
 f synchondrose intermandibulaire; symphyse mandibulaire

e sincondrosis intermandibular
d Synchondrosis intermandibularis
l synchondrosis intermandibularis

2251 intermediate carpal bone; lunar bone; semilunar bone
f os intermédiaire; os semi-lunaire
e hueso intermedio del carpo; hueso semilunar
d Os carpi intermedium
l os carpi intermedium; os lunatum

2252 intermediate ganglia
f ganglions intermédiaires
e ganglios intermedios
d Ganglia intermedia
l ganglia intermedia

2253 intermediate host
f hôte intermédiaire
e hospedador intermediario
d Zwischenwirt

2254 intermediate lobe of adenohypophysis
f lobe intermédiaire de l'adénohypophyse
e porción intermedia de la adenohipófisis
d Hypophysenzwischenlappen
l pars intermedia adenohypophysis

2255 intermediate nerve
f nerf intermédiaire (de Wrisberg)
e nervio intermedio (de Wrisberg)
d Nervus intermedius
l nervus intermedius

2256 intermediate vastus muscle
f muscle vaste intermédiaire
e músculo vasto intermedio
d Musculus vastus intermedius
l musculus vastus intermedius

2257 intermesenteric plexus
f plexus intermésentérique
e plexo intermesentérico
d Plexus intermesentericus
l plexus intermesentericus

2258 intermetacarpal articulations
f articulations intermétacarpiennes
e articulaciones intermetacarpianas
d Gelenke zwischen Vordermittelfußknochen
l articulationes intermetacarpeae

2259 intermetatarsal articulations
f articulations intermétatarsiennes
e articulaciones intermetatarsianas
d Gelenke zwischen Hintermittelfußknochen
l articulationes intermetatarseae

2260 intermuscular septum
f cloison intermusculaire
e septo intermusculoso
d Septum intermusculare
l septum intermusculare

2261 internal acoustic meatus
f méat acoustique interne; conduit auditif interne
e meato acústico interno
d innerer Gehörgang
l meatus acusticus internus

2262 internal carotid artery
f artère carotide interne
e arteria carótida interna
d Arteria carotis interna
l arteria carotis interna

2263 internal carotid nerve
f nerf carotidien interne
e nervio carotídeo interno
d Nervus caroticus internus
l nervus caroticus internus

2264 internal ear
f oreille interne
e oído interno
d inneres Ohr
l auris interna

2265 internal iliac artery
f artère iliaque interne
e arteria ilíaca interna
d Arteria iliaca interna
l arteria iliaca interna

2266 internal iliac vein
f veine iliaque interne
e vena ilíaca interna
d Vena iliaca interna
l vena iliaca interna

2267 internal iliofemoral muscle
f muscle ilio-fémoral interne
e músculo iliofemoral interno
d Musculus iliofemoralis internus
l musculus iliofemoralis internus

2268 internal jugular vein
f veine jugulaire interne
e vena yugular interna
d Vena jugularis interna
l vena jugularis interna

2269 internal layer of hoof wall
f couche interne de la paroi cornée
e estrato interno de la pared
d Blättchenschicht der Hufkapsel
l stratum internum parietis cornei

2270 internal oblique muscle of abdomen
f muscle oblique interne de l'abdomen
e músculo oblicuo interno del abdomen
d innerer schiefer Bauchmuskel
l musculus obliquus internus abdominis

2271 internal obturator muscle
f muscle obturateur interne
e músculo obturador interno
d innerer Verstopfungsmuskel
l musculus obturatorius internus

2272 internal occipital protuberance
f protubérance occipitale interne
e protuberancia occipital interna
d Protuberantia occipitalis interna
l protuberantia occipitalis interna

2273 internal opening of uterus
f orifice interne de l'utérus
e orificio interno del útero
d innerer Muttermund
l ostium uteri internum

2274 internal ophthalmic artery
f artère ophtalmique interne
e arteria oftálmica interna
d Arteria ophthalmica interna
l arteria ophthalmica interna

2275 internal ophthalmic vein
f veine ophtalmique interne
e vena oftálmica interna
d Vena ophthalmica interna
l vena ophthalmica interna

2276 internal orifice of urethra
f ostium interne de l'urètre
e orificio interno de la uretra
d innere Harnröhrenmündung
l ostium urethrae internum

* internal parasite → 1469

2277 internal pudendal artery
f artère honteuse interne
e arteria pudenda interna
d Arteria pudenda interna
l arteria pudenda interna

2278 internal sagittal crest
f crête sagittale interne
e cresta sagital interna
d Crista sagittalis interna
l crista sagittalis interna

2279 internal spiral sulcus
f sillon spiral interne
e surco espiral interno
d Sulcus spiralis internus
l sulcus spiralis internus

2280 internal thoracic artery
f artère thoracique interne
e arteria torácica interna
d Arteria thoracica interna
l arteria thoracica interna

2281 internal thoracic vein
f veine thoracique interne
e vena torácica interna
d Vena thoracica interna
l vena thoracica interna

2282 internal tunic of eyeball
f tunique interne du bulbe de l'œil
e túnica interna del globo
d innere Augenhaut
l tunica interna bulbi

2283 internal vane
f vexille interne
e estandarte interno
d Innenfahne; innere Federfahne
l vexillum internum

2284 interosseous arteries
f artères interosseuses
e arterias interóseas
d Arteriae interosseae
l arteriae interosseae

2285 interosseous ligament of forearm
f ligament interosseux antébrachial
e ligamento interóseo del antebrazo
d Ligamentum interosseum antebrachii
l ligamentum interosseum antebrachii

2286 interosseous membrane
f membrane interosseuse
e membrana interósea
d Membrana interossea
l membrana interossea

2287 interosseous muscles
f muscles interosseux
e músculos interóseos
d Zwischenknochenmuskeln
l musculi interossei

2288 interosseous veins
f veines interosseuses
e venas interóseas
d Venae interosseae
l venae interosseae

2289 interparietal bone
 f os interpariétal
 e hueso interparietal
 d Zwischenscheitelbein
 l os interparietale

2290 interposed nuclei of cerebellum
 f noyaux interposés du cervelet
 e núcleos interpuestos del cerebelo
 d Nuclei interpositi cerebelli
 l nuclei interpositi cerebelli

2291 interscapular region
 f région interscapulaire
 e región interescapular
 d Zwischenschulterblattregion
 l regio interscapularis

2292 intersex
 f intersexué
 e intersexo
 d Intersex; Zwischengeschlecht

2293 interspinal ligaments
 f ligaments interépineux
 e ligamentos interespinales
 d Ligamenta interspinalia
 l ligamenta interspinalia

2294 interspinal muscles
 f muscles interépineux
 e músculos interespinales
 d Zwischendornfortsatzmuskeln
 l musculi interspinales

2295 interstitial cell
 f cellule interstitielle
 e célula intersticial
 d Zwischenzelle; interstitiale Zelle

2296 interstitial nucleus (of Cajal)
 f noyau interstiel (de Cajal)
 e núcleo intersticial
 d Nucleus interstitialis
 l nucleus interstitialis

 * **interstitial space → 2297**

2297 interstitial tissue; interstitial space
 f tissu interstitiel
 e intersticio
 d Zwischenraum; Interstitialgewebe
 l interstitium

2298 intertarsal articulations
 f articulations intertarsiennes
 e articulaciones intertarsianas
 d Zwischenreihengelenke
 l articulationes intertarseae

2299 intertransverse ligament
 f ligament intertransversaire
 e ligamento intertransverso
 d Ligamentum intertransversarium
 l ligamentum intertransversarium

2300 intertransverse muscles
 f muscles intertransversaires
 e músculos intertransversos
 d Zwischenquerfortsatzmuskeln
 l musculi intertransversarii

2301 intervenous tubercle (*of right atrium*)
 f tubercule interveineux
 e tubérculo intervenoso
 d Muskelwulst des rechten Vorhofs
 l tuberculum intervenosum

2302 interventricular foramen
 f foramen interventriculaire
 e agujero interventricular
 d Foramen interventriculare
 l foramen interventriculare

2303 interventricular groove
 f sillon interventriculaire
 e surco interventricular
 d Herzlängsfurche
 l sulcus interventricularis

2304 interventricular septum
 f septum interventriculaire; cloison
    interventriculaire
 e septo interventricular
 d Herzkammerscheidewand
 l septum interventriculare

2305 intervertebral disk
 f disque intervertébral
 e disco intervertebral
 d Zwischenwirbelscheibe
 l discus intervertebralis

2306 intervertebral foramen
 f foramen intervertébral
 e agujero intervertebral
 d Zwischenwirbelloch
 l foramen intervertebrale

2307 intervertebral veins
 f veines intervertébrales
 e venas intervertebrales
 d Zwischenwirbelvenen
 l venae intervertebrales

2308 intestinal absorption
 f absorption intestinale
 e absorción intestinal
 d Darmresorption

2309 **intestinal glands; crypts of Lieberkühn**
f glandes intestinales (de Lieberkühn)
e glándulas intestinales
d Darmwanddrüsen; Lieberkühnsche Krypten
l glandulae intestinales

2310 **intestinal lymphatic trunk**
f tronc intestinal lymphatique
e tronco intestinal
d Truncus intestinalis
l truncus intestinalis

2311 **intestinal microorganisms; gut flora**
f flore intestinale
e flora intestinal
d Darmflora

2312 **intestinal obstruction**
f occlusion intestinale
e oclusión intestinal
d Darmverstopfung; Darmverschluß

2313 **intestinal parasite**
f parasite intestinal
e parásito intestinal
d Darmparasit

2314 **intestinal villi**
f villosités intestinales
e villosidades intestinales
d Darmzotten
l villi intestinales

2315 **intestine**
f intestin
e intestino
d Darm
l intestinum

2316 **intima; inner coat of vessels**
f intima des vaisseaux; tunique interne des vaisseaux
e íntima; túnica íntima
d Intima; Innenauskleidung der Blutgefäße
l tunica intima vasorum

2317 **intra-abdominal**
f intra-abdominal
e intraabdominal
d intraabdominal

2318 **intra-articular**
f intra-articulaire
e intraarticular
d intraartikulär

2319 **intra-articular ligament of head of rib; conjugal ligament**
f ligament intra-articulaire de la tête costale
e ligamento intraarticular de la cabeza de la costilla
d Ligamentum capitis costae intraarticularis
l ligamentum capitis costae intraarticularis

2320 **intracerebral**
f intracérébral
e intracerebral
d intrazerebral

2321 **intrachondral joints**
f articulations intrachondrales
e articulaciones intracondrales
d Rippenknorpelgelenke
l articulationes intrachondrales

2322 **intracranial**
f intracrânien
e intracraneal
d intrakranial

2323 **intradermal injection**
f injection intradermique
e inyección intradérmica
d intrakutane Injektion

2324 **intradermal test**
f épreuve intradermique
e prueba intradérmica
d Intrakutanprobe

2325 **intralaminar nuclei of thalamus**
f noyaux intralaminaires du thalamus
e núcleos intralaminares del tálamo
d Nuclei intralaminares thalami
l nuclei intralaminares thalami

2326 **intramammary**
f intramammaire
e intramamario
d intramammär

2327 **intramuscular injection**
f injection intramusculaire
e inyección intramuscular
d intramuskuläre Injektion

2328 **intraperitoneal injection**
f injection intrapéritonéale
e inyección intraperitoneal
d intraperitoneale Injektion

2329 **intratracheal injection**
f injection intratrachéale
e inyección intratraqueal
d intratracheale Injektion

2330 intra-uterine
 f intra-utérin
 e intrauterino
 d intrauterin

2331 intravenous injection
 f injection intraveineuse
 e inyección intravenosa
 d intravenöse Injektion

2332 intrinsic muscle of tongue
 f muscle lingual propre
 e músculo lingual propio
 d Musculus lingualis proprius
 l musculus lingualis proprius

2333 involuntary muscle
 f muscle involontaire
 e músculo involuntario
 d unwillkürliche Muskulatur

2334 iodine deficiency
 f carence en iode
 e carencia en yodo
 d Jodmangel

 * IPN → 2180

 * IPN virus → 2181

 * IPV → 2182

2335 iridocorneal angle
 f angle irido-cornéen
 e ángulo iridocorneal
 d Angulus iridocornealis
 l angulus iridocornealis

2336 iridovirus
 f iridovirus
 e iridovirus
 d Iridovirus

2337 iris
 f iris
 e iris
 d Iris; Regenbogenhaut
 l iris

2338 ischaemia; ischemia
 f ischémie
 e isquemia
 d Ischämie

 * ischemia → 2338

2339 ischial tuberosity; point of buttock
 f tubérosité ischiatique; pointe de la fesse

 e tuberosidad isquiática
 d Sitzbeinhöcker
 l tuber ischiadicum

 * ischiatic foramen → 4006

 * ischiatic incisure → 4008

2340 ischiatic lymphocentre
 f lymphocentre ischiatique
 e linfocentro ciático
 d Lymphocentrum ischiadicum
 l lymphocentrum ischiadicum

 * ischiatic nerve → 4007

2341 ischiocavernous muscle
 f muscle ischio-caverneux
 e músculo isquiocavernoso
 d Musculus ischiocavernosus
 l musculus ischiocavernosus

2342 ischiofemoral ligament
 f ligament ischio-fémoral
 e ligamento isquiofemoral
 d Ligamentum ischiofemorale
 l ligamentum ischiofemorale

2343 ischiofemoral muscle
 f muscle ischio-fémoral
 e músculo isquiofemoral
 d Musculus ischiofemoralis
 l musculus ischiofemoralis

2344 ischiopubic synchondrosis
 f synchondrose ischio-pubienne
 e sincondrosis isquiopúbica
 d Synchondrosis ischiopubica
 l synchondrosis ischiopubica

2345 ischiorectal fossa
 f fosse ischio-rectale
 e fosa isquiorrectal
 d Fossa ischiorectalis
 l fossa ischiorectalis

2346 ischium; pin bone
 f ischion
 e hueso isquion
 d Sitzbein
 l os ischii

 * islets of Langerhans → 3254

 * Isospora infection → 2347

2347 isosporosis; Isospora infection
 f isosporose
 e isosporosis
 d Isosporose

2348 isthmus of fauces
f isthme du gosier
e istmo de las fauces
d Rachenenge
l isthmus faucium

2349 isthmus of uterine tube
f isthme de la trompe utérine
e istmo de la trompa uterina
d Eileiterenge
l isthmus tubae uterinae

* itchiness → 3659

2350 ivermectin
f ivermectine
e ivermectina
d Ivermectin
l ivermectinum

* ixodidosis → 4581

2351 ixodid tick; hard tick
f ixode
e garrapata dura
d Schildzecke
l Ixodidae

J

2352 Japanese encephalitis virus
f virus de l'encéphalite japonaise
e virus de la encefalitis japonesa
d japanisches Enzephalitis-Virus

2353 jaundice; icterus
f ictère
e ictericia
d Ikterus; Gelbsucht

2354 jejunal arteries
f artères jéjunales
e arterias yeyunales
d Arteriae jejunales; Leerdarmarterien
l arteriae jejunales

2355 jejunal veins
f veines jéjunales
e venas yeyunales
d Venae jejunales
l venae jejunales

2356 jejunum
f jéjunum
e yeyuno
d Leerdarm
l jejunum

2357 jigger flea; sand flea; chigoe
f chique
e nigua
d Sandfloh
l Tunga penetrans

* **Johne's disease** → 3312

* **joint** → 297

2358 joint capsule; articular capsule
f capsule articulaire
e cápsula articular
d Gelenkkapsel
l capsula articularis

2359 joint cavity
f cavité articulaire
e cavidad articular
d Gelenkhöhle
l cavum articulare

* **joint disease** → 283

* **joint mice** → 1762

2360 jugular foramen
f foramen jugulaire
e agujero yugular
d Foramen jugulare
l foramen jugulare

2361 jugular fossa
f fosse jugulaire
e fosa yugular
d Fossa jugularis; Drosselgrube
l fossa jugularis

2362 jugular furrow; jugular groove
f sillon jugulaire
e surco yugular
d Drosselrinne
l sulcus jugularis

2363 jugular ganglion
f ganglion jugulaire
e ganglio yugular
d Ganglion jugulare
l ganglion jugulare

* **jugular groove** → 2362

2364 jugular incisure
f incisure jugulaire
e escotadura yugular
d Incisura jugularis
l incisura jugularis

2365 jugular nerve
f nerf jugulaire
e nervio yugular
d Nervus jugularis
l nervus jugularis

2366 jugular process
f processus jugulaire
e apófisis yugular
d Processus jugularis; Drosselfortsatz
l processus jugularis

K

2367 kanamycin
 f kanamycine
 e canamicina
 d Kanamycin
 l kanamycinum

2368 keel bone; sternal crest
 f carène sternale; crête sternale; bréchet
 e cresta esternal
 d Brustbeinkamm
 l carina sterni; crista sterni

* **keel bursa** → 4202

* **keel bursitis** → 570

2369 keratin
 f kératine
 e queratina
 d Keratin

2370 keratitis
 f kératite
 e queratitis
 d Keratitis; Hornhautentzündung

2371 keratoconjunctivitis
 f kératoconjonctive
 e queratoconjuntivitis
 d Keratokonjunktivitis

2372 keratosis
 f kératose
 e queratosis
 d Keratose

2373 ketamine
 f kétamine
 e cetamina
 d Ketamin
 l ketaminum

* **ketonuria** → 41

2374 ketosis
 f cétose

 e cetosis
 d Ketose

* **Key-Gaskell syndrome** → 1641

2375 kidney
 f rein
 e riñón
 d Niere
 l ren

2376 kidney disease; nephropathy
 f néphropathie
 e nefropatía
 d Nephropathie; Nierenerkrankung

2377 kidney failure; renal insufficiency
 f insuffisance rénale
 e insuficiencia renal
 d Niereninsuffizienz

* **kidney fat** → 74

2378 kidney function; renal function
 f fonction rénale
 e función renal
 d Nierenfunktion

* **Klossiella infection** → 2379

2379 klossiellosis; Klossiella infection
 f klossiellose
 e klosielosis
 d Klossiellose

2380 knee
 f genou
 e rodilla
 d Knie
 l genu

* **knee cap** → 3341

2381 knee region; stifle region
 f région du genou
 e región de la rodilla
 d Kniegegend
 l regio genus

2382 kyphosis
 f cyphose
 e cifosis
 d Kyphose

L

2383 labial arteries
f artères labiales
e arterias labiales
d Lippenarterien
l arteriae labiales

2384 labial frenulum
f frein de la lèvre
e frenillo del labio
d Lippenbändchen
l frenulum labii

2385 labial glands
f glandes labiales
e glándulas labiales
d Lippendrüsen
l glandulae labiales

2386 labial nerves
f nerfs labiaux
e nervios labiales
d Lippennerven
l nervi labiales

2387 labial papillae (*of ruminants*)
f papilles labiales
e papilas labiales
d Lippenpapillen
l papillae labiales

* labial tactile hairs → 4962

2388 labial veins
f veines labiales
e venas labiales
d Lippenvenen
l venae labiales

2389 laboratory animal
f animal de laboratoire
e animal de laboratorio
d Laboratoriumstier

2390 laboratory diagnosis
f diagnostic de laboratoire
e diagnóstico de laboratorio
d Labordiagnostik

2391 labyrinth
f labyrinthe

e laberinto
d Labyrinth
l labyrinthus

2392 lacerate foramen
f foramen lacerum; trou déchiré
e agujero rasgado
d Foramen lacerum
l foramen lacerum

2393 lacrimal artery
f artère lacrymale
e arteria lacrimal
d Tränendrüsenarterie
l arteria lacrimalis

2394 lacrimal bone
f os lacrymal; unguis
e hueso lacrimal
d Tränenbein
l os lacrimale

2395 lacrimal bulla
f bulle lacrymale
e bulla lacrimal
d Tränenbeinblase
l bulla lacrimalis

2396 lacrimal caruncle
f caroncule lacrymale
e carúncula lacrimal
d Tränenkarunkel
l caruncula lacrimalis

2397 lacrimal gland
f glande lacrymale
e glándula lacrimal
d Tränendrüse
l glandula lacrimalis

2398 lacrimal groove
f sillon lacrymal
e surco lacrimal
d Tränenfurche
l sulcus lacrimalis

2399 lacrimal nerve
f nerf lacrymal
e nervio lacrimal
d Nervus lacrimalis
l nervus lacrimalis

2400 lacrimal sac
f sac lacrymal
e saco lacrimal
d Tränensack
l saccus lacrimalis

2401 **lacrimal sinus**
f sinus lacrymal
e seno lacrimal
d Sinus lacrimalis
l sinus lacrimalis

2402 **lacrimation**
f larmoiement
e lacrimación
d Tränen

2403 **lactation disorder**
f trouble de la lactation
e trastorno de la lactancia
d Laktationstörung

2404 **lactiferous duct**
f conduit lactifère; canal galactophore
e conducto lactífero
d Milchgang
l ductus lactiferi

2405 **lactiferous sinus; milk cistern**
f sinus lactifère; citerne du lait
e seno lactífero
d Milchzisterne
l sinus lactiferi

2406 **lacuna vasorum**
f lacune vasculaire
e laguna vascular
d Gefäßlücke
l lacuna vasorum

2407 **lamb dysentery; Clostridium perfringens type B infection**
f dysentérie des agneaux; entérite hémorragique des agneaux; entérotoxémie à Clostridium perfringens type B
e disentería de los corderos; enteritis hemorrágica de los corderos; infección por Clostridium perfringens tipo B
d Lämmerdysenterie; Clostridium-perfringens-Typ-B-Infektion

2408 **lamellae of corium; laminae of hoof wall**
f lamelles du chorion
e laminillas coriales
d Lederhautblättchen
l lamellae coriales

2409 **lamellar corpuscle; Vater-Pacini corpuscle**
f corpuscule lamelleux (de Vater-Pacini)
e corpúsculo laminar (de Vater-Pacini)
d Lamellenkörperchen; Vater-Pacinisches Körperchen
l corpusculum lamellosum

2410 **lamellar corpuscle of birds; Herbst corpuscle**
f corpuscule lamelleux des oiseaux; corpuscule d'Herbst
e corpúsculo laminar
d Herbstsches Körperchen
l corpusculum lamellosum avium

2411 **lamellar part of nuchal ligament**
f lame du ligament nuchal
e lámina de la nuca
d Lamina nuchae; Nackenplatte
l lamina nuchae

2412 **lameness**
f boiterie
e cojera
d Lahmheit

2413 **lamina**
f lame
e lámina
d Lamina
l lamina

* **laminae of hoof wall** → 2408

2414 **lamina of vertebral arch**
f lame de l'arc vertébral
e lámina del arco de la vértebra
d Lamina arcus vertebrae
l lamina arcus vertebrae

2415 **lamina propria; mucous membrane proper**
f chorion de la membrane muqueuse
e lámina propia de la mucosa
d Eigenschicht der Schleimhaut

2416 **laminar corium; corium of hoof wall**
f chorion de la paroi; derme lamellaire
e corión de la pared; dermis de la pared
d Wandlederhaut
l corium parietis; dermis parietis

2417 **laminitis**
f fourbure
e laminitis
d Hufrehe; Klauenrehe

2418 **lancet fluke; small liver fluke**
f douve lancéolée; petite douve du foie
e distoma lanceolado; pequeña duela del hígado
d Lanzettegel; kleiner Leberegel
l Dicrocoelium dendriticum

2419 **laparotomy**
f laparotomie
e laparotomía
d Laparotomie; Bauchschnitt

2420 **large body louse of poultry**
 f Goniodes gigas
 e piojo gigante de las gallinas
 d Goniodes gigas
 l Goniodes gigas

2421 **large colon** (*of horse*)
 f grand côlon (*du cheval*)
 e colon grueso
 d großes Kolon
 l colon crassum

2422 **large intestine**
 f gros intestin
 e intestino grueso
 d Dickdarm
 l intestinum crassum

2423 **large occipital foramen**
 f foramen magnum; trou occipital
 e agujero magno
 d großes Hinterhauptsloch
 l foramen magnum

* **large pastern bone** → 1705

2424 **large round muscle**
 f muscle grand rond
 e músculo redondo mayor
 d großer runder Muskel
 l musculus teres major

2425 **laryngeal cartilages**
 f cartilages du larynx
 e cartílagos de la laringe
 d Kehlkopfknorpel
 l cartilagines laryngis

2426 **laryngeal cavity**
 f cavité laryngée
 e cavidad de la laringe
 d Kehlkopfhöhle
 l cavum laryngis

2427 **laryngeal glands**
 f glandes laryngées
 e glándulas laríngeas
 d Kehlkopfdrüsen
 l glandulae laryngeae

* **laryngeal hemiplegia** → 3902

2428 **laryngeal muscles**
 f muscles du larynx
 e músculos de la laringe
 d Kehlkopfmuskeln
 l musculi laryngis

* **laryngeal part of pharynx** → 2433

2429 **laryngeal prominence; Adam's apple**
 f proéminence laryngée; pomme d'Adam
 e prominencia laríngea
 d Prominentia laryngea; Adamsapfel
 l prominentia laryngea

2430 **laryngeal region**
 f région du larynx
 e región laríngea
 d Kehlkopfgegend
 l regio laryngea

2431 **laryngeal ventricle; ventricle of larynx**
 f ventricule du larynx
 e ventrículo de la laringe
 d Kehlkopfventrikel
 l ventriculus laryngis

2432 **laryngitis**
 f laryngite
 e laringitis
 d Laryngitis; Kehlkopfentzündung

2433 **laryngopharynx; laryngeal part of pharynx**
 f laryngopharynx; partie laryngienne du
      pharynx
 e hipofaringe
 d Kehlrachen
 l pars laryngea pharyngis

2434 **laryngotracheitis**
 f laryngotrachéite
 e laringotraqueitis
 d Laryngotracheitis

2435 **larynx**
 f larynx
 e laringe
 d Larynx; Kehlkopf
 l larynx

2436 **lasalocid**
 f lasalocide
 e lasalocido
 d Lasalocid
 l lasalocidum

2437 **latent infection**
 f infection latente
 e infección latente
 d latente Infektion

2438 **lateral abdominal region; flank region**
 f région abdominale latérale; région du flanc
 e región lateral del abdomen
 d Flankengegend; seitliche Bauchregion
 l regio abdominis lateralis

2439 **lateral angle of eye; lateral canthus**
  f angle latéral de l'œil
  e ángulo lateral del ojo
  d äußerer Winkel der Lidspalte
  l angulus oculi lateralis

2440 **lateral bicipital groove**
  f sillon bicipital latéral
  e surco bicipital lateral
  d Sulcus bicipitalis lateralis
  l sulcus bicipitalis lateralis

  * **lateral canthus** → 2439

2441 **lateral cerebellar nucleus; dentate nucleus**
  f noyau latéral du cervelet; noyau dentelé
  e núcleo lateral del cerebelo
  d Nucleus lateralis cerebelli; Nucleus dentatus
  l nucleus lateralis cerebelli; nucleus dentatus

2442 **lateral cervical nucleus**
  f noyau cervical latéral
  e núcleo cervical lateral
  d seitlicher Halskern
  l nucleus cervicalis lateralis

2443 **lateral column of spinal cord**
  f cordon latéral de la moelle épinière
  e cordón lateral del médula espinal
  d Seitenstrang des Rückenmarks
  l funiculus lateralis

2444 **lateral condyle** (*of femur*)
  f condyle latéral
  e cóndilo lateral
  d lateraler Gelenkknorren
  l condylus lateralis

2445 **lateral cutaneous nerve of thigh**
  f nerf cutané fémoral latéral
  e nervio cutáneo lateral del muslo
  d Nervus cutaneus femoris lateralis
  l nervus cutaneus femoris lateralis

2446 **lateral digital extensor muscle**
  f muscle extenseur latéral des doigts
  e músculo extensor digital lateral
  d seitlicher Zehenstrecker
  l musculus extensor digitorum lateralis

  * **lateral fissure of brain** → 4393

2447 **lateral horn of spinal cord**
  f corne latérale de la moelle épinière
  e asta lateral del médula espinal
  d Lateralhorn der grauen Substanz
  l cornu laterale

2448 **lateral intercondylar tubercle** (*of tibia*)
  f tubercule intercondylaire latéral (*du tibia*)
  e tubérculo intercondilar lateral
  d Tuberculum intercondylare laterale
  l tuberculum intercondylare laterale

2449 **lateral ligament**
  f ligament latéral
  e ligamento lateral
  d Ligamentum laterale
  l ligamentum laterale

2450 **lateral ligament of bladder**
  f ligament latéral de la vessie
  e ligamento lateral de la vejiga
  d Ligamentum vesicae laterale
  l ligamentum vesicae laterale

2451 **lateral nasal arteries**
  f artères nasales latérales
  e arterias laterales de la nariz
  d Arteriae laterales nasi; seitliche Nasenarterien
  l arteriae laterales nasi

2452 **lateral nasal gland**
  f glande nasale latérale
  e glándula nasal lateral
  d Glandula nasalis lateralis
  l glandula nasalis lateralis

2453 **lateral plantar artery**
  f artère plantaire latérale
  e arteria plantar lateral
  d laterale Sohlenarterie
  l arteria plantaris lateralis

2454 **lateral plantar nerve**
  f nerf plantaire latéral
  e nervio plantar lateral
  d Nervus plantaris lateralis
  l nervus plantaris lateralis

2455 **lateral pterygoid nerve**
  f nerf ptérygoïdien latéral
  e nervio pterigoideo lateral
  d Nervus pterygoideus lateralis
  l nervus pterygoideus lateralis

2456 **lateral recess of fourth ventricle**
  f récessus latéral du quatrième ventricule
  e receso lateral del cuarto ventrículo
  d seitlicher Zipfel des vierten Hirnventrikels
  l recessus lateralis ventriculi quarti

2457 **lateral rectus muscle** (*of eyeball*)
  f muscle droit latéral (*du bulbe de l'œil*)
  e músculo recto lateral
  d Musculus rectus lateralis

l musculus rectus lateralis

2458 lateral region of neck
f région latérale du cou
e región lateral del cuello
d laterale Halsregion
l regio colli lateralis

2459 lateral retropharyngeal lymph nodes
f nœuds lymphatiques rétro-pharyngiens latéraux
e nódulos linfáticos retrofaríngeos laterales
d laterale Retropharyngeallymphknoten
l lymphonodi retropharyngei laterales

2460 lateral ridge of frog
f branche latérale de la fourchette
e pilar lateral de la cuña
d lateraler Strahlschenkel
l crus cunei laterale

2461 lateral sacral artery
f artère sacrée latérale
e arteria sacra lateral
d laterale Kreuzbeinarterie
l arteria sacralis lateralis

2462 lateral sublingual recess
f récessus sublingual latéral
e receso sublingual lateral
d Recessus sublingualis lateralis
l recessus sublingualis lateralis

2463 lateral supracondylar tuberosity (*of femur*)
f tubérosité supracondylaire latérale (*du femur*)
e tuberosidad supracondilar lateral
d Tuberositas supracondylaris lateralis
l tuberositas supracondylaris lateralis

2464 lateral tectospinal fibres
f fibres tecto-spinales latérales
e fibras tectoespinales laterales
d Fibrae tectospinales laterales
l fibrae tectospinales laterales

2465 lateral thoracic artery
f artère thoracique latérale
e arteria torácica lateral
d Arteria thoracica lateralis
l arteria thoracica lateralis

2466 lateral thoracic nerve
f nerf thoracique latéral
e nervio torácico lateral
d seitlicher Brustkorbnerv
l nervus thoracicus lateralis

2467 lateral thoracic vein
f veine thoracique latérale
e vena torácica lateral
d lateral Brustkorbvene
l vena thoracica lateralis

* lateral ulnar muscle → 4730

2468 lateral vastus muscle
f muscle vaste latéral
e músculo vasto lateral
d Musculus vastus lateralis
l musculus vastus lateralis

2469 lateral ventricle of brain
f ventricule latéral du cerveau
e ventrículo lateral
d Seitenventrikel; laterale Gehirnkammer
l ventriculus lateralis cerebri

* lateroethmoidal bone → 1425

2470 lateropharyngeal ganglion
f ganglion latéro-pharyngien
e ganglio laterofaríngeo
d Ganglion lateropharyngeum
l ganglion lateropharyngeum

2471 lead poisoning
f saturnisme; intoxication par le plomb
e intoxicación por plomo; plumbismo
d Bleivergiftung

2472 left atrioventricular opening
f orifice atrio-ventriculaire gauche
e orificio atrioventricular izquierdo
d linke Atrioventrikularöffnung
l ostium atrioventricularis sinistrum

2473 left atrioventricular valve; bicuspid valve; mitral valve
f valve atrio-ventriculaire gauche; valve bicuspide; valve mitrale
e valva atrioventricular izquierda; valva bicúspide; valva mitral
d linke Atrioventrikularklappe; Mitralklappe; Bikuspidalklappe
l valva atrioventricularis sinistra; valva bicuspidalis; valva mitralis

2474 left atrium
f oreillette gauche du cœur
e atrio izquierdo
d linker Vorhof
l atrium sinistrum

2475 left azygous vein
f veine azygos gauche

e vena ácigos izquierda
d Vena azygos sinistra
l vena azygos sinistra

2476 left gastric artery
f artère gastrique gauche
e arteria gástrica izquierda
d Arteria gastrica sinistra; linke
Magenarterie
l arteria gastrica sinistra

2477 left gastro-epiploic artery
f artère gastro-épiploïque gauche
e arteria gastroepiplóica izquierda
d Arteria gastroepiploica sinistra
l arteria gastroepiploica sinistra

2478 left gastro-epiploic vein
f veine gastro-épiploïque gauche
e vena gastroepiplóica izquierda
d Vena gastroepiploica sinistra
l vena gastroepiploica sinistra

2479 left hepatic duct
f conduit hépatique gauche; canal hépatique
gauche
e conducto hepático izquierdo
d linker Gallengang
l ductus hepaticus sinister

2480 left lobe of liver
f lobe gauche du foie; lobe hépatique gauche
e lóbulo izquierdo del hígado
d linker Leberlappen
l lobus hepatis sinister

* left lymphatic duct → 4536

2481 left ovarian vein
f veine ovarique gauche
e vena ovárica izquierda
d linke Eierstocksvene
l vena ovarica sinistra

2482 left pulmonary artery
f artère pulmonaire gauche
e arteria pulmonar izquierda
d linke Lungenarterie
l arteria pulmonalis sinistra

2483 left ruminal artery
f artère ruminale gauche
e arteria ruminal izquierda
d linke Pansenarterie
l arteria ruminalis sinistra

2484 left testicular vein
f veine testiculaire gauche

e vena testicular izquierda
d linke Hodenvene
l vena testicularis sinistra

2485 left ventricle
f ventricule gauche
e ventrículo izquierdo
d linke Herzkammer
l ventriculus sinister

* leg → 2517

* Leishmania infection → 2486

2486 leishmaniosis; Leishmania infection
f leishmaniose
e leishmaniosis
d Leishmaniose

2487 lemniscus
f lemnisque
e lemnisco
d Schleifenbahn
l lemniscus

2488 lens
f cristallin
e cristalino
d Linse
l lens

2489 lens capsule
f capsule du cristallin
e cápsula del cristalino
d Linsenkapsel
l capsula lentis

2490 lens fibres
f fibres du cristallin
e fibras del cristalino
d Linsenfasern
l fibrae lentis

* lenticular nucleus → 2491

2491 lentiform nucleus; lenticular nucleus
f noyau lenticulaire
e núcleo lenticular
d linsenförmiger Kern
l nucleus lentiformis

2492 lentivirus; slow virus
f lentivirus; virus lent
e lentivirus
d Lentivirus

2493 leporipoxvirus
f léporipoxvirus
e leporipoxvirus
d Leporipoxvirus

2494 leptospire
f leptospire
e leptospira
d Leptospira
l Leptospira

2495 leptospirosis
f leptospirose
e leptospirosis
d Leptospirose

2496 lesion
f lésion
e lesión
d Läsion; krankhafte Veränderung
l laesio

2497 lesser omentum
f petit omentum
e omento menor
d kleines Netz
l omentum minus

2498 lesser ring of iris; inner ring of iris
f petite circonférence de l'iris
e anillo menor del iris
d innerer Irisring
l anulus iridis minor

2499 lethal
f létal
e letal
d letal

2500 lethal dose
f dose létale
e dosis letal
d Dosis letalis; tödliche Dosis
l dosis letalis

2501 lethal factor
f facteur létal
e factor letal
d Letalfaktor

* **leucaemia** → 2502

* **leucocyte** → 2503

* **Leucocytozoon infection** → 2505

* **leucosis** → 2508

2502 leukaemia; leukemia; leucaemia
f leucémie
e leucemia
d Leukämie

* **leukemia** → 2502

2503 leukocyte; leucocyte; white blood cell
f leucocyte
e leucocito; glóbulo blanco
d Leukozyt; weißes Blutkörperchen

2504 leukocytosis
f leucocytose
e leucocitosis
d Leukozytose

2505 leukocytozoonose; Leucocytozoon infection
f leucocytozoonose
e leucocitozoonosis
d Leukozytozoonose

2506 leukoderma
f leucodermie
e leucodermia
d Leukoderma

2507 leukopenia
f leucopénie
e leucopenia
d Leukopenie

2508 leukosis; leucosis
f leucose
e leucosis
d Leukose

2509 levamisole
f lévamisole
e levamisol
d Levamisol
l levamisolum

* **levarterenol** → 3023

2510 levator ani muscle
f muscle élévateur de l'anus
e músculo elevador del ano
d Heber des Afters
l musculus levator ani

2511 levator muscle of eyelid
f muscle releveur de la paupière
e músculo elevador del párpado
d Heber des Augenlids
l musculus levator palpebrae

2512 levator muscle of upper lip
f muscle releveur de la lèvre supérieure
e músculo levador del labio superior
d Heber der Oberlippe
l musculus levator labii superioris

* LH → 2609

2513 lidocaine; lignocaine
 f lidocaïne
 e lidocaína
 d Lidocain
 l lidocainum

2514 ligament
 f ligament
 e ligamento
 d Ligamentum; Band
 l ligamentum

2515 ligament of femoral head; round ligament of femur
 f ligament de la tête du fémur
 e ligamento de la cabeza del hueso fémur
 d Ligamentum capitis ossis femoris
 l ligamentum capitis ossis femoris; ligamentum teres femoris

2516 ligament of head of fibula
 f ligament de la tête fibulaire
 e ligamento de la cabeza del peroné
 d Ligamentum capitis fibulae
 l ligamentum capitis fibulae

* lignocaine → 2513

2517 limb; leg
 f membre
 e pata
 d Bein; Extremität
 l membrum

2518 limbus
 f limbe
 e limbo; eponiquio
 d Limbus; Saum
 l limbus; vallum

2519 lincomycin
 f lincomycine
 e lincomicina
 d Lincomycin
 l lincomycinum

2520 lindane; HCH; gamma benzene hexachloride
 f lindane; HCH
 e lindano
 d Lindan; HCH
 l lindanum

2521 lingual artery
 f artère linguale
 e arteria lingual
 d Zungenarterie
 l arteria lingualis

2522 lingual fossa (*of bovines*)
 f fosse linguale
 e fosa lingual
 d Futterloch
 l fossa linguae

2523 lingual frenulum
 f frein de la langue
 e frenillo de la lengua
 d Zungenbändchen
 l frenulum linguae

2524 lingual glands
 f glandes linguales
 e glándulas linguales
 d Zungendrüsen
 l glandulae linguales

* lingual muscles → 4592

2525 lingual nerve
 f nerf lingual
 e nervio lingual
 d Zungennerv
 l nervus lingualis

2526 lingual papillae; gustatory papillae
 f papilles gustatives
 e papilas linguales
 d Zungenpapillen
 l papillae linguales

2527 lingual septum; septum of tongue
 f septum lingual
 e septo de la lengua
 d Septum linguae
 l septum linguae

2528 lingual tonsil
 f tonsille linguale; amygdale linguale
 e tonsila lingual
 d Zungenmandel
 l tonsilla lingualis

2529 lingual vein
 f veine linguale
 e vena lingual
 d Zungenvene
 l vena lingualis

2530 linguatulosis; tongueworm infestation
 f linguatulose
 e linguatulosis
 d Linguatulose

2531 lingula of cerebellum
 f lingula du cervelet
 e língula del cerebelo
 d Kleinhirnzüngelchen
 l lingula cerebelli

2532 linguofacial trunk
 f tronc linguo-facial
 e tronco linguofacial
 d Truncus linguofacialis
 l truncus linguofacialis

2533 linguofacial vein
 f veine linguo-faciale
 e vena linguofacial
 d Vena linguofacialis
 l vena linguofacialis

2534 lip
 f lèvre
 e labio
 d Lippe
 l labium

2535 lipodystrophy
 f lipodystrophie
 e lipodistrofia
 d Lipodystrophie

2536 lipoprotein
 f lipoprotéine
 e lipoproteína
 d Lipoprotein

2537 lips of mouth
 f lèvres de la bouche
 e labios de la boca
 d Lippen
 l labia oris

* Listeria monocytogenes infection → 2538

2538 listeriosis; Listeria monocytogenes infection
 f listériose; infection à Listeria monocytogenes
 e listeriosis
 d Listeriose; Listeria-Infektion

2539 lithiasis
 f lithiase
 e litiasis
 d Lithiase; Steinkrankheit

2540 liver
 f foie
 e hígado
 d Leber
 l hepar

* liver cell → 1991

2541 liver cirrhosis
 f cirrhose du foie
 e cirrosis del hígado
 d Leberzirrhose

2542 liver disease
 f maladie du foie
 e enfermedad del hígado
 d Hepatopathie; Leberkrankheit

* liver enlargement → 1994

2543 liver fluke
 f douve hépatique
 e distoma hepático
 d Leberegel
 l Fasciola

* liver fluke infestation → 1625

2544 liver lobe
 f lobe hépatique
 e lóbulo del hígado
 d Leberlappen
 l lobus hepatis

2545 liver lobule
 f lobule hépatique
 e lóbulo hepático
 d Leberläppchen
 l lobulus hepatis

2546 live vaccine
 f vaccin vivant
 e vacuna viva
 d Lebendvakzine

2547 lobar bronchi
 f bronches lobaires
 e bronquios lobares
 d Lappenbronchen
 l bronchi lobares

2548 lobe
 f lobe
 e lóbulo
 d Lappen
 l lobus

2549 lobe of kidney
 f lobe rénal
 e lóbulo renal
 d Nierenlappen
 l lobus renalis

2550 lobe of mammary gland
 f lobe de la glande mammaire
 e lóbulo de la glándula mamaria
 d Milchdrüsenlappen
 l lobus glandulae mammariae

2551 lobule
 f lobule
 e lobulillo
 d Läppchen
 l lobulus

2552 lobule of epididymis
 f lobule de l'épididyme
 e lobulillo del epidídimo
 d Nebenhodenläppchen
 l lobulus epididymidis; conus epididymidis

2553 lobule of lung
 f lobule pulmonaire
 e lobulillo del pulmón
 d Lungenläppchen
 l lobulus pulmonis

2554 lobule of mammary gland
 f lobule de la glande mammaire
 e lóbulos de la glándula mamaria
 d Milchdrüsenläppchen
 l lobulus glandulae mammariae

2555 local anaesthesia
 f anesthésie locale
 e anestesia local
 d Lokalanästhesie; örtliche Betäubung

2556 local antibody
 f anticorps local
 e anticuerpo local
 d lokaler Antikörper

2557 loins
 f lombe(s)
 e lomos
 d Lende
 l lumbus

2558 long abductor of first digit
 f muscle long abducteur du pouce (doigt I)
 e músculo abductor largo del dedo 1°
 d Musculus abductor digiti I longus
 l musculus abductor digiti I longus

2559 long adductor muscle
 f muscle adducteur long
 e músculo aductor largo
 d Musculus adductor longus
 l musculus adductor longus

2560 long bone
 f os long
 e hueso largo
 d Röhrenknochen; langer Knochen
 l os longum

2561 long digital extensor muscle
 f muscle long extenseur digital des doigts
 e músculo extensor digital largo
 d langer Zehenstrecker
 l musculus extensor digitorum longus

2562 longest muscle of back
 f muscle longissimus; muscle long dorsal
 e músculo longísimo
 d langer Rückenmuskel
 l musculus longissimus (dorsi)

2563 longitudinal fissure of cerebrum
 f fissure longitudinale du cerveau
 e cisura longitudinal del cerebro
 d Längsspalt des Gehirns
 l fissura longitudinalis cerebri

2564 longitudinal groove of rumen
 f sillon longitudinal du rumen
 e surco longitudinal del rumen
 d Längsfurche des Rumens
 l sulcus longitudinalis (ruminis)

2565 longitudinal ligament
 f ligament longitudinal
 e ligamento longitudinal
 d Längsband der Wirbelsäule
 l ligamentum longitudinale

2566 longitudinal pillar of rumen
 f pilier longitudinal du rumen
 e pilar longitudinal del rumen
 d Längspfeiler des Rumens
 l pila longitudinalis ruminis

2567 longitudinal striae (of Lancisi)
 f stries longitudinales (de Lancisi)
 e estrías longitudinales
 d Striae longitudinales
 l striae longitudinales

2568 long muscle of head
 f muscle long de la tête
 e músculo largo de la cabeza
 d Musculus longus capitis
 l musculus longus capitis

2569 long muscle of neck
 f muscle long du cou
 e músculo largo del cuello
 d langer Halsmuskel
 l musculus longus colli

2570 long-nosed cattle louse
 f pou du bœuf
 e piojo azul de cabeza larga
 d langnasige Rinderlaus
 l Linognathus vituli

2571 long peroneal muscle
 f muscle long péronier
 e músculo peroneo largo
 d langer Wadenbeinmuskel
 l musculus peroneus longus

2572 long thoracic nerve
f nerf thoracique long
e nervio torácico largo
d langer Brustkorbnerv
l nervus thoracicus longus

2573 lordosis
f lordose
e lordosis
d Lordose

2574 lore
f lore
e brida
d Zügel
l lorum

2575 louping ill
f louping ill
e encefalomielitis ovina
d Louping-ill; Springkrankheit der Schafe;
schottische Schafenzephalitis

2576 louse
f pou
e piojo
d Laus

2577 louse fly
f hippobosque
e hipobóscido; hipobosca
d Lausfliege
l Hippoboscus

2578 louse infestation; mallophagosis;
anoplurosis
f phtiriase
e pediculosis
d Läusebefall

2579 lower beak
f rostre mandibulaire; bec inférieur
e pico inferior
d Unterschnabel
l rostrum mandibulare

2580 lower eyelid
f paupière inférieure
e párpado inferior
d Unterlid
l palpebra inferior

* **lower jawbone → 2663**

2581 lower leg
f jambe
e pierna
d Unterschenkel
l crus

2582 lower leg region; crural region
f région de la jambe
e región de la pierna
d Unterschenkelgegend
l regio cruris

2583 lower lip
f lèvre inférieure
e labio inferior; labio mandibular
d Unterlippe
l labium inferius; labium mandibulare

2584 lower part of neck
f col
e cuello
d Nacken
l cervix

* **LTH → 2610**

2585 lumbar arteries
f artères lombaires
e arterias lumbares
d Lendenarterien
l arteriae lumbales

2586 lumbar enlargement of spinal cord
f intumescence lombaire
e intumescencia lumbar
d Intumescentia lumbalis;
Lendenanschwellung
l intumescentia lumbalis

2587 lumbar ganglia
f ganglions lombaux/lombaires
e ganglios lumbares
d Lendenganglien
l ganglia lumbalia

2588 lumbar intertransverse articulations
f articulations intertransversaires lombaires
e articulaciones intertransversas lumbares
d Articulationes intertransversariae
lumbales
l articulationes intertransversariae lumbales

2589 lumbar lymphatic trunks
f troncs lombaires lymphatiques
e troncos lombares
d Trunci lumbales
l trunci lumbales

2590 lumbar lymphocentre
f lymphocentre lombaire
e linfocentro lumbar
d Lymphocentrum lumbale
l lymphocentrum lumbale

2591 **lumbar nerves**
f nerfs lombaires/lombaux
e nervios lumbares
d Lendennerven; Lumbalnerven
l nervi lumbales

2592 **lumbar part of pillar of diaphragm**
f pilier du diaphragme
e porción lumbar del diafragma
d Zwerchfellpfeiler
l pars lumbalis diaphragmatis

2593 **lumbar plexus**
f plexus lombaire
e plexo lumbar
d Lendengeflecht
l plexus lumbalis

2594 **lumbar region**
f région lombaire/lombale
e región lumbar
d Lendengegend; Lumbalregion
l regio lumbalis

2595 **lumbar splanchnic nerves**
f nerfs splanchniques lombaux
e nervios esplácnicos lumbares
d Nervi splanchnici lumbales
l nervi splanchnici lumbales

2596 **lumbar veins**
f veines lombaires
e venas lumbares
d Lendenvenen
l venae lumbales

2597 **lumbar vertebra**
f vertèbre lombaire
e vértebra lumbar
d Lendenwirbel
l vertebra lumbalis

2598 **lumbocostal ligament**
f ligament lombocostal
e ligamento lumbocostal
d Ligamentum lumbocostale
l ligamentum lumbocostale

2599 **lumbodiaphragmatic recess**
f récessus lombo-diaphragmatique
e receso lumbodiafragmático
d Recessus lumbodiaphragmaticus
l recessus lumbodiaphragmaticus

2600 **lumbosacral articulation**
f articulation lombo-sacrale
e articulación lumbosacra
d Junctura lumbosacralis
l junctura lumbosacralis

2601 **lumbosacral intertransverse articulation**
f articulation intertransversaire lombo-sacrale
e articulación intertransversa lumbosacra
d Articulatio intertransversaria lumbosacralis
l articulatio intertransversaria lumbosacralis

2602 **lumbosacral plexus**
f plexus lombo-sacral/lombo-sacré
e plexo lumbosacro
d Plexus lumbosacralis
l plexus lumbosacralis

2603 **lumbrical muscles of foot**
f muscles lombricaux
e músculos lumbricales del pie
d wurmförmige Muskeln
l musculi lumbricales

* **lumpy jaw of cattle** → 53

2604 **lumpy skin disease of cattle**
f dermatose nodulaire contagieuse bovine
e dermatosis nodular contagiosa
d Knotenausschlag des Rindes
l dermatosis nodularis

2605 **lumpy skin disease orthopoxvirus**
f orthopoxvirus de la dermatose nodulaire bovine
e orthopoxvirus de la dermatosis nodular bovina
d Orthopoxvirus der Lumpy skin disease

* **lunar bone** → 2251

2606 **lung**
f poumon
e pulmón
d Lunge
l pulmo

2607 **lung fluke**
f douve pulmonaire
e distoma pulmonar
d Lungenegel
l Paragonimus

2608 **lungworm; protostrongylid**
f strongle respiratoire
e estróngilo pulmonar
d Lungenwurm

2609 **luteinizing hormone; LH; lutropin**
f hormone lutéinisante; LH; lutropine
e hormona luteinisante; lutropina
d Luteinisierungshormon; LH; Lutropin

* **luteotropic hormone** → 2610

2610 luteotropin; luteotropic hormone; LTH
  f lutéotropine; hormone lutéotrope
  e luteotropina; hormona luteotrópica
  d Luteotropin; luteotropes Hormon; LTH

* **lutropin** → 2609

2611 Lyme disease; Borrelia burgdorferi infection
  f maladie de Lyme; infection à Borrelia burgdorferi
  e enfermedad de Lyme; infección por Borrelia burgdorferi
  d Lyme-Krankheit; Borrelia-burgdorferi-Infektion

2612 lymph
  f lymphe
  e linfa
  d Lymphe
  l lympha

2613 lymphadenitis
  f lymphadénite
  e linfadenitis
  d Lymphadenitis; Lymphknotenentzündung

2614 lymphangitis
  f lymphangite
  e linfangitis
  d Lymphangitis; Lymphgefäßentzündung

2615 lymphatic nodule
  f nodule lymphatique
  e nodulillo linfático
  d Lymphknötchen
  l lymphonodulus; nodulus lymphaticus

2616 lymphatic plexus
  f plexus lymphatique
  e plexo linfático
  d Lymphgefäßgeflecht
  l plexus lymphaticus

2617 lymphatic system
  f système lymphatique
  e sistema linfático
  d Lymphsystem
  l systema lymphaticum

2618 lymphatic system disease
  f maladie du système lymphatique
  e linfopatía
  d Lymphopathie

2619 lymphatic valve
  f valvule lymphatique
  e válvula linfática
  d Lymphgefäßklappe
  l valvula lymphatica

2620 lymph follicle
  f follicule lymphatique
  e folículo linfático
  d Lymphknötchen
  l folliculus lymphaticus

2621 lymph heart
  f cœur lymphatique
  e corazón linfático
  d Lymphherz
  l cor lymphaticum

2622 lymph node
  f nœud lymphatique
  e nódulo linfático
  d Lymphknoten
  l lymphonodus; nodus lymphaticus

* **lymphocenter** → 2623

2623 lymphocentre; lymphocenter
  f lymphocentre
  e linfocentro
  d Lymphzentrum
  l lymphocentrum

2624 lymphocystis disease of fish
  f maladie lympho-kystique des poissons
  e enfermedad linfocística de peces
  d Lymphocystis-Krankheit

2625 lymphocyte
  f lymphocyte
  e linfocito
  d Lymphozyt

2626 lymphocytic choriomeningitis arenavirus
  f arénavirus de la chorioméningite lymphocytaire
  e arenavirus de la coriomeningitis linfocitaria
  d lymphozytäres Choriomeningitisvirus

2627 lymphocytosis
  f lymphocytose
  e linfocitosis
  d Lymphozytose

2628 lymphoid cell
  f cellule lymphoïde
  e célula linfoida
  d Lymphoidzelle

2629 lymphoid tissue
f tissu lymphoïde
e tejido linfático
d Lymphgewebe

2630 lymphoma
f lymphome
e linfoma
d Lymphom

2631 lymphosarcoma
f lymphosarcome
e linfosarcoma
d Lymphosarkom

2632 lymph vessel
f vaisseau lymphatique
e vaso linfático
d Lymphgefäß
l vas lymphaticum

2633 lysis
f lyse
e lisis
d Lyse; Lysis

2634 lysosome
f lysosome
e lisosoma
d Lysosom

* lysotype → 3427

2635 lysozyme
f lysozyme
e lisozima
d Lysozym

M

2636 macrolide antibiotic
f antibiotique macrolide
e antibiótico macrolido
d Makrolidantibiotikum

2637 macrophage
f macrophage
e macrófago
d Makrophage

2638 macroscopic
f macroscopique
e macroscópico
d makroskopisch

2639 maedi; progressive interstitial pneumonia of sheep
f maedi; pneumonie progressive interstitielle du mouton
e maedi; neumonía progresiva de las ovejas
d Maedi; chronische Pneumopathie des Schafes

2640 magnesium deficiency
f carence en magnésium; déficit magnésien
e carencia en magnesio
d Magnesiummangel

2641 main pancreatic duct
f conduit pancréatique; canal de Wirsung
e conducto pancreático principal
d Hauptausführungsgang des exokrinen Pankreas
l ductus pancreaticus

2642 major histocompatibility complex
f système majeur d'histocompatibilité
e sistema mayor de histocompatibilidad
d Haupthistokompatibilitätskomplex

2643 major petrosal nerve
f nerf grand pétreux
e nervio petroso mayor
d Nervus petrosus major
l nervus petrosus major

2644 major vestibular gland
f glande vestibulaire majeure; glande de Bartholin
e glándula vestibular mayor; glándula de Bartholino
d große Vorhofsdrüse
l glandula vestibularis major

2645 malabsorption
f malabsorption
e absorción defectuosa
d Malabsorption

2646 malar artery
f artère malaire
e arteria malar
d Wangenarterie
l arteria malaris

2647 malar muscle
f muscle malaire
e músculo malar
d Musculus malaris
l musculus malaris

2648 male genital disease
f maladie de l'appareil génital mâle
e enfermedad de los órganos genitales masculinos
d Andropathie

2649 male genital organs
f organes génitaux masculins
e órganos genitales masculinos
d männliche Geschlechtsorgane
l organa genitalia masculina

* **malformation** → 1028

* **malignant catarrhal fever** → 536

2650 malignant catarrhal fever herpesvirus; alcelaphine herpesvirus type 1
f virus du coryza gangréneux des bovins
e virus de la fiebre catarral maligna
d Virus des bösartigen Katarrhalfiebers

* **malignant foulbrood** → 139

2651 malignant hyperthermia of pigs
f hyperthermie maligne du porc
e hipertermia maligna del cerdo
d maligne Hyperthermie des Schweins

2652 malignant oedema; Clostridium novyi infection
f œdème malin; infection à Clostridium novyi
e edema maligno; infección por Clostridium novyi
d malignes Ödem; Pararauschbrand; Clostridium-novyi-Infektion

2653 **mallein; Pseudomonas mallei allergen**
f malléine
e maleina
d Mallein

2654 **malleus; hammer**
f marteau
e martillo
d Hammer
l malleus

* **mallophagosis** → 2578

* **malpighian corpuscle** → 3810

2655 **mamillary body**
f corps mamillaire
e cuerpo mamilar
d Corpus mamillare
l corpus mamillare

2656 **mamillary process**
f processus mamillaire; apophyse mamillaire
e apófisis mamilar
d Zitzenfortsatz
l processus mamillaris

2657 **mamillitis; mammillitis; thelitis**
f mamillite; thélite
e mamilitis
d Mamillitis; Zitzenentzündung

2658 **mamillohypothalamic fascicle**
f faisceau mamillo-hypothalamique
e fascículo mamilohipotalámico
d Fasciculus mamillohypothalamicus
l fasciculus mamillohypothalamicus

* **mammalian adenovirus** → 2690

2659 **mammary gland**
f mamelle; glande mammaire
e mama; glándula mamaria
d Milchdrüse
l mamma; glandula mammaria

2660 **mammary gland disease**
f maladie de la mamelle
e enfermedad de la glándula mamaria
d Erkrankung der Milchdrüsen

2661 **mammary lymph nodes**
f nœuds lymphatiques mammaires
e nódulos linfáticos mamarios
d Euterlymphknoten; Gesäugelymphknoten
l lymphonodi mammarii

2662 **mammary oedema; udder oedema**
f œdème mammaire
e edema mamario
d Euterödem

* **mammillitis** → 2657

2663 **mandible; lower jawbone**
f mandibule
e mandíbula
d Unterkiefer
l mandibula

2664 **mandibular alveolar nerve**
f nerf alvéolaire mandibulaire
e nervio alveolar mandibular
d Nervus alveolaris mandibularis
l nervus alveolaris mandibularis

2665 **mandibular articulation; temporomandibular articulation**
f articulation temporo-mandibulaire
e articulación temporomandibular
d Kiefergelenk
l articulatio temporomandibularis

2666 **mandibular depressor muscle**
f muscle abaisseur de la mandibule
e músculo depresor de la mandíbula
d Musculus depressor mandibulae
l musculus depressor mandibulae

2667 **mandibular duct; duct of mandibular salivary gland; Wharton's duct**
f conduit submandibulaire/sous-mandibulaire; canal de Wharton
e conducto mandibular; conducto submaxilar; conducto de Wharton
d Ausführungsgang der Unterkieferspeicheldrüse; Wharton-Gang
l ductus mandibularis

2668 **mandibular foramen**
f foramen mandibulaire
e agujero mandibular
d Foramen mandibulae
l foramen mandibulae

2669 **mandibular fossa; glenoid fossa**
f fosse mandibulaire
e fosa mandibular
d Fossa mandibularis
l fossa mandibularis

2670 **mandibular ganglion; submandibular ganglion**
f ganglion mandibulaire; ganglion submandibulaire
e ganglio mandibular

d Ganglion mandibulare
l ganglion mandibulare; ganglion
  submandibulare

2671 mandibular gland
f glande mandibulaire
e glándula mandibular
d Unterkieferdrüse
l glandula mandibularis

2672 mandibular lymph nodes
f nœuds lymphatiques mandibulaires
e nódulos linfáticos mandibulares
d Kehlgangslymphknoten
l lymphonodi mandibulares

2673 mandibular lymphocentre
f lymphocentre mandibulaire
e linfocentro mandibular
d Lymphocentrum mandibulare
l lymphocentrum mandibulare

2674 mandibular nerve
f nerf mandibulaire
e nervio mandibular
d Unterkiefernerv
l nervus mandibularis

2675 mandibular notch
f incisure mandibulaire
e escotadura de la mandíbula
d Incisura mandibulae
l incisura mandibulae

* **mandibular symphysis** → 2249, 2250

2676 mane
f crinière
e crinera
d Mähne
l juba

2677 mange; scabies
f gale
e sarna
d Räude

2678 mange mite; sarcoptiform mite
f sarcopte; sarcoptide
e ácaro de la sarna
d Räudemilbe
l Sarcoptiformes

2679 manica flexoria
f manica flexoria
e manguito de los flexores
d Manica flexoria
l manica flexoria

2680 mannosidosis
f mannosidose
e manosidosis
d Mannosidose

2681 manubrial cartilage; cariniform cartilage
f cartilage du manubrium sternal
e cartílago del mango o manubrio
d Habichtsknorpel
l cartilago manubrii

2682 manubrium; presternum
f poignée du sternum; manche du sternum;
  manubrium
e mango del esternón
d Manubrium
l manubrium sterni

* **MAP** → 2745

**2683 Marek's disease; avian
  neurolymphomatosis**
f maladie de Marek; neurolymphomatose
  aviaire
e enfermedad de Marek; neurolinfomatosis
  aviar
d Mareksche Krankheit

**2684 Marek's disease herpesvirus; gallid
  herpesvirus type 1**
f herpèsvirus de la maladie de Marek;
  herpèsvirus des gallinacées type 1
e virus de la enfermedad de Marek;
  herpesvirus de las gallináceas tipo 1
d Marek-Virus; Gallid-Herpesvirus Typ 1

* **Marteilia infection** → 2685

2685 marteilosis; Marteilia infection
f infection à Marteilia
e marteilosis
d Marteilia-Infektion

2686 masseteric artery
f artère massétérique
e arteria masetérica
d Arteria masseterica
l arteria masseterica

2687 masseteric nerve
f nerf massétérique
e nervio masetérico
d Nervus massetericus
l nervus massetericus

2688 masseteric region
f région massétérine
e región masetérica

d Massetergegend
l regio masseterica

2689 **masseter muscle**
f muscle masséter
e músculo masetero
d äußerer Kaumuskel
l musculus masseter

2690 **mastadenovirus; mammalian adenovirus**
f mastadénovirus
e mastadenovirus
d Mastadenovirus

* **mast cell** → 2693

* **masticatory surface** → 3069

2691 **mastitis**
f mammite
e mastitis
d Mastitis; Euterentzündung

2692 **mastitis-metritis-agalactiae syndrome; MMA syndrome**
f syndrome mammite-métrite-agalactie; syndrome MMA
e síndrome metritis-mastitis-agalactia; síndrome MMA
d Mastitis-Metritis-Agalaktie-Syndrom; MMA-Komplex

2693 **mastocyte; mast cell**
f mastocyte
e mastocito
d Mastozyt; Mastzelle

2694 **mastoid foramen**
f foramen mastoïdien
e agujero mastoideo
d Foramen mastoideum
l foramen mastoideum

2695 **maternal placenta**
f placenta maternel
e porción uterina; placenta maternal
d mütterlicher Anteil der Plazenta
l pars uterina placentae

2696 **maxilla; upper jawbone**
f maxillaire
e maxilar
d Maxilla; Oberkieferbein
l maxilla

2697 **maxillary artery**
f artère maxillaire
e arteria maxilar
d Oberkieferarterie
l arteria maxillaris

2698 **maxillary gland**
f glande maxillaire
e glándula maxilar
d Oberkeiferdrüse
l glandula maxillaris

2699 **maxillary nerve**
f nerf maxillaire
e nervio maxilar
d Oberkiefernerv
l nervus maxillaris

2700 **maxillary recess** (*of carnivores*)
f récessus maxillaire
e receso maxilar
d Recessus maxillaris
l recessus maxillaris

2701 **maxillary sinus**
f sinus maxillaire
e seno maxilar
d Kieferhöhle
l sinus maxillaris

2702 **maxillary vein**
f veine maxillaire
e vena maxilar
d Oberkiefervene
l vena maxillaris

2703 **maxillopalatine aperture**
f aperture maxillo-palatine
e abertura maxilopalatina
d Oberkiefer-Gaumenbein-Öffnung
l apertura maxillopalatina

2704 **meat hygiene**
f hygiène de la viande
e higiene de carnes
d Fleischhygiene

2705 **meat inspection**
f inspection des viandes
e inspección de carnes
d Fleischbeschau

2706 **meat inspector**
f inspecteur des viandes
e inspector de la carne
d Fleischbeschauer

2707 **meatus**
f méat
e meato
d Meatus; Verbindungsgang
l meatus

2708 mebendazole
 f mébendazole
 e mebendazol
 d Mebendazol
 l mebendazolum

2709 Meckel's diverticulum
 f diverticule de Meckel
 e divertículo de Meckel
 d Meckelsches Divertikel
 l diverticulum ilei verum; diverticulum vitelli

2710 meconium
 f méconium
 e meconio
 d Mekonium; Darmpech

2711 media; middle coat of vessels
 f média; tunique moyenne des vaisseaux
 e media; túnica media
 d Media; mittlere Schicht der Gefäßwand
 l tunica media vasorum

2712 medial angle of eye; medial canthus
 f angle médial de l'œil
 e ángulo medial del ojo
 d innerer Winkel der Lidspalte
 l angulus oculi medialis

 * **medial angle of ilium** → 3968

2713 medial bicipital groove
 f sillon bicipital médial
 e surco bicipital medial
 d Sulcus bicipitalis medialis
 l sulcus bicipitalis medialis

 * **medial canthus** → 2712

2714 medial condyle (*of femur*)
 f condyle médial
 e cóndilo medial
 d medialer Gelenkknorren
 l condylus medialis

2715 medial cutaneous nerve of forearm
 f nerf cutané antébrachial médial
 e nervio cutáneo medial del antebrazo
 d Nervus cutaneus antebrachii medialis
 l nervus cutaneus antebrachii medialis

2716 medial eminence; median eminence
 f éminence médiale
 e eminencia medial
 d Eminentia medialis
 l eminentia medialis

2717 medial fascicle of telencephalon
 f faisceau médial du télencéphale
 e fascículo medial del telencéfalo
 d Fasciculus medialis telencephali
 l fasciculus medialis telencephali

2718 medial intercondylar tubercle (*of tibia*)
 f tubercule intercondylaire médial (*du tibia*)
 e tubérculo intercondilar medial
 d Tuberculum intercondylare mediale
 l tuberculum intercondylare mediale

2719 medial longitudinal fascicle
 f faisceau longitudinal médial
 e fascículo longitudinal medio
 d Fasciculus longitudinalis medialis
 l fasciculus longitudinalis medialis

2720 medial plantar artery
 f artère plantaire médiale
 e arteria plantar medial
 d mediale Sohlenarterie
 l arteria plantaris medialis

2721 medial plantar nerve
 f nerf plantaire médial
 e nervio plantar medial
 d Nervus plantaris medialis
 l nervus plantaris medialis

2722 medial pterygoid nerve
 f nerf ptérygoïdien médial
 e nervio terigoideo medial
 d Nervus pterygoideus medialis
 l nervus pterygoideus medialis

2723 medial rectus muscle (*of eyeball*)
 f muscle droit médial (*du bulbe de l'œil*)
 e músculo recto medial
 d Musculus rectus medialis
 l musculus rectus medialis

2724 medial retropharyngeal lymph nodes
 f nœuds lymphatiques rétro-pharyngiens médiaux
 e nódulos linfáticos retrofaríngeos mediales
 d mittlere Retropharyngeallymphknoten
 l lymphonodi retropharyngei mediales

2725 medial ridge of frog
 f branche médiale de la fourchette
 e pilar medial de la cuña
 d medialer Strahlschenkel
 l crus cunei mediale

2726 medial sacral artery
 f artère sacrée médiane
 e arteria sacra media
 d mediane Kreuzbeinarterie
 l arteria sacralis mediana

2727 **medial supracondylar tuberosity** (*of femur*)
f tubérosité supracondylaire médiale (*du fémur*)
e tuberosidad supracondilar medial
d Tuberositas supracondylaris medialis
l tuberositas supracondylaris medialis

2728 **medial vastus muscle**
f muscle vaste médial
e músculo vasto medio
d Musculus vastus medialis
l musculus vastus medialis

2729 **median artery**
f artère médiane (*de l'avant bras*)
e arteria mediana
d Arteria mediana
l arteria mediana

2730 **median cubital vein**
f veine médiane du coude
e vena mediana del codo
d Vena mediana cubiti
l vena mediana cubiti

* **median eminence** → 2716

2731 **median fissure of spinal cord**
f fissure médiane de la moelle épinière
e cisura media de la médula espinal
d Medianspalt des Rückenmarks
l fissura mediana medullae spinalis

2732 **median ligament of bladder; median vesical ligament**
f ligament médian de la vessie
e ligamento medio de la vejiga
d Ligamentum vesicae medianum
l ligamentum vesicae medianum

2733 **median nerve**
f nerf médian
e nervio mediano
d Nervus medianus
l nervus medianus

2734 **median sulcus of tongue**
f sillon médian de la langue
e surco medio de la lengua
d Längsfurche der Zunge
l sulcus medianus linguae

2735 **median vein**
f veine médiane
e vena mediana
d Vena mediana
l vena mediana

* **median vesical ligament** → 2732

2736 **mediastinal lymph nodes**
f nœuds lymphatiques médiastinaux
e nódulos linfáticos mediastínicos
d Mediastinallymphknoten
l lymphonodi mediastinales

2737 **mediastinal lymphocentre**
f lymphocentre médiastinal
e linfocentro mediastínico
d Lymphocentrum mediastinale
l lymphocentrum mediastinale

2738 **mediastinal pleura**
f plèvre médiastinale
e pleura mediastínica
d Mittelfell
l pleura mediastinalis

2739 **mediastinal recess** (*of pleural cavity*)
f récessus médiastinal
e receso del mediastino
d Recessus mediastini
l recessus mediastini

2740 **mediastinal veins**
f veines médiastinales
e venas mediastínicas
d Mittelfellvenen
l venae mediastinales

2741 **mediastinum**
f médiastin
e mediastino
d Mediastinum
l mediastinum

2742 **medicated feed**
f aliment médicamenteux
e alimento medicado
d Medizinalfutter

2743 **medication**
f médication
e medicación
d Medikation

2744 **medicinal plant**
f plante médicinale
e planta medicinal
d Arzneipflanze; Heilpflanze

* **medicinal treatment** → 1392

2745 **medroxyprogesterone; MAP**
f médroxyprogestérone
e medroxiprogesterona
d Medroxyprogesteron
l medroxyprogesteronum

2746 medulla oblongata
 f moelle allongé; bulbe rachidien
 e médula oblongada
 d Medulla oblongata; verlängertes
   Rückenmark
 l medulla oblongata

 * **medulla of kidney → 3815**

2747 medullary cavity
 f cavité médullaire
 e cavidad medular
 d Markhöhle
 l cavum medullare

2748 medullary cone; terminal cone
 f cône médullaire; cône terminal
 e cono medular
 d Conus medullaris; Endkegel
 l conus medullaris

2749 medullary lamina
 f lame médullaire
 e lámina medular
 d Lamina medullaris
 l lamina medullaris

2750 megestrol
 f mégestrol
 e megestrol
 d Megestrol
 l megestrolum

 * **meibomian glands → 4445**

 * **Meissner's plexus → 4283**

2751 melanocyte
 f mélanocyte
 e melanocito
 d Melanozyt

2752 melanocyte-stimulating hormone; MSH
 f hormone mélanotrope; mélanostimuline;
   MSH
 e melanostimulina
 d melanozytstimulierendes Hormon; MSH

2753 melanoma
 f mélanome
 e melanoma
 d Melanom

2754 melanosis
 f mélanose
 e melanosis
 d Melanose

2755 melatonin
 f mélatonine
 e melatonina
 d Melatonin

2756 melioidosis; Pseudomonas pseudomallei infection
 f mélioïdose; infection à Pseudomonas
   pseudomallei
 e melioidosis; infección por Pseudomonas
   pseudomallei
 d Melioidose; Pseudomonas-pseudomallei-
   Infektion

 * **Melissococcus pluton infection → 1557**

2757 membrane
 f membrane
 e membrana
 d Membran
 l membrana

2758 membranous labyrinth
 f labyrinthe membraneux
 e laberinto membranoso
 d häutiges Labyrinth
 l labyrinthus membranaceus

 * **membranous semicircular canals → 4035**

2759 meningeal arteries
 f artères méningées
 e arterias meníngeas
 d Arteriae meningeae; Hirnhautarterien
 l arteriae meningeae

2760 meninges
 f méninges
 e meninges
 d Meningen; Hirnhäute
 l meninges

2761 meningitis
 f méningite
 e meningitis
 d Meningitis

2762 meningo-encephalitis
 f méningo-encéphalite
 e meningoencefalitis
 d Meningoenzephalitis

 * **meniscal ligament → 4641**

2763 meniscofemoral ligament
 f ligament ménisco-fémoral
 e ligamento meniscofemoral
 d Ligamentum meniscofemorale
 l ligamentum meniscofemorale

2764 mental foramen
f foramen mentonnier
e agujero mentoniano
d Foramen mentale
l foramen mentale

2765 mental nerves
f nerfs mentonniers
e nervios mentonianos
d Nervi mentales
l nervi mentales

* **mental organ** → 859

2766 mentolabial sulcus
f sillon mento-labial
e surco mentolabial
d Sulcus mentolabialis
l sulcus mentolabialis

2767 mercury poisoning
f hydrargyrisme
e intoxicación por mercurio
d Quecksilbervergiftung

2768 Merkel's tactile cells/disks
f disques de Merkel
e meniscos táctilos
d Merkelsche Tastzellen
l menisci tactus

2769 merozoite
f mérozoïte
e merozoito
d Merozoit

2770 mesencephalic tegmentum
f tegmen mésencéphalique
e tegmento del mesencéfalo
d Mittelhirnhaube
l tegmentum mesencephali

2771 mesencephalon; midbrain
f mésencéphale; cerveau moyen
e mesencéfalo
d Mesencephalon; Mittelhirn
l mesencephalon

2772 mesenchyme
f mésenchyme
e mesénquima
d Mesenchym

2773 mesenteric plexus
f plexus mésentérique
e plexo mesentérico
d Plexus mesentericus
l plexus mesentericus

2774 mesentery
f mésentère
e mesenterio
d Mesenterium; Gekröse
l mesenterium

2775 mesethmoidal bone
f os mésethmoïde
e hueso mesetmoides
d Os mesethmoidale
l os mesethmoidale

2776 mesocolon; colonic mesentery
f mésocolon
e mesocolon
d Grimmdarmgekröse
l mesocolon

2777 mesoduodenum
f mésoduodénum
e mesoduodeno
d Duodenalgekröse
l mesoduodenum

2778 mesometrium; uterine mesentery
f mésomètre
e mesometrio
d Mesometrium; Gebärmuttergekröse
l mesometrium

* **mesonephric duct** → 4978

2779 mesonephros
f mésonéphros; corps de Wolff
e mesonefros
d Mesonephros; Urniere
l mesonephros

2780 mesorchium; testicular mesentery
f mésorchium
e mesorquio
d Mesorchium; Hodengekröse
l mesorchium

2781 mesorectum
f mésorectum
e mesorrecto
d Mastdarmgekröse
l mesorectum

2782 mesosalpinx
f mésosalpinx
e mesosalpinx
d Mesosalpinx; Eileitergekröse
l mesosalpinx

2783 mesovarium
f mésovarium

e mesovario
d Mesovarium; Eierstockgekröse
l mesovarium

2784 metabolic disorder
f trouble du métabolisme
e trastorno metabólico
d Stoffwechselstörung

2785 metacarpal arteries
f artères métacarpiennes
e arterias metacarpianas
d Metakarpalarterien
l arteriae metacarpeae

2786 metacarpal bones I-V
f os métacarpiens
e huesos metacarpianos 1°-5°
d Metakarpalien I-V
l ossa metacarpalia I-V

* **metacarpal exostosis** → 4174

2787 metacarpal ligaments
f ligaments métacarpiens
e ligamentos metacarpianos
d Ligamenta metacarpea
l ligamenta metacarpea

2788 metacarpal pad
f coussinet métacarpien
e almohadilla metacarpiana
d Metakarpalballen
l torus metacarpeus

2789 metacarpal regions
f régions métacarpiennes
e regiones del metacarpo
d Metakarpalregionen
l regiones metacarpi

2790 metacarpal veins
f veines métacarpiennes
e venas metacarpianas
d Metakarpalvenen
l venae metacarpeae

* **metacarpophalangeal articulations** → 1673

2791 metacarpophalangeal region; fetlock of ruminants
f région métacarpo-phalangienne; boulet des ongulés
e región metacarpofalangiana
d Fesselgelenksgegend
l regio metacarpophalangea

2792 metacarpus
f métacarpe
e metacarpo
d Metacarpus; Mittelhand
l metacarpus

2793 metacercaria
f métacercaire
e metacercaria
d Metazerkarie

2794 metallibure; methallibure
f métallibure
e metalibura
d Metallibur
l metalliburum

2795 metapatagium
f métapatagium
e metapatagio
d Metapatagium
l metapatagium

2796 metaphysis
f métaphyse
e metáfisis
d Metaphyse
l metaphysis

2797 metastrongylosis; Metastrongylus infestation
f métastrongylose
e metastrongilosis
d Metastrongylose

* **Metastrongylus infestation** → 2797

2798 metatarsal arteries
f artères métatarsiennes
e arterias metatarsianas
d Metatarsalarterien
l arteriae metatarseae

2799 metatarsal bones I-V
f os métatarsiens I-V
e huesos metatarsianos 1°-5°
d Metatarsalien I-V
l ossa metatarsalia I-V

2800 metatarsal ligaments
f ligaments métatarsiens
e ligamentos metatarsianos
d Ligamenta metatarsea
l ligamenta metatarsea

2801 metatarsal pad
f coussinet métatarsien
e almohadilla metatarsiana
d Metatarsalballen
l torus metatarseus

2802 **metatarsal regions**
f régions métatarsiennes
e regiones del metatarso
d Metatarsalregionen
l regiones metatarsi

* **metatarsal spur** → 4187

2803 **metatarsal veins**
f veines métatarsiennes
e venas metatarsianas
d Metatarsalvenen
l venae metatarseae

2804 **metatarsophalangeal articulations; fetlock joint**
f articulations métatarso-phalangiennes; articulation du boulet
e articulaciones metatarsofalangianas
d Articulationes metatarsophalangeae; Fesselgelenk
l articulationes metatarsophalangeae

2805 **metatarsophalangeal region; fetlock**
f région métatarso-phalangienne; boulet
e región metatarsofalangiana
d Fesselgelenksgegend (der Beckgliedmaße)
l regio metatarsophalangea

2806 **metatarsus**
f métatarse
e metatarso
d Metatarsus; Mittelfuß
l metatarsus

2807 **metencephalon**
f métencéphale
e metencéfalo
d Hinterhirn
l metencephalon

2808 **methadone**
f méthadone
e metadona
d Methadon
l methadonum

2809 **methaemoglobin; methemoglobin**
f méthémoglobine
e methemoglobina
d Methämoglobin

* **methallibure** → 2794

* **methemoglobin** → 2809

2810 **metomidate**
f métomidate

e metomidato
d Metomidat
l metomidatum

2811 **metrifonate; trichlorfon**
f métrifonate; trichlorphon
e metrifonato
d Metrifonat; Trichlorphon
l metrifonatum

2812 **metritis**
f métrite
e metritis
d Metritis; Gebärmutterentzündung

2813 **metronidazole**
f métronidazole
e metronidazol
d Metronidazol
l metronidazolum

2814 **Meynert's fascicle**
f faisceau rétroflexe de Meynert
e fascículo retroflexo
d Fasciculus retroflexus; Meynert-Bündel
l fasciculus retroflexus

2815 **microbial**
f microbien
e microbiano
d mikrobiell

2816 **microbiology**
f microbiologie
e microbiología
d Mikrobiologie

2817 **microcephaly**
f microcéphalie
e microcefalía
d Mikrozephalie

2818 **micrococcus**
f microcoque
e micrococo
d Mikrokokk
l Micrococcus

2819 **microgamete**
f microgamète
e microgameto
d Mikrogamet

2820 **microgamont**
f microgamonte
e microgamonto
d Mikrogamont

2821 micro-organism
f micro-organisme; microbe
e microorganismo
d Mikroorganismus; Mikrobe

2822 microphthalmia; microphthalmos
f microphtalmie
e microftalmía
d Mikrophthalmie; Augapfelverkleinerung

* **microphthalmos** → 2822

2823 microscopy
f microscopie
e microscopia
d Mikroskopie

2824 microsporidia
f microsporidies
e microsporidios
d Mikrosporidien
l Microsporidia

* **midbrain** → 2771

2825 midcarpal joint; middle carpal articulation
f articulation médio-carpienne
e articulación mediocarpiana
d Zwischenreihengelenk
l articulatio mediocarpea

2826 middle abdominal region; epigastrium
f région abdominale centrale; région épigastrique
e región media del abdomen
d mittlere Bauchregion
l regio abdominis media

* **middle carpal articulation** → 2825

2827 middle cervical ganglion
f ganglion cervical moyen
e ganglio cervical medio
d mittlerer sympathischer Halsnervenknoten
l ganglion cervicale medium

2828 middle clunial nerves
f nerfs cluniaux moyens
e nervios medios de los nalgas
d Nervi clunium medii
l nervi clunium medii

* **middle coat of vessels** → 2711

2829 middle ear
f oreille moyenne
e oído medio
d Mittelohr
l auris media

2830 middle gluteal muscle
f fessier moyen
e músculo glúteo medio
d mittlere Kruppenmuskel
l musculus gluteus medius

2831 middle layer of hoof wall
f couche moyenne de la paroi cornée
e estrato medio de la pared
d Mittelschicht der Hufplatte; Hauptschicht der Hufplatte
l stratum medium parietis cornei

2832 middle lobe of lung
f lobe moyen du poumon
e lóbulo medio
d Mittellappen der Lunge
l lobus medius pulmonalis dextri

2833 middle nasal concha; middle turbinate
f cornet nasal moyen
e concha nasal media
d mittlere Nasenmuschel
l concha nasalis media

2834 middle phalangeal region; pastern region
f région de la phalange moyenne; région du pâturon
e región de la falange media
d Kronbeinregion
l regio phalangis mediae

2835 middle phalanx; second phalanx; small pastern bone
f phalange moyenne
e falange media; hueso coronal
d Kronbein
l phalanx media; os coronale

2836 middle sacral vein
f veine sacrale moyenne
e vena sacra media
d Vena sacralis mediana
l vena sacralis mediana

* **middle turbinate** → 2833

* **midge** → 1868

* **milk cistern** → 2405

* **milkers' nodule** → 3660

* **milk fever** → 3333

2837 milk hygiene
f hygiène du lait
e higiene de la leche
d Milchhygiene

* milk teeth → 1220

* milk vein → 4321

2838 mineral deficiency
 f carence minérale
 e deficiencia mineral
 d Mineralstoffmangel

2839 mineral metabolism disorder
 f trouble du métabolisme des minéraux
 e trastorno del metabolismo mineral
 d Störung im Mineralstoffwechsel

2840 mineralocorticoid
 f minéralocorticoïde
 e mineralocorticoide
 d Mineralokortikoid

2841 minimum lethal dose; MLD
 f dose létale minimale; DLM
 e dosis letal mínima
 d minimal letale Dosis; Dosis letalis minima;
   DLM
 l dosis letalis minima

2842 minor petrosal nerve
 f nerf petit pétreux
 e nervio petroso.menor
 d Nervus petrosus minor
 l nervus petrosus minor

2843 minor sublingual ducts
 f conduits sublinguaux mineurs
 e conductos sublinguales menores
 d kleine Ausführungsgänge der
   Unterzungendrüse
 l ductus sublinguales minores

2844 minor vestibular glands
 f glandes vestibulaires mineures
 e glándulas vestibulares menores
 d kleine Vorhofsdrüsen
 l glandulae vestibulares minores

2845 miracidium
 f miracidium
 e miracidio
 d Mirazidium

2846 mite
 f acarien
 e ácaro
 d Milbe
 l Acari

2847 mitochondria
 f mitochondries
 e mitocondria
 d Mitochondrien

* mitral valve → 2473

* MLD → 2841

* MMA syndrome → 2692

2848 modiolus; central pillar of cochlea
 f modiolus
 e modiolo
 d Modiolus; Schneckenspindel
 l modiolus

2849 molar glands (*of cat*)
 f glandes molaires
 e glándulas molares
 d Glandulae molares
 l glandulae molares

2850 molar teeth
 f dents molaires
 e dientes molares
 d Molaren; hintere Backenzähne
 l dentes molares

2851 molecular biology
 f biologie moléculaire
 e biología molecular
 d Molekularbiologie

2852 molecular layer of cerebellum
 f stratum moléculaire cérébelleux
 e estrato molecular del cerebelo
 d Molekularschicht der Kleinhirnrinde
 l stratum moleculare cerebelli

2853 molybdenosis; molybdenum poisoning
 f molybdénose
 e molibdenosis
 d Molybdänose; Molybdän-Vergiftung

* molybdenum poisoning → 2853

2854 monensin
 f monensin
 e monensina
 d Monensin
 l monensinum

* Moniezia infestation → 2855

2855 monieziosis; Moniezia infestation
 f moniéziose
 e monieziosis
 d Monieziose

2856 monoclonal antibody
 f anticorps monoclonal
 e anticuerpo monoclonal
 d monoklonaler Antikörper

2857 monoclonal gammopathy
f dysglobulinémie monoclonale
e gammapatía monoclonal
d monoklonale Gammopathie

2858 morantel
f morantel
e morantel
d Morantel
l morantelum

* **Moraxella bovis infection** → 533

2859 morbillivirus
f morbillivirus
e morbillivirus
d Morbillivirus

2860 morphological
f morphologique
e morfológico
d morphologisch

2861 morphology
f morphologie
e morfología
d Morphologie

2862 mortality
f mortalité
e mortalidad
d Mortalität; Sterblichkeit

2863 mortality rate
f taux de mortalité
e tasa de mortalidad; porcentaje de mortalidad
d Mortalitätsziffer; Verlustquote

2864 mosquito
f moustique
e mosquito
d Stechmücke
l Culicidae

2865 motor nerve
f nerf moteur
e nervio motor
d motorischer Nerv

2866 motor nucleus
f noyau moteur
e núcleo motor
d motorischer Kern
l nucleus motorius

2867 motor root of trigeminal nerve
f racine motrice du nerf trijumeau

e raíz motora del nervio trigémino
d motorische Wurzel des Nervus trigeminus
l radix motoria nervi trigemini

* **mouse pox** → 1429

2868 mouth
f bouche
e boca
d Mund
l os; oris

2869 mouth region
f région de la bouche
e región oral
d Mundregion
l regio oralis

* **MSH** → 2752

2870 mucormycosis
f mucormycose
e mucormicosis
d Mukormykose

2871 mucosa; mucous membrane
f muqueuse; tunique muqueuse
e mucosa; tunica mucosa
d Schleimhaut
l tunica mucosa

2872 mucosal disease of cattle
f maladie des muqueuses
e enfermedad de las mucosas
d Mucosal disease des Rindes

* **mucosal disease virus** → 528

2873 mucous gland
f glande muqueuse
e glándula mucosa
d Schleimdrüse
l glandula mucosa

* **mucous membrane** → 2871

* **mucous membrane proper** → 2415

2874 mucus
f mucosité
e moco
d Schleim
l mucus

* **Muelleria infestation** → 2875

* **muellerian duct** → 3286

2875 **muelleriosis; Muelleria infestation**
  f muellériose
  e mueleriosis
  d Mülleriose

2876 **multifidous muscles**
  f muscles multifides (*du rachis*)
  e músculos multífidos (*del raquis*)
  d vielästige Muskeln
  l musculi multifidi

2877 **multiplex placenta**
  f placenta multiplex
  e placenta múltiple
  d Placenta mutiplex
  l placenta multiplex

2878 **mummified fetus; fetal mummification**
  f foetus momifié
  e feto atrófico/strofiado
  d mumifizierte Frucht

2879 **muscle**
  f muscle
  e músculo
  d Muskel
  l musculus

2880 **muscle fibre**
  f fibre musculaire
  e fibra muscular
  d Muskelfaser

2881 **muscle relaxant**
  f myorelaxant
  e miorelajante
  d Muskelrelaxans; Myorelaxans

2882 **muscles of the head**
  f muscles de la tête
  e músculos de la cabeza
  d Kopfmuskeln
  l musculi capitis

2883 **muscle tissue**
  f tissu musculaire
  e tejido muscular
  d Muskelgewebe

  * **muscular atrophy** → 155

2884 **muscular disease; myopathy**
  f myopathie
  e miopatía
  d Myopathie; Muskelerkrankung

2885 **muscular dystrophy**
  f dystrophie musculaire
  e distrofia muscular
  d Muskeldystrophie

2886 **muscular lacuna**
  f lacune musculaire
  e laguna muscular
  d Muskellücke
  l lacuna musculorum

2887 **muscular layer of mucous membrane**
  f lame musculaire de la muqueuse
  e lámina muscular de la mucosa
  d Muskelschicht der Schleimhaut
  l lamina muscularis mucosae

  * **muscular part of stomach** → 1838

2888 **muscular process** (*of temporal bone*)
  f processus musculaire
  e apófisis muscular
  d Processus muscularis
  l processus muscularis

2889 **muscular ridges of ventricles**
  f trabécules charnues; colonnes charnues
  e trabéculas carnosas
  d Trabeculae carneae cordis
  l trabeculae carneae cordis

2890 **musculature**
  f musculature
  e musculatura
  d Muskulatur

2891 **musculocutaneous nerve**
  f nerf musculo-cutané
  e nervio musculocutáneo del brazo
  d Nervus musculocutaneus
  l nervus musculocutaneus

2892 **musculophrenic artery**
  f artère musculo-phrénique
  e arteria musculofrénica
  d Arteria musculophrenica
  l arteria musculophrenica

2893 **musculophrenic vein**
  f veine musculo-phrénique
  e vena musculofrénica
  d Vena musculophrenica
  l vena musculophrenica

2894 **mutagenicity**
  f mutagénicité
  e mutagenicidad
  d Mutagenität

  * **muzzle** → 2942, 2949

2895 **mycobacterium**
  f mycobactérie

e micobacteria
d Mykobakterie
l Mycobacterium

* **Mycobacterium avium infection** → 364

* **Mycobacterium bovis infection** → 551

* **Mycobacterium farcinogenes infection** →
531

* **Mycobacterium paratuberculosis infection**
→ 3312

2896 mycological
f mycologique
e micológico
d mykologisch

2897 mycology
f mycologie
e micología
d Mykologie

2898 mycoplasma
f mycoplasma
e micoplasma
d Mykoplasma
l Mycoplasma

* **Mycoplasma agalactiae infection** → 1040

* **Mycoplasma hyopneumoniae infection** →
3555

* **Mycoplasma mycoides infection** → 1041

* **Mycoplasma mycoides var. capri infection**
→ 1042

* **Mycoplasma synoviae infection** → 354

2899 mycoplasmosis
f mycoplasmose
e micoplasmosis
d Mykoplasmose

2900 mycosis; fungal disease
f mycose
e micosis
d Mykose; Pilzkrankheit

* **mycotic dermatitis** → 1269

2901 mycotic mastitis
f mammite mycosique
e mastitis micótica
d Pilzmastitis; Hefemastitis

2902 mycotoxicosis
f mycotoxicose
e micotoxicosis
d Mykotoxikose

2903 mycotoxin
f mycotoxine
e micotoxina
d Mykotoxin

2904 myelencephalon
f myélencéphale
e mielencéfalo
d Myelencephalon
l myelencephalon

2905 myeloblastosis
f myéloblastose
e mieloblastosis
d Myeloblastose

2906 myelocytomatosis
f myélocytomatose
e mielocitomatosis
d Myelozytomatose

2907 myenteric plexus; Auerbach's plexus
f plexus myentérique; plexus d'Auerbach
e plexo mientérico (de Auerbach)
d Plexus myentericus
l plexus myentericus

2908 myiasis; fly strike; calliphoridosis
f myiase cutanée
e miasis cutánea
d Myiase

2909 mylohyoid muscle
f muscle mylo-hyoïdien
e músculo milohioideo
d Musculus mylohyoideus
l musculus mylohyoideus

2910 mylohyoid nerve
f nerf mylo-hyoïdien
e nervio milohioideo
d Nervus mylohyoideus
l nervus mylohyoideus

2911 myocarditis
f myocardite
e miocarditis
d Myokarditis

2912 myocardium; heart muscle
f myocarde; muscle cardiaque
e miocardio; músculo cardíaco
d Myokard; Herzmuskel
l myocardium

2913 myoclonia
f myoclonie
e mioclonía
d Myoklonie

2914 myofibrillar hypoplasia
f hypoplasie myofibrillaire
e hipoplasia miofibrilar
d myofibrilläre Hyoplasie

2915 myoglobin
f myoglobine
e mioglobina
d Myoglobin

2916 myoglobinuria
f myoglobinurie
e mioglobinuria
d Myoglobinurie

2917 myometrium; uterine muscle
f myomètre
e miometrio
d Myometrium; Muskulatur der
　Gebärmutter
l myometrium

* myopathy → 2884

2918 myositis
f myosite
e miositis
d Myositis; Muskelentzündung

2919 myxobolosis; Myxobolus pfeifferii infection
f myxobolose; infection par Myxobolus
　pfeifferii

e mixobolosis; infección por Myxobolus
　pfeifferii
d Myxobolose; Myxobolus-pfeifferii-
　Infektion

* Myxobolus pfeifferii infection → 2919

2920 myxomatosis; rabbit myxomatosis
f myxomatose
e mixomatosis
d Myxomatose
l myxomatosis

* myxomatosis virus → 2921

2921 myxoma virus; myxomatosis virus
f virus de la myxomatose
e virus de la mixomatosis
d Kaninchenpestvirus; Myxomatose-Virus

* Myxosoma cerebralis infection → 2922

**2922 myxosomosis; Myxosoma cerebralis
infection; whirling disease of trout**
f myxosomose; infection par Myxosoma
　cerebralis
e mixosomosis; infección por Myxosoma
　cerebralis
d Myxosomose; Myxosoma-cerebralis-
　Infektion

* myxosporidial infection → 2923

2923 myxosporidiosis; myxosporidial infection
f myxosporidiose
e mixosporidiosis
d Myxosporidiose

N

2924 naftalofos; naphthalophos
f naphtalofos
e naftalofos
d Naftalafos
l naftalofosum

* nagana → 95

2925 nail
f ongle
e uña
d Nagel
l unguis

2926 Nairobi sheep disease
f maladie dc Nairobi
e enfermedad de Nairobi
d Nairobi sheep disease

2927 nalidixic acid
f acide nalidixique
e ácido nalidíxico
d Nalidixsäure
l acidum nalidixicum

2928 nandrolone
f nandrolone
e nandrolona
d Nandrolon
l nandrolonum

2929 nape of neck
f nuque
e nuca
d Nacken
l nucha

* naphthalophos → 2924

2930 narasin
f narasine
e narasina
d Narasin
l narasinum

2931 nasal arteries
f artères nasales
e arterias nasales
d Nasenarterien
l arteriae nasales

2932 nasal bone
f os nasal
e hueso nasal
d Nasenbein
l os nasale

2933 nasal cartilage
f cartilage du nez
e cartílago de la nariz
d Nasenknorpel
l cartilago nasi

2934 nasal cavity
f cavité nasale
e cavidad de la nariz
d Nasenhöhle
l cavum nasi

2935 nasal concha; turbinate
f cornet nasal
e cornete nasal
d Nasenmuschel
l concha nasalis

2936 nasal discharge
f jetage; écoulement nasal
e flujo nasal
d Nasenausfluß

2937 nasal diverticulum (*horse*)**; false nostril**
f diverticule nasal
e divertículo de la nariz
d Nasentrompete; falsches Nasenloch (*Pferd*)
l diverticulum nasi

2938 nasal glands
f glandes nasales
e glándulas nasales
d Nasendrüsen
l glandulae nasales

2939 nasal meatus
f méat nasal
e meato nasal
d Nasengang
l meatus nasi

2940 nasal mucosa
f muqueuse nasale
e túnica mucosa de la nariz
d Nasenschleimhaut
l tunica mucosa nasi

2941 nasal operculum
f opercule nasal
e opérculo nasal
d Operculum nasale
l operculum nasale

* nasal part of pharynx → 2954

2942 nasal plane; muzzle
f plan nasal; museau
e plano nasal
d Nasenspiegel
l planum nasale

2943 nasal process
f processus nasal
e apófisis nasal
d Nasenfortsatz
l processus nasalis

2944 nasal septum
f cloison nasale
e septo de la nariz
d Nasenscheidewand
l septum nasi

2945 nasal septum cartilage
f cartilage de la cloison nasale
e cartílago del septo nasal
d Nasenscheidewandknorpel
l cartilago septi nasi

2946 nasociliary nerve
f nerf naso-ciliaire
e nervio nasociliar
d Nervus nasociliaris
l nervus nasociliaris

2947 nasogastric tube; stomach tube
f sonde naso-œsophagienne
e sonda nasogástrica
d Nasenschlundsonde

2948 nasolabial glands
f glandes naso-labiales
e glándulas nasolabiales
d Flotzmauldrüsen
l glandulae plani nasolabiales

2949 nasolabial plane; muzzle
f plan naso-labial; mufle
e plano nasolabial
d Nasenlippenspiegel; Flotzmaul
l planum nasolabiale

2950 nasolacrimal canal
f canal naso-lacrymal
e canal nasolacrimal
d Tränennasenkanal
l canalis nasolacrimalis

2951 nasolacrimal duct
f conduit naso-lacrymal
e conducto nasolacrimal
d Tränennasengang
l ductus nasolacrimalis

2952 nasomaxillary aperture
f aperture naso-maxillaire
e abertura nasomaxilar
d Nasen-Oberkiefer-Öffnung
l apertura nasomaxillaris

2953 nasopalatine nerve
f nerf naso-palatin
e nervio nasopalatino
d Nervus nasopalatinus
l nervus nasopalatinus

2954 nasopharynx; rhinopharynx; nasal part of pharynx
f nasopharynx; rhinopharnyx; partie nasale du pharynx
e nasofaringe
d Atmungsrachen
l pars nasalis pharyngis

2955 natural antibody
f anticorps naturel
e anticuerpo natural
d natürlicher Antikörper

2956 naturally-acquired immunity
f immunité naturelle
e inmunidad natural
d natürlich erworbene Immunität

2957 navel; umbilicus
f ombilic
e ombligo
d Nabel
l umbilicus

* navicular bone → 797, 1341

2958 navicular disease; podotrochlitis
f maladie naviculaire; podotrochlite aseptique chronique
e enfermedad navicular
d Podotrochlose; Strahlbeinerkrankung; Hufrollenentzündung

2959 neck
f cou
e cuello
d Hals
l collum

2960 neck muscles
f muscles du cou
e músculos del cuello
d Halsmuskeln
l musculi colli

2961 neck of bladder
  f col vésical
  e cuello de la vejiga
  d Harnblasenhals
  l cervix vesicae

2962 neck of femur
  f col du fémur
  e cuello del hueso fémur
  d Schenkelhals
  l collum ossis femoris

2963 neck of humerus
  f col de l'humérus
  e cuello del húmero
  d Humerushals
  l collum humeri

2964 neck of omasum
  f col du feuillet
  e cuello del omaso
  d Psalterhals
  l collum omasi

  * **neck of uterus** → 840

  * **necrobacillosis** → 2965

**2965 necrobacteriosis; necrobacillosis;
  Fusobacterium necrophorum infection**
  f nécrobacillose; infection à Fusobacterium
    necrophorum
  e necrobacilosis; infección por
    Fusobacterium necrophorum
  d Nekrobazillose; Fusobacterium-
    necrophorum-Infektion

  * **necropsy** → 3569

2966 necrosis
  f nécrose
  e necrosis
  d Nekrose

2967 necrotic
  f nécrotique; nécrosé
  e necrótico
  d nekrotisch

2968 necrotic enteritis
  f entérite nécrosante
  e enteritis necrótica
  d nekrotische Enteritis
  l enteritis necrotica

  * **necrotic hepatitis** → 460

2969 necrotic stomatitis
  f stomatite nécrotique
  e estomatitis necrótica
  d nekrotische Stomatitis
  l stomatitis necrotica

2970 nematode; roundworm
  f nématode
  e nematodo
  d Nematode
  l Nematoda

  * **nematode infestation** → 2971

2971 nematodosis; nematode infestation
  f nématodose
  e nematodosis
  d Nematodose

2972 neomycin
  f néomycine
  e neomicina
  d Neomycin
  l neomycinum

2973 neonatal; newborn
  f néonatal; nouveau-né
  e recién nacido
  d neugeboren

2974 neopallium
  f néopallium
  e neopalio
  d Neopallium
  l neopallium

2975 neoplasm; tumour; tumor
  f néoplasme; tumeur
  e neoplasma; tumor
  d Neoplasma; Geschwulst; Tumor

  * **Neorickettsia helminthoeca infection** →
    3982

2976 nephritis
  f néphrite
  e nefritis
  d Nephritis; Nierenentzündung

  * **nephrolith** → 3807

2977 nephrolithiasis
  f néphrolithiase; lithiase rénale
  e nefrolitiasis
  d Nephrolithiase

2978 nephron
  f néphron
  e nefrón
  d Nephron

* nephropathy → 2376

2979 nequinate
 f néquinate
 e nequinato
 d Nequinat
 l nequinatum

2980 nerve
 f nerf
 e nervio
 d Nerv
 l nervus

2981 nerve block; conduction anaesthesia
 f anesthésie par blocage nerveux
 e bloqueo nervioso
 d Nervenblock; Leitungsanästhesie

2982 nerve endings
 f terminaisons nerveuses
 e terminaciones nerviosas
 d Nervenendigungen
 l terminationes nervorum

* nerve fiber → 2983

2983 nerve fibre; nerve fiber
 f fibre nerveuse
 e fibra nerviosa
 d Nervenfaser

2984 nervous disorder
 f désordre nerveux; trouble nerveux
 e trastorno nervioso
 d nervöse Störung; Nervenstörung

2985 nervous system
 f système nerveux
 e sistema nervioso
 d Nervensystem
 l systema nervosum

2986 nervous system disease; neuropathy
 f maladie du système nerveux; neuropathie
 e neuropatía; enfermedad del sistema nervioso
 d Neuropathie; Nervensystemerkrankung

2987 netobimin
 f nétobimine
 e netobimina
 d Netobimin
 l netobiminum

* nettlerash → 4780

2988 neural transmission
 f transmission nerveuse
 e transmisión neural
 d nervöse Übermittlung

2989 neuraminidase
 f neuraminidase
 e neuraminidasa
 d Neuraminidase

2990 neuraxial oedema
 f œdème névraxial
 e edema neuraxial
 d neuraxiales Ödem

2991 neurectomy
 f neurectomie; névrectomie
 e neurectomia
 d Neurektomie

2992 neuritis
 f névrite
 e neuritis
 d Neuritis; Nervenentzündung

2993 neuroepithelial layer of retina
 f stratum neuro-épithélial de la rétine
 e estrato neuroepitelial de la retina
 d Neuroepithelschicht der Retina
 l stratum neuroepitheliale retinae

2994 neurofibroma
 f neurofibrome
 e neurofibroma
 d Neurofibrom

2995 neurohypophyseal recess; infundibular recess
 f récessus neurohypophysaire; récessus de l'infundibulum
 e receso neurohipofisario; receso del infundíbulo
 d Recessus neurohypophysialis
 l recessus neurohypophysialis; recessus infundibuli

2996 neurohypophysis; posterior lobe of pituitary gland
 f neurohypophyse; lobe postérieur de la glande pituitaire
 e neurohipófisis; lóbulo posterior de la hipófisis
 d Neurohypophysis; Hypophysenhinterlappen
 l neurohypophysis; lobus posterior hypophyseos

2997 neuroleptic
 f neuroleptique
 e neuroléptico
 d Neuroleptikum

2998 neuromuscular
f neuromusculaire
e neuromuscular
d neuromuskulär

2999 neuron
f neurone
e neurona
d Neuron

* **neuropathy** → 2986

3000 neurotropic
f neurotrope
e neurotrópico
d neurotrop

3001 neurotropic drug
f médicament neurotrope
e fármaco neurotrópico
d neurotrope Substanz

3002 neutropenia
f neutropénie
e neutropenia
d Neutropenie

3003 neutrophil granulocyte
f granulocyte neutrophile
e granulocito neutrófilo
d neutrophiler Granulozyt

3004 neutrophilia
f neutrophilie
e neutrofilia
d Neutrophilie

* **newborn** → 2973

3005 Newcastle disease
f maladie de Newcastle
e enfermedad de Newcastle
d Newcastle-Krankheit; Pseudogeflügelpest

3006 Newcastle disease paramyxovirus; avian paramyxovirus-1
f paramyxovirus de la maladie de Newcastle
e virus de la enfermedad de Newcastle
d Paramyxovirus der Newcastle-Krankheit

3007 niclofolan
f niclofolan
e niclofolan
d Niclofolan
l niclofolanum

3008 niclosamide
f niclosamide
e niclosamida
d Niclosamid
l niclosamidum

3009 nicotinic acid
f acide nicotinique
e ácido nicotínico
d Nikotinsäure
l acidum nicotinicum

* **nictitating membrane** → 4526

3010 nifursol
f nifursol
e nifursol
d Nifursol
l nifursolum

3011 nitrate poisoning
f intoxication par les nitrates
e intoxicación por nitratos
d Nitratvergiftung

3012 nitrofural; nitrofurazone
f nitrofural; nitrofurazone
e nitrofural; nitrofurazona
d Nitrofural; Nitrofurazon
l nitrofuralum

* **nitrofurazone** → 3012

3013 nitroscanate
f nitroscanate
e nitroscanato
d Nitroscanat
l nitroscanatum

3014 nitrous oxide
f protoxyde d'azote
e óxido nitroso
d Lachgas; Stickstoffoxydul

3015 nitroxinil; nitroxynil
f nitroxinil
e nitroxinilo
d Nitroxinil
l nitroxinilum

* **nitroxynil** → 3015

3016 node
f nœud
e nódulo
d Knoten
l nodus

3017 nodular worm
f œsophagostome
e gusano nodular

  d Knötchenwurm
  l Oesophagostomum

* **nodular worm disease** → 3081

3018 nodule
  f nodule
  e nodulillo
  d Knötchen
  l nodulus

3019 noninfectious disease
  f maladie non infectieuse
  e enfermedad no infecciosa
  d nichtinfektiöse Krankheit

3020 nonpathogenic
  f non pathogène
  e no patógeno
  d apathogen

3021 nonspecific immunostimulation;
    paraimmunization
  f immunostimulation non spécifique
  e inmunoestimulación no específica
  d unspezifische Immunstimulierung;
    Paramunisierung

3022 nontoxic
  f non toxique
  e atóxico
  d ungiftig; unschädlich

* **noradrenaline** → 3023

3023 norepinephrine; noradrenaline;
    levarterenol
  f norépinéphrine; noradrénaline
  e norepinefrina; noradrenalina
  d Norepinephrin; Noradrenalin
  l norepinephrinum

3024 norgestomet
  f norgestomet
  e norgestomet
  d Norgestomet
  l norgestometum

3025 North American blastomycosis;
    Blastomyces dermatitidis infection
  f blastomycose nord-américaine; infection à
    Blastomyces dermatitidis
  e blastomicosis norteamericana; infección
    por Blastomyces dermatitidis
  d nordamerikanische Blastomykose;
    Blastomyces-dermatitidis-Infektion

3026 nose
  f nez
  e nariz
  d Nase
  l nasus

* **nosebleed** → 1508

* **Nosema infection** → 3027

* **nosematosis** → 3027

3027 nosemosis; nosematosis; Nosema infection
  f nosémose
  e nosemosis
  d Nosematose

3028 nostril
  f narine
  e orificio de la nariz
  d Nasenöffnung; Nasenloch
  l naris

3029 notarium; dorsal bone
  f notarium; os dorsal
  e notario
  d Notarium; Os dorsale
  l notarium; os dorsale

* **notch** → 2161

3030 notifiable disease
  f maladie à déclaration obligatoire
  e enfermedad de declaración obligatoria
  d anzeigepflichtige Seuche; meldepflichtige
    Krankheit

3031 notoedric mange; notoedrosis
  f gale notoédrique
  e sarna notédrica
  d Notoedresräude der Katze

* **notoedrosis** → 3031

3032 novobiocin
  f novobiocine
  e novobiocina
  d Novobiocin
  l novobiocinum

3033 nuchal crest
  f crête de la nuque
  e cresta de la nuca
  d Genickkamm
  l crista nuchae

3034 nuchal fascia
  f fascia nuchal

e fascia de la nuca
d Fascia nuchae
l fascia nuchae

3035 nuchal ligament
f ligament nuchal; ligament cervical
e ligamento de la nuca
d Nackenband
l ligamentum nuchae

3036 nuclei of cranial nerves
f noyaux des nerfs crâniens
e núcleos de los nervios craneales
d Kerne der Hirnnerven
l nuclei nervorum cranialum

3037 nuclei of lateral lemniscus
f noyaux du lemnisque latéral
e núcleos del lemnisco lateral
d Nuclei lemnisci lateralis
l nuclei lemnisci lateralis

3038 nuclei of origin
f noyaux d'origine
e núcleos de origen
d Ursprungskerne
l nuclei originis

* **nuclei of pons → 3545**

* **nuclei of thalamus → 4509**

3039 nucleoprotein
f nucléoprotéine
e nucleoproteína
d Nukleoprotein

3040 nucleotide
f nucléotide
e nucleótido
d Nukleotid

3041 nucleus
f noyau
e núcleo
d Nukleus; Kern
l nucleus

3042 nucleus gracilis
f noyau gracile
e núcleo grácil
d Nucleus gracilis
l nucleus gracilis

3043 nucleus of lens
f noyau du cristallin

e núcleo del cristalino
d Linsenkern
l nucleus lentis

3044 nucleus pulposus
f noyau pulpeux
e núcleo pulposo
d Gallertkern
l nucleus pulposus

3045 nutrient arteries (*of bone*)
f artères nourricières (*de l'os*)
e arterias nutricias
d Arteriae nutriciae
l arteriae nutriciae

3046 nutrient canal of bone; haversian canal
f canal nourricier de l'os; canal de Havers
e canal nutricio
d Ernährungskanal; Haversscher Kanal
l canalis nutricius

3047 nutrient foramen of bone
f trou nourricier de l'os
e agujero nutricio
d Ernährungsloch
l foramen nutricium

3048 nutritional deficiency
f carence; déficience nutritionnelle
e deficiencia nutritiva
d Nährstoffmangel; Ernährungsmangel

3049 nutritional disorder
f trouble nutritionnel
e trastorno nutritivo
d Ernährungsstörung

3050 nymphomania
f nymphomanie
e ninfomania
d Nymphomanie; Brüllerkrankheit;
Dauerbrunst

3051 nystagmus
f nystagmus
e nistagmo
d Nystagmus; Augenzittern

3052 nystatin
f nystatine
e nistatina
d Nystatin
l nystatinum

O

3053 obesity
f obésité
e obesidad; adiposidad
d Adipositas; Fettsucht
l adipositas

3054 obstetrics
f obstétrique
e obstetricia
d Obstetrik; Geburtshilfe

3055 obstruction; blockage
f obstruction; blocage
e obstrucción
d Verlegung; Verschluß

3056 obturator artery
f artère obturatrice
e arteria obturadora
d Arteria obturatoria
l arteria obturatoria

3057 obturator foramen
f foramen obturateur; trou obturateur
e agujero obturador
d Hüftbeinloch
l foramen obturatum

3058 obturator nerve
f nerf obturateur
e nervio obturador
d Nervus obturatorius
l nervus obturatorius

3059 obturator paralysis
f paralysie du nerf obturateur
e parálisis del nervio obturador
d Lähmung des Nervus obturatorius

3060 obturator vein
f veine obturatrice
e vena obturadora
d Vena obturatoria
l vena obturatoria

3061 occipital artery
f artère occipitale
e arteria occipital
d Arteria occipitalis
l arteria occipitalis

3062 occipital bone
f os occipital
e hueso occipital
d Hinterhauptsbein
l os occipitale

3063 occipital condyle
f condyle de l'occipital
e cóndilo occipital
d Gelenkfortsatz des Hinterhauptbeins
l condylus occipitalis

3064 occipital crest
f crête occipitale
e cresta occipital
d Crista occipitalis
l crista occipitalis

3065 occipital lobe
f lobe occipital
e lóbulo occipital
d Hinterhauptlappen
l lobus occipitalis

3066 occipital vein
f veine occipitale
e vena occipital
d Hinterhauptvene
l vena occipitalis

3067 occipitohyoid muscle
f muscle occipito-hyoïdien
e músculo occipitohioideo
d Musculus occipitohyoideus
l musculus occipitohyoideus

3068 occiput; back of head; poll
f occiput; partie postérieure de la tête
e occipucio
d Okziput; Hinterhaupt
l occiput

3069 occlusal surface; masticatory surface
f face occlusale de la dent; surface triturante
e cara oclusal; superficie triturante
d Kaufläche; Reibefläche
l facies occlusalis dentis

3070 occupational disease
f maladie professionnelle
e enfermedad profesional
d Berufskrankheit

3071 ochratoxin; Penicillium viridicatum toxin
f ochratoxine
e ocratoxina; toxina de Penicillium viridicatum
d Ochratoxin

3072 oculomotor nerve (*3rd cranial*)
f nerf oculo-moteur
e nervio oculomotor
d Nervus oculomotorius
l nervus oculomotorius

3073 Oddi's sphincter; sphincter of hepatopancreatic ampulla
f sphincter d'Oddi
e músculo esfínter de la ampolla
d Oddi-Sphinkter
l musculus sphincter ampullae hepatopancreaticae

* **odontoid process → 4599**

3074 oedema; edema
f œdème
e edema
d Ödem; Wassersucht

3075 oedematous; edematous
f œdémateux
e edematoso
d ödematös

3076 oesophageal glands; esophageal glands
f glandes œsophagiennes
e glándulas esofágicas
d Speiseröhrendrüsen
l glandulae (o)esophageae

* **oesophageal groove → 1808**

3077 oesophageal hiatus; esophageal hiatus
f hiatus œsophagien
e hiato esofágico
d Speiseröhrenschlitz
l hiatus (o)esophageus

3078 oesophageal impression; esophageal impression
f empreinte œsophagienne
e impresión esofágica
d Impressio oesophagea
l impressio (o)esophagea

3079 oesophageal obstruction; esophageal obstruction
f obstruction de l'œsophage
e obstrucción del esófago
d Schlundverstopfung; Ösophagusverstopfung

3080 oesophageal veins; esophageal veins
f veines œsophagiennes
e venas esofágicas
d Speiseröhrenvenen
l venae (o)esophageae

3081 oesophagostomosis; nodular worm disease
f œsophagostomose
e esofagostomosis
d Ösophagostomose; Knötchenwurmkrankheit

3082 oesophagus; esophagus
f œsophage
e esófago
d Speiseröhre
l oesophagus; esophagus

3083 oestradiol; estradiol
f estradiol; œstradiol
e estradiol
d Estradiol
l estradiolum

3084 oestrid fly
f œstridés
e éstrido
d Dasselfliege
l Oestridae

3085 oestrogen; estrogen
f œstrogène
e estrógeno
d Östrogen

3086 oestrone; estrone
f estrone; œstrone
e estrona
d Estron
l estronum

3087 oestrus synchronization; estrus synchronization
f synchronisation de l'œstrus
e sincronización del estro
d Brunstsynchronisation

3088 ofloxacin
f ofloxacine
e ofloxacino
d Ofloxacin
l ofloxacinum

3089 olaquindox
f olaquindox
e olaquindox
d Olaquindox
l olaquindoxum

3090 olecranal fossa
f fosse olécrânienne
e fosa del olécranon
d Fossa olecrani
l fossa olecrani

3091 **olecranal ligament**
  f ligament olécrânien
  e ligamento del olécranon
  d Ligamentum olecrani
  l ligamentum olecrani

3092 **olecranal tuber**
  f tubérosité de l'olécrâne
  e tuberosidad del olécranon
  d Tuber olecrani
  l tuber olecrani

3093 **olecranon**
  f olécrâne
  e olecranón
  d Ellbogenhöcker
  l olecranon

3094 **olfactory bulb**
  f bulbe olfactif
  e bulbo olfatorio
  d Riechkolben
  l bulbus olfactorius

* **olfactory cortex** → 126

3095 **olfactory glands; Bowman's glands**
  f glandes olfactives
  e glándulas olfatorias
  d Riechdrüsen
  l glandulae olfactoriae

3096 **olfactory nerves**
  f nerfs olfactifs
  e nervios olfatorios
  d Riechnerven
  l nervi olfactorii

3097 **olfactory organ; organ of smell**
  f organe olfactif
  e órgano del olfato
  d Geruchsorgan
  l organum olfactus

3098 **olfactory region** (*of nasal mucosa*)
  f région olfactive (*de la muqueuse du nez*)
  e región olfatoria
  d Riechgegend
  l regio olfactoria

3099 **olfactory tract**
  f tractus olfactif
  e tracto olfatorio
  d Riechbahn
  l tractus olfactorius

3100 **oligodontia**
  f oligodontie
  e oligodoncia
  d Oligodontie

3101 **oliguria**
  f oligurie
  e oliguria
  d Oligurie

3102 **olivary nucleus**
  f noyau olivaire
  e núcleo olivar
  d Nucleus olivarius
  l nucleus olivarius

3103 **omasal canal**
  f canal omasique
  e canal del omaso
  d Psalterkanal
  l canalis omasi

* **omasal fold** → 3105

3104 **omasal groove**
  f sillon omasal
  e surco del omaso
  d Sulcus omasi; Psalterrinne
  l sulcus omasi

3105 **omasal lamina; omasal fold**
  f lame omasique
  e lámina del omaso
  d Psalterblatt
  l lamina omasi

3106 **omasal papilla**
  f papille omasale
  e papila del omaso
  d Psalterpapille
  l papilla omasi

3107 **omaso-abomasal opening**
  f orifice omaso-abomasique
  e orificio omasoabomásico
  d Psalterlabmagenöffnung
  l ostium omasoabomasicum

3108 **omasum**
  f omasum; feuillet
  e omaso; librillo
  d Omasum; Psalter; Blättermagen
  l omasum

3109 **omental bursa**
  f bourse omentale
  e bolsa omental
  d Netzbeutel
  l bursa omentalis

3110 **omobrachial vein**
  f veine omo-brachiale
  e vena omobraquial
  d Vena omobrachialis
  l vena omobrachialis

3111 omohyoid muscle
f muscle omo-hyoïdien
e músculo omohioideo
d Musculus omohyoideus
l musculus omohyoideus

3112 omotransverse muscle
f muscle omo-transversaire
e músculo omotransverso
d Schulter-Halsmuskel
l musculus omotransversarius

3113 omphalitis
f omphalite
e onfalitis
d Nabelentzündung

3114 omphalomesenteric duct; vitelline duct
f canal omphalo-mésentérique; canal vitellin
e conducto onfalomesentérico; conducto
vitelino
d Dottergang; Dottersackstiel
l ductus omphalomesentericus

* Onchocerca infestation → 3115

3115 onchocercosis; Onchocerca infestation
f onchocercose
e oncocercosis
d Onchozerkose

3116 oncogenic
f oncogène
e oncogénico
d onkogen

3117 oncogenicity
f oncogénicité; pouvoir oncogène
e oncogenicidad
d Onkogenität

3118 oncogenic virus
f virus oncogène
e virus oncogénico
d onkogenes Virus

3119 oncosphere
f oncosphère
e oncosfera
d Onkosphäre

3120 oncovirus type C
f oncovirus type C
e oncovirus tipo C
d Oncovirus Typ C

3121 Ondiri disease; bovine infectious petechial fever; Cytoecetes ondiri infection
f maladie d'Ondiri; infection à Cytoecetes
ondiri
e enfermedad de Ondiri; fiebre petequial
infecciosa bovina; infección por Cytoecetes
ondiri
d Ondiri-Krankheit; infektiöses
Petechialfieber der Rinder; Cytoecetes-
ondiri-Infektion

3122 oocyst
f oocyste
e ooquiste; oocisto
d Oozyste

3123 oocyte
f oocyte; ovocyte
e ovocito
d Oozyt
l ovocytus

* open fracture → 1019

3124 opening of pulmonary trunk
f ostium du tronc pulmonaire
e orificio del tronco pulmonar
d Ostium trunci pulmonalis
l ostium trunci pulmonalis

3125 opening of vena cava
f ostium de la veine cave
e orificio de la vena cava
d Ostium venae cavae
l ostium venae cavae

3126 ophthalmic nerve
f nerf ophtalmique
e nervio oftálmico
d Nervus ophthalmicus
l nervus ophthalmicus

3127 opisthorchiosis; Opisthorchis infestation
f opisthorchose
e opistorquiosis
d Opisthorchiose

* Opisthorchis infestation → 3127

3128 opsonin
f opsonine
e opsonina
d Opsonin

3129 optical part of retina
f partie optique de la rétine
e porción óptica de la retina
d lichtempfindlicher Teil der Netzhaut;
Hauptteil der Netzhaut
l pars optica retinae

3130 optic chiasm
 f chiasma optique
 e quiasma óptico
 d Sehnervenkreuzung
 l chiasma opticum

3131 optic disk; optic papilla
 f disque du nerf optique; papille du nerf optique
 e disco del nervio óptico
 d Sehnervenpapille
 l discus nervi optici

3132 optic nerve
 f nerf optique
 e nervio óptico
 d Nervus opticus; Sehnerv
 l nervus opticus

 * **optic papilla** → 3131

3133 optic radiation
 f radiation optique de Gratiolet
 e radiación óptica
 d Sehstrahlung (von Gratiolet)
 l radiatio optica

3134 optic recess
 f récessus optique
 e receso óptico
 d Recessus opticus
 l recessus opticus

3135 optic tract
 f tractus optique
 e tracto óptico
 d Sehbahn
 l tractus opticus

3136 oral administration
 f administration orale
 e administración oral
 d orale Verabreichung
 l per os

 * **oral aperture** → 3138

3137 oral cavity
 f cavité buccale; cavité orale
 e cavidad de la boca
 d Mundhöhle
 l cavum oris

3138 oral cleft; oral aperture
 f orifice buccal
 e hendidura de la boca
 d Mundspalte
 l rima oris

3139 oral glands; glands of the mouth
 f glandes orales
 e glándulas de la boca
 d Munddrüsen
 l glandulae oris

3140 oral mucosa
 f muqueuse orale
 e túnica mucosa de la boca
 d Mundschleimhaut
 l tunica mucosa oris

 * **oral part of pharynx** → 3164

3141 oral vaccination
 f vaccination orale
 e vacunación oral
 d orale Vakzination

3142 orbicular muscle
 f muscle orbiculaire
 e músculo orbicular
 d Ringmuskel
 l musculus orbicularis

3143 orbicular muscle of eye
 f muscle orbiculaire del l'œil
 e músculo orbicular del ojo
 d Augenringmuskel
 l musculus orbicularis oculi

3144 orbicular muscle of mouth
 f orbiculaire des lèvres de la bouche
 e músculo orbicular de la boca
 d Mundringmuskel
 l musculus orbicularis oris

3145 orbit; orbital cavity
 f orbite; cavité orbitaire
 e orbita
 d Augenhöhle
 l orbita

 * **orbital arch** → 4352

 * **orbital cavity** → 3145

3146 orbital fascia
 f fascia orbitaire
 e fascia orbital
 d Augenfaszie
 l fascia orbitalis

3147 orbital fissure
 f fente orbitaire; fissure orbitaire
 e cisura orbitaria
 d Fissura orbitalis
 l fissura orbitalis

* **orbital gland** → 5006

3148 orbital muscle
f muscle orbitaire
e músculo orbitario
d Musculus orbitalis
l musculus orbitalis

3149 orbital part of frontal bone
f partie orbitale de l'os frontal
e porción orbitaria del hueso frontal
d Augenhöhlenplatte
l pars orbitalis ossis frontalis

3150 orbital region
f région orbitaire
e región orbitaria
d Augenbogengegend
l regio orbitalis

* **orbital wing of sphenoid bone** → 4974

3151 orbitorotund foramen
f foramen orbito-rond
e agujero orbitorredondo
d Foramen orbitorotundum
l foramen orbitorotundum

3152 orbitosphenoidal bone
f os orbito-sphénoïde
e hueso orbitoesfenoides
d Os orbitosphenoidale
l os orbitosphenoidale

3153 orbivirus
f orbivirus
e orbivirus
d Orbivirus

* **orchidectomy** → 704

3154 orchitis
f orchite
e orquitis
d Orchitis; Hodenentzündung

* **orf** → 1044

3155 organic disease
f maladie organique
e enfermedad orgánica
d Organkrankheit

3156 organochlorine compound
f composé organochloré
e compuesto organoclórico
d organische Chlorverbindung

* **organ of Corti** → 4157

* **organ of smell** → 3097

* **organ of taste** → 1912

3157 organ of vision
f organe de la vue
e órgano de vista
d Sehorgan
l organum visus

3158 organophosphorus compound
f composé organophosphoré
e compuesto organofosfórico
d organische Phosphorverbindung;
Posphorsäureester

3159 organs of sense; sensory organs
f organes des sens
e órganos de los sentidos
d Sinnesorgane
l organa sensuum

3160 orifice of vagina
f orifice du vagin
e orificio de la vagina
d Scheideneingang
l ostium vaginae

3161 ormetoprim
f ormétoprime
e ormetoprima
d Ormetoprim
l ormetoprimum

3162 ornithosis; psittacosis; Chlamydia psittaci infection
f ornithose; psittacose; infection à Chlamydia psittaci
e ornitosis; psitacosis; infección por Chlamydia psittaci
d Ornithose; Psittakose; Chlamydia-psittaci-Infektion

3163 orobasal organ; Ackerknecht's organ
f organe orobasal (d'Ackerknecht)
e órgano orobasal
d Ackerknechtsches Organ
l organum orobobasale

3164 oropharynx; oral part of pharynx
f oro-pharynx; partie orale du pharynx
e orofaringe
d Mundrachen
l pars oralis pharyngis

3165 **orthomyxovirus**
  f orthomyxovirus
  e orthomyxovirus
  d Orthomyxovirus

3166 **orthopoxvirus**
  f orthopoxvirus
  e orthopoxvirus
  d Orthopoxvirus

3167 **osseous labyrinth; bony labyrinth**
  f labyrinthe osseux
  e laberinto óseo
  d knöchernes Labyrinth
  l labyrinthus osseus

3168 **ossification**
  f ossification
  e osificación
  d Ossifikation; Verknöcherung

3169 **osteitis**
  f ostéite
  e osteítis
  d Ostitis; Knochenentzündung

3170 **osteoblast**
  f ostéoblaste
  e osteoblasto
  d Osteoblast

3171 **osteochondritis**
  f ostéochondrite
  e osteocondritis
  d Osteochondritis

3172 **osteoclast**
  f ostéoclaste
  e osteoclasto
  d Osteoklast

3173 **osteocyte; bone cell**
  f ostéocyte; cellule osseuse
  e osteocito; célula ósea
  d Osteozyt; Knochenzelle

3174 **osteodystrophy**
  f ostéodystrophie
  e osteodistrofia
  d Osteodystrophie

  * **osteogenesis** → 512

3175 **osteogenesis imperfecta**
  f ostéogenèse imparfaite
  e osteogénesis imperfecta
  d Osteogenesis imperfecta
  l osteogenesis imperfecta

3176 **osteomalacia; bone softening**
  f ostéomalacie
  e osteomalacia
  d Osteomalazie; Knochenerweichung

3177 **osteomyelitis**
  f ostéomyélite
  e osteomielitis
  d Osteomyelitis; Knochenmarkentzündung

3178 **osteopetrosis**
  f ostéopétrose
  e osteopetrosis
  d Osteopetrose

3179 **osteophyte; bony outgrowth**
  f ostéophyte; excroissance osseuse
  e osteofito
  d Osteophyt; Knochenauswuchs

3180 **osteoporosis**
  f ostéoporose
  e osteoporosis
  d Osteoporose; Knochenschwund

  * **osteosynthesis** → 1760

  * **Ostertagia infestation** → 3181

3181 **ostertagiosis; Ostertagia infestation**
  f ostertagiose
  e ostertagiosis
  d Ostertagiose

3182 **otic bone**
  f os otique
  e hueso ótico
  d Os otica
  l os otica

3183 **otic ganglion; Arnold's ganglion**
  f ganglion otique; ganglion d'Arnold
  e ganglio ótico
  d Ganglion oticum
  l ganglion oticum

3184 **otitis**
  f otite
  e otitis
  d Otitis

3185 **otitis externa**
  f otite externe
  e otitis externa
  d Otitis externa; Entzündung des äußeren
    Gehörgangs
  l otitis externa

3186 otitis media
 f otite moyenne
 e otitis media
 d Otitis media; Mittelohrentzündung
 l otitis media

3187 otodectic mange; ear mange; otodectosis
 f gale otodectique
 e sarna otodéctica; sarna de la oreja
 d Otodektesräude; Ohrräude

* otodectosis → 3187

* outer ring of iris → 1898

3188 oval foramen of heart
 f foramen ovale; trou de Botal
 e agujero oval del corazón
 d Foramen ovale cordis
 l foramen ovale cordis

3189 oval fossa of heart
 f fosse ovale du cœur
 e fosa oval del corazón
 d Fossa ovalis cordis
 l fossa ovalis cordis

3190 oval window on middle ear; vestibular fenestra
 f fenêtre ovale; fenêtre du vestibule
 e ventana del vestíbulo; ventana oval
 d Vorhofsfenster
 l fenestra vestibuli; fenestra ovalis

3191 ovarian artery
 f artère ovarique
 e arteria ovárica
 d Eierstockarterie
 l arteria ovarica

3192 ovarian bursa
 f bourse ovarienne
 e bolsa ovárica
 d Eierstocktasche
 l bursa ovarica

3193 ovarian cumulus
 f cumulus oophorus
 e acúmulo ooforo
 d Eihügel
 l cumulus oophorus

3194 ovarian cyst
 f kyste ovarien
 e quiste ovárico
 d Zystovar; Eierstockzyste

3195 ovarian disease
 f maladie de l'ovaire
 e enfermedad del ovario
 d Eierstockerkrankung

3196 ovarian fimbria
 f frange ovarienne
 e fimbria ovárica
 d Eierstockfranse
 l fimbria ovarica

3197 ovarian follicles
 f follicules ovariques
 e folículos ováricos
 d Eierstockfollikel
 l folliculi ovarici

* ovarian ligament → 4792

3198 ovarian plexus
 f plexus ovarique
 e plexo ovárico
 d Eierstockgeflecht
 l plexus ovaricus

3199 ovariectomy; spaying
 f ovariectomie
 e ovariectomía
 d Ovarektomie

3200 ovary
 f ovaire
 e ovario
 d Ovar; Eierstock
 l ovarium

3201 overdosage
 f surdosage
 e dosis excesiva
 d Überdosierung

3202 overfeeding
 f suralimentation
 e sobrealimentación
 d Überfütterung

3203 overmilking
 f surtraite
 e ordeño excesivo
 d Übermelken

3204 oviduct; uterine tube; fallopian tube
 f oviducte; trompe utérine (de Fallope)
 e oviducto; trompa uterina
 d Ovidukt; Eileiter
 l tuba uterina

3205 **ovine babesiosis**
  f babésiose ovine
  e babesiosis ovina
  d Babesiose des Schafes

3206 **ovine brucellosis**
  f brucellose ovine
  e brucelosis ovina
  d Schafbrucellose

3207 **ovine enterotoxaemia; struck; Clostridium perfringens type C infection**
  f entérotoxémie hémorragique à Clostridium perfringens type C
  e enterotoxemia hemorrágica ovina; infección por Clostridium perfringens tipo C
  d Enterotoxämie des Schafes durch Clostridium perfringens Typ C

  * **ovine lentiviral meningoencephalitis →4934**

  * **ovine orbivirus infection → 496**

  * **ovine pestivirus infection → 520**

  * **ovine poxvirus → 4073**

3208 **ovulation**
  f ovulation
  e ovulación
  d Ovulation

3209 **ovum; egg**
  f ovule; œuf
  e huevo
  d Eizelle; Ei
  l ovum

3210 **oxacillin**
  f oxacilline
  e oxacilina
  d Oxacillin
  l oxacillinum

3211 **oxfendazole**
  f oxfendazole

  e oxfendazol
  d Oxfendazol
  l oxfendazolum

3212 **oxibendazole**
  f oxibendazole
  e oxibendazol
  d Oxibendazol
  l oxibendazolum

3213 **oxolinic acid**
  f acide oxolinique
  e ácido oxolínico
  d Oxolinsäure
  l acidum oxolinicum

3214 **oxyclozanide**
  f oxyclozanide
  e oxiclozanida
  d Oxyclozanid
  l oxyclozanidum

3215 **oxygen consumption**
  f consommation d'oxygène
  e consumo de oxígeno
  d Sauerstoffverbrauch

3216 **oxytetracycline**
  f oxytétracycline
  e oxitetraciclina
  d Oxytetracyclin
  l oxytetracyclinum

3217 **oxytocin**
  f oxytocine
  e oxitocina
  d Oxytocin
  l oxytocinum

3218 **oxyuriosis; Oxyuris infestation**
  f oxyurose
  e oxiurosis
  d Oxyuriose

  * **Oxyuris infestation → 3218**

P

* pacchionian granulations → 256

3219 **pad**
f coussinet; torus
e almohadilla
d Ballen
l torus

3220 **pad of fat**
f corps adipeux
e cuerpo adiposo
d Fettpolster
l corpus adiposum

* **Paecilomyces infection** → 3221

3221 **paecilomycosis; Paecilomyces infection**
f pæcilomycose
e pecilomicosis; infección por Paecilomyces spp.
d Paecilomykose

3222 **palate**
f palais
e paladar
d Gaumen
l palatum

3223 **palatine arteries**
f artères palatines
e arterias palatinas
d Gaumenarterien
l arteriae palatinae

3224 **palatine bone**
f os palatin
e hueso palatino
d Gaumenbein
l os palatinum

3225 **palatine fissure**
f fissure palatine
e cisura palatina
d Fissura palatina
l fissura palatina

3226 **palatine foramen**
f foramen palatin
e agujero palatino
d Foramen palatinum
l foramen palatinum

3227 **palatine glands**
f glandes palatines
e glándulas palatinas
d Gaumendrüsen
l glandulae palatinae

3228 **palatine muscle**
f muscle palatin
e músculo palatino
d Musculus palatinus
l musculus palatinus

3229 **palatine nerves**
f nerfs palatins
e nervios palatinos
d Gaumennerven
l nervi palatini

3230 **palatine process of maxilla**
f processus palatin de l'os maxillaire
e apófisis palatina
d Gaumenfortsatz des Oberkiefers
l processus palatinus maxillae

3231 **palatine sinus**
f sinus palatin
e seno palatino
d Gaumenhöhle
l sinus palatinus

3232 **palatine tonsil**
f amygdale palatine; tonsille palatine
e amígdala palatina; tonsila palatina
d Gaumenmandel
l tonsilla palatina

3233 **palatine veins**
f veines palatines
e venas palatinas
d Gaumenvenen
l venae palatinae

3234 **palatoglossal arch**
f arc palato-glosse
e arco palatogloso
d Arcus palatoglossus
l arcus palatoglossus

3235 **palatoglossal muscle**
f muscle palato-glosse
e músculo palatogloso
d Musculus palatoglossus
l musculus palatoglossus

3236 **palatopharyngeal arch**
f arc palato-pharyngien
e arco palatofaríngeo
d Arcus palatopharyngeus
l arcus palatopharyngeus

3237 **palatopharyngeal muscle**
  f muscle palato-pharyngien
  e músculo palatofaríngeo
  d Musculus palatopharyngeus
  l musculus palatopharyngeus

* **palatoschisis** → 934

3238 **pale, soft exudative meat; PSE meat**
  f viande pâle, molle et exsudative; viande PSE
  e carne pálida, blanda y exsudativa
  d blasses, weiches, wäßriges, exsudatives Fleisch; PSE-Fleisch

3239 **pallium**
  f pallium
  e palio
  d Pallium; Gehirnmantel
  l pallium

3240 **palmar common digital arteries**
  f artères digitales communes palmaires
  e arterias digitales palmares comunes
  d Hauptmittelfußarterien; mediale Palmararterien
  l arteriae digitales palmares communes

3241 **palmar digital nerves**
  f nerfs digitaux palmaires
  e nervios digitales palmares
  d Nervi digitales palmares
  l nervi digitales palmares

3242 **palmar ligaments**
  f ligaments palmaires
  e ligamentos palmares
  d Ligamenta palmaria
  l ligamenta palmaria

3243 **palmar metacarpal nerves**
  f nerfs métacarpiens palmaires
  e nervios metacarpianos palmares
  d Nervi metacarpei palmares
  l nervi metacarpei palmares

3244 **palm of the hand**
  f paume de la main
  e palma de la mano
  d Hohlhand; Handteller
  l palma manus

3245 **palpation**
  f palpation
  e palpación
  d Palpation; Betastung

3246 **palpebral arteries**
  f artères palpébrales
  e arterias palpebrales
  d Lidarterien
  l arteriae palpebrales

3247 **palpebral conjunctiva**
  f conjonctive palpébrale
  e túnica conjuntiva de los párpados
  d Lidbindehaut
  l tunica conjunctiva palpebrarum

3248 **palpebral fissure**
  f fente palpébrale
  e hendidura de los párpados
  d Lidspalte
  l rima palpebrarum

3249 **palpebral reflex**
  f réflexe palpébral
  e reflejo palpebral
  d Lidschlußreflex

3250 **palpebral veins**
  f veines palpébrales
  e venas palpebrales
  d Lidvenen
  l venae palpebrales

3251 **pampiniform venous plexus**
  f plexus pampiniforme (veineux)
  e plexo pampiniforme
  d Plexus pampiniformis
  l plexus pampiniformis

3252 **pancreas**
  f pancréas
  e páncreas
  d Pankreas; Bauchspeicheldrüse
  l pancreas

3253 **pancreatic disease**
  f maladie du pancréas
  e enfermedad del páncreas
  d Pankreaserkrankung

3254 **pancreatic islets; islets of Langerhans**
  f îlots pancréatiques (de Langerhans)
  e islotes pancreáticos (de Langerhans)
  d Pankreasinseln; Langerhanssche Inseln
  l insulae pancreaticae

3255 **pancreatic juice**
  f suc pancréatique
  e jugo pancreático
  d Pankreassaft
  l succus pancreaticus

3256 **pancreaticoduodenal veins**
 f veines pancréatico-duodénales
 e venas pancréaticoduodenales
 d Venae pancreaticoduodenales
 l venae pancreaticoduodenales

3257 **pancreatic plexus**
 f plexus pancréatique
 e plexo pancreático
 d Plexus pancreaticus
 l plexus pancreaticus

3258 **pancreatic ring; portal ring**
 f anneau du pancréas
 e anillo del páncreas
 d Bauchspeicheldrüsenring; Pfortaderring
 l anulus pancreaticus

3259 **pancreatic veins**
 f veines pancréatiques
 e venas pancréaticas
 d Venae pancreaticae
 l venae pancreaticae

3260 **pancreatitis**
 f pancréatite
 e pancreatitis
 d Pankreatitis

3261 **panzootic**
 f panzootie
 e panzootia
 d Panzootie

3262 **papilla**
 f papille
 e papila
 d Papille
 l papilla

3263 **papillae of proventriculus**
 f papilles proventriculaires
 e papilas del proventrículo
 d Papillen der Drüsenmagen;
    Drüsenknospen
 l papillae proventriculares

3264 **papillae of reticulum**
 f papilles réticulaires
 e papilas del retículo
 d Haubenpapillen
 l papillae reticuli

 * **papillary duct of teat** → 4457

3265 **papillary ducts of kidney**
 f conduits papillaires du rein
 e conductillos papilares
 d Ductus papillares renis
 l ductus papillares renis

3266 **papillary foramina of kidney**
 f foramens papillaires du rein
 e agujeros papilares
 d Foramina papillaria; Harnporen
 l foramina papillaria renis

3267 **papillary layer of corium**
 f stratum papillare; derme papillaire
 e estrato papilar
 d Stratum papillare; Papillarkörper
 l stratum papillare corii

3268 **papillary muscles of heart**
 f muscles papillaires du cœur
 e músculos papilares
 d Papillarmuskeln; Warzenmuskeln des
    Herzens
 l musculi papillares

3269 **papillary process of liver**
 f processus papillaire du foie
 e apófisis papilar del hígado
 d Processus papillaris hepatis
 l processus papillaris hepatis

3270 **papilloma**
 f papillome
 e papiloma
 d Papillom

3271 **papillomavirus**
 f papillomavirus
 e papillomavirus
 d Papillomavirus

3272 **papovavirus**
 f papovavirus
 e papovavirus
 d Papovavirus

3273 **papular**
 f papuleux
 e papular
 d papulös

3274 **papular stomatitis**
 f stomatite papuleuse
 e estomatitis papular
 d Stomatitis papulosa
 l stomatitis papulosa

3275 **papule**
 f papule
 e pápula
 d Papel
 l papula

3276 **parabronchus**
 f parabronche
 e parabronquia
 d Lungenpfeife
 l parabronchus

3277 **paracentesis**
 f paracentèse
 e paracentesis
 d Parazentese

3278 **paradidymis**
 f paradidyme
 e paradídimo
 d Paradidymis
 l paradidymis

 * **paraglossal bone** → 1480

3279 **paragonimosis; Paragonimus infestation**
 f paragonimose
 e paragonimosis
 d Paragonimose

 * **Paragonimus infestation** → 3279

3280 **parahippocampal gyrus; hippocampal gyrus**
 f gyrus parahippocampal
 e giro parahipocampal; giro del hipocampo
 d Gyrus parahippocampalis
 l gyrus parahippocampalis

 * **paraimmunization** → 3021

3281 **parainfluenza**
 f parainfluenza
 e parainfluenza
 d Parainfluenza

3282 **parakeratosis**
 f parakératose
 e paraqueratosis
 d Parakeratose

 * **paralumbar fossa** → 2023

3283 **paralysis**
 f paralysie
 e parálisis
 d Paralyse; Lähmung

3284 **paralytic**
 f paralytique
 e paralítico
 d paralytisch

3285 **paralytic myoglobinuria; equine azoturia**
 f myopathie myoglobinurique paralytique
 e mioglobinuria paralítica
 d paralytische Myoglobinurie des Pferdes

3286 **paramesonephric duct; muellerian duct**
 f canal paramésonéphrotique (de Mueller)
 e conducto paramesonéfrico (de Müller)
 d Müllerscher Gang
 l ductus paramesonephricus

3287 **paramphistome; rumen fluke**
 f paramphistome
 e paranfistoma
 d Pansenegel
 l Paramphistomum

3288 **paramphistomosis**
 f paramphistomose
 e paranfistomiasis
 d Paramphistomose; Pansenegelbefall

3289 **paramyxovirus**
 f paramyxovirus
 e paramyxovirus
 d Paramyxovirus

 * **paranal sinus** → 166

3290 **paranasal sinuses**
 f sinus paranasaux
 e senos paranasales
 d Nasennebenhöhlen
 l sinus paranasales

3291 **paraplegia; posterior paralysis**
 f paraplégie
 e parálisis posterior
 d Paraplegie; Nachhandlähmung

3292 **parapoxvirus**
 f parapoxvirus
 e parapoxvirus
 d Parapoxvirus

3293 **paraproteinaemia**
 f paraprotéinémie
 e paraproteinemia
 d Paraproteinämie

3294 **pararectal fossa**
 f fosse pararectale
 e fosa pararrectal
 d Fossa pararectalis
 l fossa pararectalis

3295 **parascariosis; Parascaris equorum infestation**
 f ascaridose des équidés

e parascariosis; ascaridosis de los équidos
d Paraskaridose; Spulwurmbefall der Pferde

* **Parascaris equorum infestation** → 3295

3296 parasitaemia
f parasitémie
e parasitemia
d Parasitämie

3297 parasite
f parasite
e parásito
d Parasit

3298 parasitic
f parasitaire
e parasítico; parasitario
d parasitär

* **parasitic disease** → 3304

3299 parasiticide
f parasiticide
e parasiticido
d Parazitizid

3300 parasitic infestation
f infestation parasitaire
e infestación parasitaria
d Parasitenbefall

3301 parasitism
f parasitisme
e parasitismo
d Parasitismus

3302 parasitological
f parasitologique
e parasitológico
d parasitologisch

3303 parasitology
f parasitologie
e parasitología
d Parasitologie

3304 parasitosis; parasitic disease
f parasitose; maladie parasitaire
e parasitosis; enfermedad parasitaria
d Parasitose; parasitäre Krankheit

3305 parasphenoidal bone
f os parasphénoïde
e hueso paraesfenoides
d Os parasphenoidale
l os parasphenoidale

3306 parasympathetic nervous system
f système nerveux parasympathique; partie
parasympathique
e sistema nervioso parasimpático; porción
parasimpática
d parasympathisches Nervensystem
l pars parasympathica systematis nervosi
autonomici

3307 parasympathetic nucleus
f noyau parasympathique
e núcleo parasimpático
d parasympathischer Kern
l nucleus parasympathicus

3308 parasympatholytic
f parasympatholytique
e parasimpaticolítico
d Parasympatholytikum

3309 parasympathomimetic
f parasympathomimétique
e parasimpaticomimético
d Parasympathomimetikum

* **parathormone** → 3311

3310 parathyroid gland
f glande parathyroïde
e glándula paratiroidea
d Parathyreoidea; Nebenschilddrüse;
Epithelkörperchen
l glandula parathyroidea

3311 parathyroid hormone; parathormone
f hormone parathyroïdienne
e hormona paratiroides
d Parathormon

**3312 paratuberculosis; Johne's disease;
Mycobacterium paratuberculosis infection**
f paratuberculose bovine; infection à
Mycobacterium paratuberculosis
e paratuberculosis; enfermedad de Johne;
infección por Mycobacterium
paratuberculosis
d Paratuberkulose; Mycobacterium-
paratuberculosis-Infektion
l enteritis paratuberculosa

* **paratyphoid** → 3981

3313 paraventricular nucleus
f noyau paraventriculaire
e núcleo paraventricular
d Nucleus paraventricularis
l nucleus paraventricularis

3314 **parbendazole**
  f parbendazole
  e parbendazol
  d Parbendazol
  l parbendazolum

3315 **parenchyma**
  f parenchyme
  e parénquima
  d Parenchym
  l parenchyma

3316 **parenteral**
  f parentéral
  e parenteral
  d parenteral

3317 **parenteral administration**
  f administration parentérale
  e administración parenteral
  d parenterale Verabreichung

3318 **paresis; partial paralysis**
  f parésie; paralysie partielle
  e parálisis parcial
  d Parese

3319 **parietal bone**
  f os pariétal
  e hueso parietal
  d Scheitelbein
  l os parietale

3320 **parietal layer of pericardium**
  f lame pariétale du péricarde
  e lámina parietal del pericardio
  d Lamina parietalis pericardii
  l lamina parietalis pericardii

3321 **parietal peritoneum**
  f péritoine pariétal
  e peritoneo parietal
  d parietales Blatt des Bauchfells
  l peritoneum parietale

3322 **parietal pleura**
  f plèvre pariétale
  e pleura parietal
  d Pleura parietalis
  l pleura parietalis

3323 **parietal region**
  f région pariétale
  e región parietal
  d Scheitelregion
  l regio parietalis

3324 **paroophoron**
  f paroophoron
  e paraovario; paroóforo
  d Paroophoron
  l paroophoron

3325 **parotid artery**
  f artère parotidienne
  e arteria parotídea
  d Ohrspeicheldrüsenarterie
  l arteria parotidea

3326 **parotid duct**
  f conduit parotidien; canal de Sténon
  e conducto parotídeo; conducto de Estenón
  d Parotisgang; Stenonscher Gang
  l ductus parotideus

3327 **parotid gland**
  f glande parotide
  e glándula parótida
  d Parotis; Ohrspeicheldrüse
  l glandula parotis

3328 **parotid lymphocentre**
  f lymphocentre parotidien
  e linfocentro parotídeo
  d Lymphocentrum parotideum
  l lymphocentrum parotideum

3329 **parotido-auricular muscle**
  f muscle parotido-auriculaire
  e músculo parotidoauricular
  d Musculus parotidoauricularis
  l musculus parotidoauricularis

3330 **parotid papilla**
  f papille parotidienne
  e papila parotídea
  d Parotispapille
  l papilla parotidea

3331 **parotid plexus**
  f plexus parotidien
  e plexo parotídeo
  d Parotisgeflecht
  l plexus parotideus

3332 **parotid region**
  f région parotidienne
  e región parotídea
  d Ohrspeicheldrüsenregion
  l regio parotidea

* **partial paralysis** → 3318

3333 **parturient paresis; milk fever**
  f fièvre vitulaire; fièvre du lait
  e fiebre de la leche; paresis puerperal
  d Festliegen; Gebärparese

3334 parturition
f parturition; mise bas
e parto; parición
d Partus; Geburt

3335 parturition complication
f complication du part
e complicación del parto
d Geburtsstörung

3336 parvovirus
f parvovirus
e parvovirus
d Parvovirus

3337 paspalum staggers; Claviceps paspali intoxication
f intoxication à Claviceps paspali
e intoxicación por Claviceps paspali
d Claviceps-paspali-Intoxikation

3338 passalurosis; Passalurus infestation
f passalurose
e pasalurosis; infestación por Passalurus sp.
d Passalurose

* **Passalurus infestation → 3338**

3339 passive immunity
f immunité passive
e inmunidad pasiva
d passive Immunität

* **pastern joint → 3655**

* **pastern region → 2834**

* **Pasteurella infection → 3340**

* **Pasteurella multocida infection → 1939**

3340 pasteurellosis; Pasteurella infection
f pasteurellose; infection à Pasteurella
e pasteurelosis; infección por Pasteurella spp.
d Pasteurellose; Pasteurella-Infektion

3341 patella; knee cap
f rotule
e rótula
d Patella; Kniescheibe
l patella

3342 patellar ligaments
f ligaments patellaires
e ligamento rotuliano
d Ligamenta patellae
l ligamenta patellae

3343 patellar luxation
f luxation de la rotule
e luxación de la rótula
d Patella-Luxation

3344 patellar reflex
f réflexe rotulien
e reflejo rotuliano
d Patellarsehnenreflex

3345 pathogen
f agent pathogène
e agente patógeno
d Infektionserreger; Krankheitserreger

3346 pathogenesis
f pathogenèse
e patogénesis
d Pathogenese

3347 pathogenic
f pathogène
e patógeno
d pathogen

3348 pathogenicity
f pouvoir pathogène
e patogenicidad
d Pathogenität

3349 pathological change
f altération pathologique
e alteración patológica
d krankhafte Veränderung; Verletzung

3350 pathology
f anatomie pathologique
e patología
d Pathologie

* **paunch → 3935**

3351 pectinate muscles (*of right atrium*)
f trabécules charnues de l'oreillette droite
e músculos pectinados
d Musculi pectinati
l musculi pectinati

3352 pectineal line; cranial edge of pubis
f pecten du pubis; surface pectinéale
e pecten del hueso pubis
d Schambeinkamm
l pecten ossis pubis

3353 pectineal muscle
f muscle pectiné
e músculo pectíneo
d Kammuskel
l musculus pectineus

3354 **pectoral fascia**
f fascia pectoral
e fascia pectoral
d Fascia pectoralis
l fascia pectoralis

3355 **pectoral nerves**
f nerfs pectoraux
e nervios pectorales
d Nervi pectorales
l nervi pectorales

3356 **pectoral regions**
f régions de la poitrine
e regiones del tórax
d Brustregionen
l regiones pectoris

3357 **peduncle**
f pédoncule
e pedúnculo
d Stiel; Füßchen
l pedunculus

3358 **pelvic axis**
f axe du bassin
e eje de la pelvis
d Beckenachse
l axis pelvis

* **pelvic bone** → 2010

3359 **pelvic cavity**
f cavité pelvienne
e cavidad pelviana
d Beckenhöhle
l cavum pelvis

3360 **pelvic diaphragm**
f diaphragme pelvien
e diafragma de la pelvis
d Muskeln des Beckenausgangs
l diaphragma pelvis

3361 **pelvic fascia**
f fascia pelvien
e fascia de la pelvis
d Beckenfaszie
l fascia pelvis

3362 **pelvic ganglia**
f ganglions pelviens
e ganglios pélvicos
d Beckenganglien
l ganglia pelvina

3363 **pelvic girdle**
f ceinture du membre pelvien
e cinturón del miembro pelviano
d Beckengürtel
l cingulum membri pelvini

3364 **pelvic inlet; cranial aperture of pelvis**
f détroit crânial du bassin; ouverture
craniale du pelvis
e abertura craneal de la pelvis
d Beckeneingang
l apertura pelvis cranialis

3365 **pelvic limb; hindleg**
f membre pelvique; membre postérieur
e miembro pelviano
d Beckengliedmaße; Hinterextremität
l membrum pelvinum

3366 **pelvic nerves**
f nerfs pelviens
e nervios pélvicos
d Nervi pelvini
l nervi pelvini

3367 **pelvic outlet; caudal aperture of pelvis**
f détroit caudal du bassin; ouverture caudale
du pelvis
e abertura caudal de la pelvis
d Beckenausgang
l apertura pelvis caudalis

3368 **pelvic part of ureter**
f partie pelvienne de l'uretère
e porción pelviana del uréter
d Beckenstück des Harnleiters
l pars pelvina ureteris

3369 **pelvic plexus**
f plexus pelvien
e plexo pélvico
d Beckengeflecht
l plexus pelvinus

3370 **pelvic recess** (*of kidney*)
f récessus pelvique
e receso de la pelvis
d Recessus pelvis
l recessus pelvis

3371 **pelvic regions**
f régions pelviennes
e regiones de la pelvis
d Beckengegenden
l regiones pelvis

3372 **pelvic symphysis**
f symphyse pelvienne
e sínfisis pélvica
d Beckensymphyse
l symphysis pelvina

* pelvic vertebrae → 4417

3373 **pelvis**
   f bassin
   e pelvis
   d Becken
   l pelvis

3374 **penicillamine**
   f pénicillamine
   e penicilamina
   d Penicillamin
   l penicillaminum

* **penicillin G** → 431

3375 **penicillins**
   f pénicillines
   e penicilinas
   d Penizilline

* **penicillin V** → 3449

* **Penicillium viridicatum toxin** → 3071

3376 **penile artery**
   f artère du pénis
   e arteria del pene
   d Penisarterie
   l arteria penis

3377 **penile bone**
   f os pénien
   e hueso del pene
   d Penisknochen
   l os penis

3378 **penile veins**
   f veines du pénis
   e venas del pene
   d Penisvenen
   l venae penis

3379 **penis**
   f pénis; verge
   e pene
   d Penis; männliches Glied
   l penis

3380 **pennaceous part of vexillum**
   f partie pennacée du vexillum
   e porción penácea
   d Pars pennacea
   l pars pennacea

3381 **pentamidine**
   f pentamidine
   e pentamidina
   d Pentamidin
   l pentamidinum

3382 **pentastome**
   f pentastome
   e pentastoma; linguatula
   d Zungenwurm; Pentastomide
   l Pentastoma

3383 **pentobarbital; pentobarbitone**
   f pentobarbital
   e pentobarbital
   d Pentobarbital
   l pentobarbitalum

* **pentobarbitone** → 3383

3384 **peracute**
   f suraigu
   e sobreagudo
   d perakut

3385 **percutaneous**
   f percutané
   e percutáneo
   d perkutan

3386 **perforating branch**
   f rameau perforant
   e rama perforante
   d perforierender Ast
   l ramus perforans

3387 **perforating veins**
   f veines perforantes
   e venas perforantes
   d Venae perforantes
   l venae perforantes

3388 **pericardiacophrenic artery**
   f artère péricardiaco-phrénique
   e arteria pericardiacofrénica
   d Arteria pericardiacophrenica
   l arteria pericardiacophrenica

3389 **pericardial cavity**
   f cavité péricardique
   e cavidad pericárdica
   d Herzbeutelhöhle
   l cavum pericardii

3390 **pericarditis**
   f péricardite
   e pericarditis
   d Perikarditis; Herzbeutelentzündung

3391 **pericardium**
   f péricarde
   e pericardio
   d Perikard; Herzbeutel
   l pericardium

3392 perilymph
f périlymphe
e perilinfa
d Perilymphe
l perilympha

3393 perilymphatic duct
f conduit périlymphatique
e conducto perilínfatico
d Ductus perilymphaticus
l ductus perilymphaticus

3394 perimetrium
f périmétrium
e perimetrio
d Perimetrium; Serosa der Gebärmutter
l perimetrium

3395 perinatal
f périnatal
e perinatal
d perinatal

* perineal body → 798

3396 perineal hernia
f hernie périnéale
e hernia perineal
d Perinealhernie; Dammbruch
l hernia perinealis

3397 perineal muscles
f muscles du périnée
e músculos del periné
d Dammuskeln
l musculi perinei

3398 perineal nerves
f nerfs périnéaux
e nervios perineales
d Dammnerven
l nervi perineales

3399 perineal region
f région périnéale
e región perineal
d Dammregion
l regio perinealis

3400 perineal veins
f veines périnéales
e venas perineales
d Venae perineales
l venae perineales

3401 perineum
f périnée
e periné
d Perineum; Damm
l perineum

* periodic ophthalmia → 3789

3402 periodontium
f parodonte
e periodontio
d Zahnwurzelhaut
l periodontium

3403 periople; epidermis of limbus
f périople
e periople; epidermis del limbo
d Perioplum; Saumhorn
l perioplum; epidermis limbi

3404 perioplic corium; corium of limbus
f chorion limbique; derme périoplique
e corión del limbo
d Saumlederhaut
l corium limbi; dermis limbi

3405 perioral glands
f glandes périorales
e glándulas circumorales
d Zirkumoraldrüse
l glandulae circumorales

3406 periorbit
f périorbite
e periórbita
d Periorbita
l periorbita

3407 periosteum
f périoste
e periosteo
d Periost; Knochenhaut
l periosteum

3408 periostitis
f périostite
e periostitis
d Periostitis; Knochenhautentzündung

3409 peripheral nerve
f nerf périphérique
e nervio periférico
d peripherer Nerv

3410 peripheral nervous system
f système nerveux périphérique
e sistema nervioso periférico
d peripheres Nervensystem
l systema nervosum peripherium

3411 peritoneal cavity
f cavité péritonéale
e cavidad peritoneal
d Peritonealhöhle
l cavum peritonei

3412 peritoneum
 f péritoine
 e peritoneo
 d Peritonäum; Bauchfell
 l peritoneum

3413 peritonitis
 f péritonite
 e peritonitis
 d Peritonitis; Bauchfellentzündung

3414 perivascular
 f périvasculaire
 e perivascular
 d perivaskulär

3415 periventricular fibres
 f fibres périventriculaires
 e fibras periventriculares
 d Fibrae periventriculares
 l fibrae periventriculares

3416 permanent teeth
 f dents permanentes; dents de
 remplacement
 e dientes permanentes; dientes de
 reemplazo
 d bleibende Zähne
 l dentes permanentes

3417 permethrin
 f perméthrine
 e permetrina
 d Permethrin
 l permethrinum

3418 perosis
 f pérose
 e perosis
 d Perosis

3419 perpendicular plate of palatine bone
 f lame perpendiculaire de l'os palatin
 e lámina perpendicular del hueso palatino
 d Sagittalteil des Gaumenbeins
 l lamina perpendicularis ossis palatini

3420 pessulus
 f pessulus
 e pessulus
 d Steg
 l pessulus

3421 pesticide
 f pesticide
 e pesticida
 d Pestizid; Schädlingsbekämpfungsmittel

3422 pestivirus
 f pestivirus
 e pestivirus
 d Pestivirus

3423 petechia
 f pétéchie
 e petequia
 d Petechie; punktförmige Blutung

3424 petechial
 f pétéchial
 e petequial
 d petechial; punktförmig

3425 petrotympanic fissure
 f fissure pétro-tympanique
 e cisura petrotimpánica
 d Fissura petrotympanica
 l fissura petrotympanica

3426 petrous part of temporal bone
 f partie pétreuse de l'os temporal; rocher
 e porción petrosa del temporal
 d Felsenteil des Schläfenbeins
 l pars petrosa ossis temporalis

 * **Peyer's patches → 103**

 * **PFU → 3503**

 * **phage → 398**

3427 phage type; lysotype
 f lysotype; type phagique
 e lisotipo
 d Lysotyp

3428 phagocyte
 f phagocyte
 e fagocito
 d Phagozyt

3429 phagocytic
 f phagocytaire
 e fagocitario
 d Phagozyten-

3430 phagocytosis
 f phagocytose
 e fagocitosis
 d Phagozytose

 * **phalangeal exostosis → 3900**

3431 phalanx; finger bone; toe bone
 f phalange; os d'un doigt; os d'un orteil
 e falange
 d Phalanx; Zehenglied; Fingerglied
 l phalanx

3432 **phallus**
 f phallus
 e falo
 d Begattungsorgan
 l phallus

3433 **pharmacodynamics**
 f pharmacodynamie
 e farmacodinamia
 d Pharmakodynamik

3434 **pharmacokinetics**
 f pharmacocinétique
 e farmacocinética
 d Pharmakokinetik

3435 **pharmacology**
 f pharmacologie
 e farmacología
 d Pharmakologie

3436 **pharmacy**
 f pharmacie
 e farmacia
 d Pharmazie; Arzneimittelkunde

3437 **pharyngeal cavity**
 f cavité pharyngée
 e cavidad faríngea
 d Rachenhöhle
 l cavum pharyngis

3438 **pharyngeal diverticulum of pig**
 f diverticule pharyngien du porc
 e divertículo faríngeo
 d Backentasche des Schweins
 l diverticulum pharyngeum

3439 **pharyngeal glands**
 f glandes pharyngiennes
 e glándulas faríngeas
 d Speicheldrüsen des Rachens
 l glandulae pharyngea

3440 **pharyngeal opening of auditory tube**
 f ostium pharyngien de la trompe auditive
 e orificio faríngeo de la trompa auditiva
 d Rachenhöhlenmündung der Ohrtrompete
 l ostium pharyngeum tubae auditivae

3441 **pharyngeal plexus**
 f plexus pharyngé
 e plexo faríngeo
 d Rachengeflecht
 l plexus pharyngeus

3442 **pharyngeal recess; Rosenmüller's cavity**
 f récessus pharyngien
 e receso faríngeo

 d Recessus pharyngeus; Rosenmüllersche
 Grube
 l recessus pharyngeus

3443 **pharyngeal tonsil**
 f amygdale pharyngienne; tonsille
 pharyngienne
 e amígdala faríngea; tonsila faríngea
 d Rachenmandel
 l tonsilla pharyngea

3444 **pharyngeal veins**
 f veines pharyngiennes
 e venas faríngeas
 d Venae pharyngeae
 l venae pharyngeae

3445 **pharyngitis**
 f pharyngite
 e faringitis
 d Pharyngitis; Rachenentzündung

3446 **pharynx**
 f pharynx
 e faringe
 d Rachen; Schlundkopf
 l pharynx

3447 **phenobarbital; phenobarbitone**
 f phénobarbital
 e fenobarbital
 d Phenobarbital
 l phenobarbitalum

* **phenobarbitone** → 3447

3448 **phenothiazine**
 f phénothiazine
 e fenotiazina
 d Phenothiazin
 l phenothiazinum

3449 **phenoxymethylpenicillin; penicillin V**
 f phénoxyméthylpénicilline; pénicilline V
 e fenoximetilpenicilina; penicilina V
 d Phenoxymethylpenicillin; Penicillin V
 l phenoxymethylpenicillinum

3450 **phenylbutazone**
 f phénylbutazone
 e fenilbutazona
 d Phenylbutazon
 l phenylbutazonum

3451 **phimosis**
 f phimosis
 e fimosis
 d Phimose

3452 phlebitis
  f phlébite
  e flebitis
  d Phlebitis; Venenentzündung

3453 phosphatase
  f phosphatase
  e fosfatasa
  d Phosphatase

3454 phospholipid
  f phospholipide
  e fosfolipida
  d Phospholipid

3455 phosphoprotein
  f phosphoprotéine
  e fosfoproteína
  d Phosphoprotein

3456 phosphorus deficiency; aphosphorosis
  f aphosphorose
  e carencia cn fósforo
  d Phosphormangel

3457 photosensitization
  f photosensibilisation
  e fotosensibilización
  d Photosensibilisierung

3458 phrenic ganglia
  f ganglions phréniques
  e ganglios frénicos
  d Ganglia phrenica
  l ganglia phrenica

3459 phrenic nerve
  f nerf phrénique
  e nervio frénico
  d Nervus phrenicus; Zwerchfellsnerv
  l nervus phrenicus

3460 phrenicopericardiac ligament
  f ligament phrénico-péricardique
  e ligamento frenicopericárdico
  d Ligamentum phrenicopericardiacum
  l ligamentum phrenicopericardiacum

3461 phycomycosis
  f phycomycose
  e ficomicosis
  d Phykomykose

3462 physiological
  f physiologique
  e fisiológico
  d physiologisch

3463 physiology
  f physiologie
  e fisiología
  d Physiologie

3464 physocephalosis; Physocephalus infestation
  f physocéphalose
  e fisocefalosis
  d Physocephalose

* Physocephalus infestation → 3464

3465 phytohaemagglutinin
  f phytohémagglutinine
  e fitohemoaglutinina
  d Phytohämagglutinin

3466 pica; depraved appetite
  f pica; allotriophagie; aberration du goût
  e pica
  d Lecksucht; Parorexie; Allotriophagie

3467 picornavirus
  f picornavirus
  e picornavirus
  d Picornavirus

3468 pig disease; swine disease
  f maladie des porcins
  e enfermedad de los porcinos
  d Schweinekrankheit

3469 pigeon pox
  f variole du pigeon
  e viruela del palomo
  d Taubenpocken

3470 piglet anaemia
  f anémie des porcelets
  e anemia de los lechones
  d Ferkelanämie

3471 pig louse
  f pou du porc
  e piojo del cerdo
  d Schweinelaus
  l Haematopinus suis

3472 pigmentation disorder
  f trouble de la pigmentation
  e trastorno de la pigmentación
  d Störung der Pigmentierung

* pigmented epithelium → 3473

3473 pigmented layer of retina; pigmented epithelium
  f stratum pigmentaire de la rétine;
     épithélium pigmentaire

e estrato pigmentario de la retina
d Pigmentepithelschicht der Retina
l stratum pigmentosum retinae

* **pig pox** → 4389

3474 pig tapeworm; armed tapeworm
f ténia du porc; ténia armé
e tenia armada
d Schweinebandwurm des Menschen
l Taenia solium

3475 pillar
f pilier
e pilar
d Pfeiler
l pila

* **pin bone** → 2346

3476 pineal gland
f glande pinéale; épiphyse cérébrale
e glándula pineal; epífisis cerebral
d Epiphyse; Zirbeldrüse
l glandula pinealis

3477 pineal recess
f récessus pinéal
e receso pineal
d Recessus pinealis
l recessus pinealis

3478 pinna; ear flap; auricle of ear
f pavillon de l'oreille; auricule de l'oreille
e aurícula; orejuela; pabellón de la oreja
d Ohrmuschel
l auricula

3479 piperazine
f pipérazine
e piperazina
d Piperazin
l piperazinum

3480 piriform lobe; pyriform lobe
f lobe piriforme
e lóbulo piriforme
d Lobus piriformis
l lobus piriformis

3481 piriform muscle
f muscle piriforme
e músculo piriforme
d Musculus piriformis
l musculus piriformis

3482 piriform recess
f récessus piriforme

e receso piriforme
d Recessus piriformis
l recessus piriformis

* **piroplasmosis in cattle** → 526

* **piroplasmosis of horses** → 1517

* **Pithomyces chartarum toxin** → 4180

3483 pituitary; hypophyseal
f pituitaire
e pituitaria
d Hypophysen-

* **pituitary fossa** → 2091

* **pituitary gland** → 2092

3484 pituitary stalk; infundibulum of hypothalamus
f tige pituitaire; infundibulum de l'hypothalamus
e infundíbulo de la hipófisis; porción proximal de la neurohipófisis
d Hypophysenstiel
l infundibulum hypothalami; pars proximalis neurohypophysis

3485 pivot joint; trochoid articulation
f articulation trochoïde; articulation pivotante
e articulación trocoidea; articulación en pivote
d Drehgelenk; Zapfengelenk
l articulatio trochoidea

3486 placebo
f placebo
e placebo
d Placebo

3487 placenta
f placenta
e placenta
d Plazenta
l placenta

3488 placental barrier
f barrière placentaire
e barrera placentaria
d Plazentarbarriere; Plazentarschranke

3489 placental retention; retained placenta
f rétention placentaire; non-délivrance
e retención de la placenta
d Retentio secundinarum; Nachgeburtsverhaltung
l retentio secundinarum

3490 placentome
f placentome
e placentomo
d Plazentom
l placentomus

3491 plagiorchiosis; Plagiorchis infestation
f plagiorchiose
e plagiorciosis
d Plagiorchiose; Eileiteregelbefall

* **Plagiorchis infestation → 3491**

3492 plague of small ruminants
f peste des petits ruminants; PPR
e peste de pequeños ruminantes
d Pestivirose des Schafes und der Ziege

3493 plague of small ruminants virus; caprine morbillivirus
f virus de la peste des petits ruminants
e virus de la peste de los pequeños ruminantes
d ovines und caprines Morbillivirus

3494 plane articulation; gliding joint; arthrodia
f arthrodie
e articulación plana
d Schiebegelenk
l articulatio plana

3495 plantar arch
f arc plantaire
e arco plantar
d Plantarbogen
l arcus plantaris

3496 plantar cushion; ball of sole
f coussinet plantaire
e almohadilla plantar
d Sohlenballen
l pulvinus metatarsalis

3497 plantar digital nerves
f nerfs digitaux plantaires
e nervios digitales plantares
d Nervi digitales plantares
l nervi digitales plantares

3498 plantar ligaments
f ligaments plantaires
e ligamentos plantares
d Ligamenta plantaria
l ligamenta plantaria

3499 plantar metatarsal nerves
f nerfs métatarsiens plantaires
e nervios metatarsianos plantares
d Nervi metatarsei plantares
l nervi metatarsei plantares

3500 plantar talofibular ligament
f ligament talo-fibulaire plantaire
e ligamento taloperoneo plantar
d Ligamentum talofibulare plantare
l ligamentum talofibulare plantare

3501 plant poisoning
f intoxication par les plantes
e intoxicación por plantas
d Pflanzenvergiftung

3502 plaque
f plage
e placa
d Plaque

3503 plaque-forming unit; PFU
f unité formant plages; UFP
e unidad formando las placas
d plaquebildende Einheit

3504 plasma cell; plasmocyte
f cellule plasmatique; plasmocyte
e célula plasmática; plasmocito
d Plasmazelle

* **plasmacytosis of mink → 115**

3505 plasma protein
f protéine plasmatique
e proteína plasmática
d Plasmaprotein

3506 plasma volume
f volume plasmatique
e volumen plasmático
d Plasmavolumen

* **plasmocyte → 3504**

3507 platysma; cutaneous muscle of neck and face
f platysma; muscle cutané du cou et de la face
e músculo platisma; músculo cutáneo del cuello e de la cara
d Platysma; platte Hautmuskel des Halses
l platysma

3508 plerocercoid
f plérocercoïde
e plerocercoida
d Plerozerkoid

3509 pleura
f plèvre
e pleura
d Pleura; Brustfell
l pleura

3510 **pleural cavity**
 f cavité pleurale
 e cavidad pleural
 d Pleuralhöhle
 l cavum pleurae

3511 **pleural recesses**
 f récessus pleuraux
 e recesos pleurales
 d Brustfelltaschen
 l recessus pleurales

3512 **pleurisy; pleuritis**
 f pleurésie; pleurite
 e pleuresía
 d Pleuritis; Brustfellentzündung

* **pleuritis** → 3512

3513 **pleuroesophageal muscle**
 f muscle pleuro-œsophagien
 e músculo pleuroesofágico
 d Musculus pleuroesophageus
 l musculus pleuroesophageus

3514 **pleuropneumonia**
 f pleuropneumonie; péripneumonie
 e pleuroneumonía
 d Pleuropneumonie

3515 **plexus**
 f plexus
 e plexo
 d Plexus; Nervengeflecht
 l plexus

* **plexuses of bladder** → 4882

3516 **plumaceous part of vexillum**
 f partie plumacée du vexillum
 e porción plumácea
 d Pars plumacea
 l pars plumacea

3517 **plumage**
 f plumage
 e plumaje
 d Gefieder; Federkleid
 l pennae

3518 **plumed crest**
 f crête plumée; huppe
 e cresta plumosa
 d gefiederter Kamm
 l crista pennarum

3519 **plumes; down feathers**
 f plumes duveteuses; duvet

 e plumones
 d Dunen; Flaumfedern
 l pluma

* **PMS** → 4058

3520 **pneumatic bone**
 f os pneumatique
 e hueso neumático
 d pneumatischer Knochen
 l os pneumaticum

3521 **pneumocapillaries; respiratory tubules; air capillaries**
 f capillaires aériens
 e capilares aéreos
 d Luftkapillaren
 l pneumocapillares

3522 **pneumonia**
 f pneumonie
 e neumonía
 d Pneumonie; Lungenentzündung

3523 **pneumostrongylosis; Pneumostrongylus infestation**
 f pneumostrongylose
 e pneumostrongilosis
 d Pneumostrongylose

* **Pneumostrongylus infestation** → 3523

3524 **pneumothorax**
 f pneumothorax
 e neumotórax
 d Pneumothorax

3525 **pneumovirus**
 f pneumovirus
 e pneumovirus
 d Pneumovirus

3526 **pododermatitis**
 f pododermatite
 e pododermatitis
 d Pododermatitis

3527 **podotheca; foot covering** (*of birds*)
 f podothèque
 e podoteca
 d Podotheca
 l podotheca

3528 **podotrochlear bursa**
 f bourse podotrochléaire
 e bolsa podotroclear
 d Hufrollenschleimbeutel
 l bursa podotrochlearis

* podotrochlitis → 2958

* point of buttock → 2339

* point of hip → 1121

3529 point of hip region; coxal tuberosity region
  f région de l'angle de la hanche
  e región de la tuberosidad coxal
  d Hüfthöckergegend
  l regio tuberis coxae

* point of hock → 625

3530 poison
  f poison
  e tóxico; veneno
  d Gift

3531 poisoning; toxicosis
  f intoxication; toxicose
  e intoxicación; toxicosis
  d Vergiftung; Toxikose

3532 poisonous plant
  f plante toxique
  e planta tóxica
  d Giftpflanze

* polioencephalomalacia → 819

* poll → 3068

3533 polyarthritis
  f polyarthrite
  e poliartritis
  d Polyarthritis

3534 polydactylia; polydactyly; supernumerary
     digits
  f polydactylie
  e polidactilia
  d Polydaktylia

* polydactyly → 3534

3535 polydipsia
  f polydipsie
  e polidipsia
  d Polydipsie

3536 polymastia; supernumerary mammary
     glands
  f polymastie
  e polimastia
  d Polymastie; Hypermastie

3537 polymelia; supernumerary limbs
  f polymélie
  e polimelia
  d Polymelie; Hypermelie

3538 polymorph; polymorphonuclear
     granulocyte
  f granulocyte polynucléaire neutrophile
  e granulocito polimorfonuclear
  d polymorphkerniger Granulozyt

* polymorphonuclear granulocyte → 3538

3539 polymyxin B
  f polymyxine B
  e polimixina B
  d Polymyxin B
  l polymyxinum B

3540 polyomavirus
  f polyomavirus
  e polyomavirus; virus del polioma
  d Polyomavirus

3541 polyserositis
  f polysérosite
  e poliserositis
  d Polyserositis

3542 polythelia; supernumerary teats
  f polythélie
  e politelia
  d Polythelie; Hyperthelie

3543 polyuria
  f polyurie
  e poliuria
  d Polyurie

* polyvalent vaccine → 1000

3544 pons
  f pont (de Varole)
  e puente (de Varolio)
  d Pons; Brücke
  l pons

3545 pontine nuclei; nuclei of pons
  f noyaux du pont
  e núcleos del puente
  d Brückenkerne
  l nuclei pontis

3546 popliteal artery
  f artère poplitée
  e arteria poplítea
  d Kniekehlarterie
  l arteria poplitea

3547 popliteal lymph nodes
 f nœuds lymphatiques poplités
 e nódulos linfáticos poplíteos
 d Lymphonodi poplitei
 l lymphonodi poplitei

3548 popliteal lymphocentre
 f lymphocentre poplité
 e linfocentro poplíteo
 d Lymphocentrum popliteum
 l lymphocentrum popliteum

3549 popliteal muscle
 f muscle poplité
 e músculo poplíteo
 d Kniekehlmuskel
 l musculus popliteus

3550 popliteal notch
 f échancrure poplitée
 e escotadura poplítea
 d Kniekehlausschnitt
 l incisura poplitea

3551 popliteal vein
 f veine poplitée
 e vena poplítea
 d Vena poplitea
 l vena poplitea

3552 porcine brucellosis; Brucella suis infection
 f brucellose porcine; infection à Brucella
 suis
 e brucelosis porcina; infección por Brucella
 suis
 d Schweinebrucellose; Brucella-suis-
 Infektion

**3553 porcine encephalomyelitis enterovirus;
 Teschen virus**
 f entérovirus de l'encéphalomyélite porcine
 e virus de la enfermedad de Teschen
 d porzines Enzephalomyelitis-Enterovirus;
 Teschen-Virus

3554 porcine enterovirus
 f entérovirus porcin
 e enterovirus porcino
 d Schweine-Enterovirus

**3555 porcine enzootic pneumonia; Mycoplasma
 hyopneumoniae infection**
 f pneumonie enzootique du porc; infection
 à Mycoplasma hyopneumoniae
 e neumonía enzoótica porcina; neumonía
 vírica; infección por Mycoplasma
 hyopneumoniae
 d enzootische Pneumonie des Schweines;
 Mycoplasma-hyopneumoniae-Infektion

**3556 porcine haemagglutinating
 encephalomyelitis; vomiting and wasting
 disease**
 f encéphalomyélite hémagglutinante
 porcine
 e encefalomielitis hemoaglutinante porcina;
 enfermedad de los vómitos y marasmo de
 los porcinos
 d hämagglutinierende Enzephalomyelitis der
 Ferkel

3557 porcine herpesvirus
 f herpèsvirus porcin
 e herpesvirus porcino
 d porcines Herpesvirus

 ***** **porcine herpesvirus type 1 → 332**

3558 porcine parvovirus
 f parvovirus porcin
 e parvovirus porcino
 d porcines Parvovirus

**3559 porcine pleuropneumonia; Actinobacillus
 pleuropneumoniae infection**
 f pleuropneumonie infectieuse porcine;
 infection à Actinobacillus
 pleuropneumoniae
 e pleuroneumonía porcina; infección por
 Actinobacillus pleuropneumoniae
 d infektiöse Pleuropneumonie des
 Schweines; Actinobacillus-
 pleuropneumoniae-Infektion

**3560 porcine polyarthritis/polyserositis;
 Glässer's disease; Haemophilus parasuis
 infection**
 f polysérosite infectieuse du porc; infection
 à Haemophilus parasuis
 e poliartritis/poliserositis porcina;
 enfermedad de Glässer; infección por
 Haemophilus parasuis
 d Glässersche Krankheit; Polyserositis und
 Polyarthritis des Schweines;
 Haemophilus-parasuis-Infektion

 ***** **porcine viral encephalomyelitis → 4495**

 ***** **pore of sweat duct → 4382**

3561 porphyria
 f porphyrie
 e porfiria
 d Porphyrie

 ***** **portal fissure → 1988**

 ***** **portal ring → 3258**

3562 portal vein
f veine porte
e vena porta
d Pfortader
l vena portae

3563 postacetabular wing of ilium
f aile postacétabulaire de l'ilium
e ala postacetabular del ílion
d Ala postacetabularis ilii
l ala postacetabularis ilii

3564 posterior chamber of eye
f chambre postérieure du bulbe de l'œil
e cámara posterior del globo
d hintere Augenkammer
l camera posterior bulbi

3565 posterior limiting lamina; Descemet's membrane
f lame limitante interne; membrane de Descemet
e lámina limitante posterior
d hintere Grenzschicht der Hornhaut
l lamina limitans posterior

* posterior lobe of pituitary gland → 2996

3566 posterior nares
f choanes
e coanas
d Choanen
l choanae

* posterior paralysis → 3291

3567 posthitis
f posthite
e postitis
d Posthitis; Vorhautentzündung

3568 post mortem change
f altération post-mortem
e alteración post mortem
d postmortale Veränderung

3569 post mortem examination; autopsy; necropsy
f autopsie; nécropsie
e autopsia; necropsia
d Sektion; Obduktion; Autopsie

3570 post mortem finding
f résultat d'autopsie
e resulta de la autopsia
d Obduktionsbefund

* post mortem rigidity → 3897

3571 postnatal development
f développement post-natal
e desarrollo postnatal
d postnatale Entwicklung

3572 postnatal period
f période post-natale
e período postnatal
d postnatale Periode

3573 postoperative
f post-opératoire
e postoperativo
d postoperativ

* postparturient period → 3688

3574 postpatagium; caudal wing web
f postpatagium; membrane alaire caudale
e postpatagio
d Postpatagium; hintere Flughaut
l postpatagium

3575 postvaccinal
f post-vaccinal
e postvacunal
d postvakzinal

3576 poultry body louse
f Eomenacanthus stramineus
e piojo del cuerpo de las gallinas
d Federling des Huhnes
l Eomenacanthus stramineus

3577 poultry disease
f maladie des volailles
e enfermedad de las aves
d Geflügelkrankheit

3578 poultry fluff louse
f Goniocotes gallinae
e piojo del plumón
d Goniocotes gallinae
l Goniocotes gallinae

3579 poultry head louse
f Cuclotogaster heterographus
e piojo oscuro de las gallinas
d Cuclotogaster heterographus
l Cuclotogaster heterographus

3580 poultry shaft louse
f ménopon
e piojo pequeño de las gallinas; menopón de las gallinas
d Schaftlaus; Menopon gallinae
l Menopon gallinae

* poultry tick → 1757

3581 poultry wing louse
f Lipeurus caponis
e piojo del ala
d Lipeurus caponis
l Lipeurus caponis

3582 pour-on application
f application sur le dos
e aplicación sobre el dorso
d Aufgießverfahren

3583 powder down feather
f duvet poudreux
e pulvipluma
d Pulvipluma
l pulvipluma

3584 poxvirus
f poxvirus
e poxvirus
d Poxvirus

3585 pralidoxime iodide
f iodure de pralidoxime
e ioduro de pralidoxima
d Pralidoxim-Jodid
l pralidoximi iodidum

3586 praziquantel
f praziquantel
e praziquantel
d Praziquantel
l praziquantelum

3587 preacetabular wing of ilium
f aile préacétabulaire de l'ilium
e ala preacetabular del ílion
d Ala preacetabularis ilii
l ala preacetabularis ilii

3588 prearticular bone
f os préarticulaire
e hueso prearticular
d Os prearticulare
l os prearticulare

3589 precipitation test
f réaction de précipitation
e prueba de precipitación
d Präzipitationsreaktion

3590 precipitin
f précipitine
e precipitina
d Präzipitin

3591 predisposing factor
f facteur favorisant; facteur prédisposant
e factor de predisposición
d prädisponierender Faktor

3592 predisposition to disease
f prédisposition aux maladies
e predisposición
d Prädisposition; Disposition

3593 prednisolone
f prednisolone
e prednisolona
d Prednisolon
l prednisolonum

3594 prednisone
f prednisone
e prednisona
d Prednison
l prednisolum

* preen gland → 4779

* prefemoral lymph nodes → 4275

3595 prefrontal bone
f os préfrontal
e hueso prefrontal
d Os prefrontale
l os prefrontale

3596 pregnancy; gestation
f gestation
e gestación
d Trächtigkeit; Gravidität

3597 pregnancy complication
f complication de la gestation
e complicación de la gestación
d Graviditätsstörung

3598 pregnancy diagnosis
f diagnostic de gestation
e diagnóstico de la gestación
d Trächtigkeitsdiagnose

3599 pregnancy toxaemia
f toxémie de gestation
e toxemia de gestación
d Graviditätstoxikose

3600 pregnant
f gravide; gestante
e grávida; gestante
d trächtig

* premaxilla → 2156

3601 premolar teeth
 f dents prémolaires
 e dientes premolares
 d Prämolaren; vordere Backenzähne
 l dentes premolares

3602 premunition
 f prémunition
 e preinmunidad
 d Prämunität

 * **prenasal bone** → 3914

3603 preoptic nucleus
 f noyau préoptique
 e núcleo preóptico
 d Nucleus praeopticus
 l nucleus pr(a)eopticus

3604 prepuce; sheath
 f prépuce; fourreau
 e prepucio
 d Vorhaut; Präputium
 l preputium; praeputium

3605 preputial diverticulum
 f diverticule préputial
 e divertículo prepucial
 d Präputialbeutel
 l diverticulum pr(a)eputiale

3606 preputial frenulum
 f frein préputial
 e frenillo del prepucio
 d Vorhautbändchen
 l frenulum pr(a)eputii

3607 preputial glands
 f glandes préputiales
 e glándulas prepuciales
 d Vorhautdrüsen
 l glandulae pr(a)eputiales

3608 preputial muscles
 f muscles préputiaux
 e músculos prepuciales
 d Vorhautmuskeln
 l musculi pr(a)eputiales

3609 preputial orifice
 f ostium préputial
 e orificio prepucial
 d Präputialöffnung
 l ostium pr(a)eputiale

3610 preputial ring (*of stallion*)
 f anneau du fourreau (*du cheval*)
 e anillo prepucial
 d Ringfalte der inneren Vorhaut (*Hengst*)
 l anulus pr(a)eputialis

3611 prescapular region
 f région préscapulaire
 e región preescapular
 d Vorderschulterregion
 l regio pr(a)escapularis

3612 presphenoid bone
 f os présphénoïde
 e hueso presfenoides
 d vorderes Keilbein
 l os pr(a)esphenoidale

3613 presternal region
 f région présternale
 e región presternal
 d Vorderbrustgegend
 l regio pr(a)esternalis

 * **presternum** → 2682

3614 prevalence rate
 f taux de prévalence
 e tasa de prevalencia
 d Durchseuchungsrate

 * **preventive** → 3634

 * **prevertebral fascia** → 3615

3615 prevertebral lamina; prevertebral fascia
 f lame prévertébrale
 e lámina prevertebral
 d Lamina pr(a)evertebralis
 l lamina pr(a)evertebralis

3616 primary infection
 f infection primaire
 e infección primaria
 d Primärinfektion

3617 principal sublingual duct
 f conduit sublingual majeur
 e conducto sublingual mayor
 d großer Ausführungsgang der
   Unterzungendrüse; Bartholinischer Gang
 l ductus sublingualis major

3618 probang
 f sonde œsophagienne
 e sonda esofágica
 d Schlundsonde

3619 probiotic
 f probiotique
 e probiótico
 d Probiotikum

3620 procaine
f procaïne
e procaína
d Procain; Novokain
l procainum

3621 proctodeum
f proctodéum
e proctodeo
d Proctodeum; Endraum
l proctodeum

* **progestational hormone → 3622**

3622 progesterone; progestational hormone
f progestérone; hormone lutéale
e progesterona
d Progesteron
l progesteronum

3623 prognathia
f prognathie
e prognatismo
d Prognathie

3624 prognosis
f pronostic
e pronóstico
d Prognose

* **progressive interstitial pneumonia of sheep → 2639**

3625 prolactin
f prolactine
e prolactina
d Prolaktin

3626 prolapse
f prolapsus
e prolapso
d Vorfall
l prolapsus

* **prolapse of uterus → 4789**

* **prolapse of vagina → 4809**

3627 proligestone
f proligestone
e proligestona
d Proligeston
l proligestonum

3628 promazine
f promazine
e promazina
d Promazin
l promazinum

3629 promethazine
f prométhazine
e prometazina
d Promethazin
l promethazinum

3630 propatagial elastic ligament
f ligament élastique propatagial
e ligamento elástico propatagial
d Ligamentum elasticum propatagiale
l ligamentum elasticum propatagiale

3631 propatagial tensor muscle
f muscle tenseur du propatagium
e músculo tensor propatagial
d Musculus tensor propatagialis
l musculus tensor propatagialis

3632 propatagium; alar membrane; cranial wing web
f propatagium; membrane alaire crânienne
e propatagio
d vordere Flughaut
l propatagium

* **proper coat → 4704**

3633 properdin
f properdine
e properdina
d Properdin

* **proper tunic → 4704**

3634 prophylactic; preventive
f prophylactique
e profiláctico
d prophylaktisch; vorbeugend

3635 prophylaxis
f prophylaxie
e profilaxis
d Prophylaxe; Verhütung

3636 propiolactone; beta-propiolactone
f propiolactone
e propiolactona
d Propiolakton; Betapropiolakton
l propiolactonum

3637 prosencephalon; forebrain
f prosencéphale; cerveau antérieur
e prosencéfalo
d Prosenzephalon; Vorderhirn
l prosencephalon

3638 prostacyclin
f prostacycline
e prostaciclina
d Prostazyklin

3639 prostaglandin
f prostaglandine
e prostaglandina
d Prostaglandin

3640 prostate
f prostate
e próstata
d Prostata; Vorstehdrüse
l prostata

3641 prostatic venous plexus
f plexus prostatique (veineux)
e plexo prostático
d Plexus prostaticus
l plexus prostaticus

3642 prosthogonimosis; Prosthogonimus infestation
f prosthogonimose
e prostogonimosis
d Prosthogonimose

* Prosthogonimus infestation → 3642

3643 protease; proteolytic enzyme
f protéase; protéinase; enzyme protéolytique
e proteasa; proteinasa
d Protease; proteolytisches Enzym

3644 protein
f protéine
e proteína
d Protein; Eiweiß

3645 protein metabolism disorder
f trouble du métabolisme protéique
e trastorno del metabolismo de proteína
d Störung des Proteinstoffwechsels

3646 proteinuria
f protéinurie
e proteinuria
d Proteinurie

* proteolytic enzyme → 3643

3647 prothrombin
f prothrombine
e protrombina
d Prothrombin

* protostrongylid → 2608

3648 protostrongylosis; Protostrongylus infestation
f protostrongylose
e protostrongilosis
d Protostrongylose

* Protostrongylus infestation → 3648

3649 protozoa
f protozoaires
e protozoos; protozoarios
d Protozoen
l Protozoa

3650 protozoal
f protozoaire
e protozoario
d protozoal

* protozoal infection → 3651

3651 protozoosis; protozoal infection
f protozoose
e protozoosis; infección protozoaria
d protozoäre Infektion; Protozoeninfektion

3652 protractor muscle (*of prepuce*)
f muscle protracteur (*du fourreau*)
e músculo protractor
d Musculus protractor
l musculus protractor

3653 protuberance
f protubérance
e protuberancia
d Protuberantia; Knorren; Hervorragung
l protuberantia

3654 proventriculus; forestomach
f proventricule; pré-estomac
e proventrículo
d Vormagen
l proventriculus

* proventriculus → 1845

3655 proximal interphalangeal articulations; pastern joint
f articulations interphalangiennes proximales
e articulaciones interfalangianas proximales
d zweite Zehengelenke; Krongelenk
l articulationes interphalangeae proximales

3656 proximal interphalangeal region
f région interphalangienne proximale
e región interfalangiana proximal
d Krongelenkgegend
l regio interphalangea proximalis

3657 proximal phalangeal region; fetlock
f région de la phalange proximale
e región de la falange proximal
d Fesselbeingegend
l regio phalangis proximalis

* **proximal phalanx** → 1705

3658 proximal sesamoid bones
f os sésamoïdes proximaux
e huesos sesamoideos proximales
d Sesambeine des Fesselgelenks;
Gleichbeine
l ossa sesamoidea proximalia

3659 pruritus; itchiness
f prurit; démangeaison
e prurito
d Pruritus; Juckreiz; Hautjucken

* **PSE meat** → 3238

3660 pseudocowpox; milkers' nodule
f nodules des trayeurs; infection à parapoxvirus
e nódulos de los ordeñadores; seudoviruela
d Melkerknoten; Parapoxvirusinfektion;
Pseudopocken

* **Pseudomonas mallei allergen** → 2653

* **Pseudomonas mallei infection** → 1841

* **Pseudomonas pseudomallei infection** → 2756

3661 pseudopregnancy; false pregnancy
f pseudogestation; gestation imaginaire
e seudogestación; gestación nerviosa
d Pseudoträchtigkeit; Scheinträchtigkeit

* **pseudorabies** → 331

* **pseudorabies virus** → 332

3662 pseudotuberculosis; Yersinia pseudotuberculosis infection
f pseudotuberculose; infection à Yersinia pseudotuberculosis
e seudotuberculosis; infección por Yersinia pseudotuberculosis
d Pseudotuberkulose; Yersinia-pseudotuberculosis-Infektion

* **psittacosis** → 3162

3663 psoroptic mange; psoroptosis
f gale psoroptique
e sarna psoróptica
d Psoroptesräude

* **psoroptosis** → 3663

3664 pterygoid bone
f os ptérygoïde
e hueso terigoideo
d Flügelbein
l os pterygoideum

3665 pterygoid hamulus
f hamulus ptérygoïdien
e ganchillo terigoideo
d Hamulus des Flügelbeins
l hamulus pterygoideus

3666 pterygoid muscle
f muscle ptérygoïdien
e músculo terigoideo
d innerer Kaumuskel
l musculus pterygoideus

3667 pterygoid process
f processus ptérygoïdien
e apófisis terigoidea
d Flügelfortsatz
l processus pterygoideus

3668 pterygoid venous plexus
f plexus ptérygoïdien (veineux)
e plexo terigoideo
d Plexus pterygoideus
l plexus pterygoideus

3669 pterygomandibular fold
f pli ptérygo-mandibulaire
e pliegue terigomandibular
d Kieferfalte
l plica pterygomandibularis

3670 pterygomandibular ligament
f ligament ptérygo-mandibulaire
e ligamento terigomandibular
d Ligamentum pterygomandibulare
l ligamentum pterygomandibulare

3671 pterygopalatine fossa
f fosse ptérygo-palatine
e fosa terigopalatina
d Flügelgaumengrube
l fossa pterygopalatina

3672 pterygopalatine ganglion
f ganglion ptérygo-palatin
e ganglio terigopalatino
d Ganglion pterygopalatinum
l ganglion pterygopalatinum

3673 pterygopalatine nerve
f nerf ptérygo-palatin
e nervio terigopalatino
d Nervus pterygopalatinus
l nervus pterygopalatinus

3674 **pterygopharyngeal muscle**
 f muscle ptérygo-pharyngien
 e músculo terigofaríngeo
 d Musculus pterygopharyngeus
 l musculus pterygopharyngeus

3675 **pubic region**
 f région pubienne
 e región púbica
 d Schamgegend
 l regio pubica

3676 **pubic symphysis**
 f symphyse pubienne
 e sínfisis púbica
 d Schambeinfuge
 l symphysis pubica

3677 **pubis**
 f pubis; os pubien
 e hueso pubis
 d Schambein
 l os pubis

3678 **pubocaudal muscle; pubococcygeal muscle**
 f muscle pubo-coccygien
 e músculo pubocaudal
 d Musculus pubocaudalis; Musculus
   pubococcygeus
 l musculus pubocaudalis; musculus
   pubococcygeus

 * **pubococcygeal muscle → 3678**

3679 **pubofemoral ligament**
 f ligament pubo-fémoral
 e ligamento pubofemoral
 d Ligamentum pubofemorale
 l ligamentum pubofemorale

3680 **pubo-ischio-femoral muscle**
 f muscle pubo-ischio-fémoral
 e músculo puboisquiofemoral
 d Musculus puboischiofemoralis
 l musculus puboischiofemoralis

3681 **pubovesical ligament**
 f ligament pubo-vésical
 e ligamento pubovesical
 d Ligamentum pubovesicale
 l ligamentum pubovesicale

3682 **pubovesical muscle**
 f muscle pubo-vésical
 e músculo pubovesical
 d Musculus pubovesicalis
 l musculus pubovesicalis

3683 **pubovesical pouch**
 f cul-de-sac pubo-vésical
 e excavación pubovesical
 d Excavatio pubovesicalis
 l excavatio pubovesicalis

3684 **pudendal nerve**
 f nerf honteux
 e nervio pudendo
 d Nervus pudendus
 l nervus pudendus

3685 **pudendal veins**
 f veines honteuses
 e venas pudendas
 d Venae pudendae
 l venae pudendae

3686 **pudendo-epigastric trunk**
 f tronc pudendo-épigastrique
 e tronco pudendoepigástrico
 d Truncus pudendoepigastricus
 l truncus pudendoepigastricus

3687 **puerperal disorder**
 f trouble puerpéral
 e trastorno puerperal
 d Puerperalstörung

3688 **puerperal period; postparturient period**
 f période puerpérale
 e período puerperal
 d Puerperium

3689 **pulmonary adenomatosis of sheep**
 f adénomatose pulmonaire ovine
 e adenomatosis pulmonar ovina
 d Lungenadenomatose des Schafes

3690 **pulmonary alveoli**
 f alvéoles pulmonaires
 e alveolos del pulmón
 d Lungenalveolen
 l alveoli pulmonis

3691 **pulmonary emphysema**
 f emphysème pulmonaire
 e enfisema pulmonar
 d Lungenemphysem
 l emphysema pulmonum

 * **pulmonary hilus → 2005**

3692 **pulmonary lymph nodes**
 f nœuds lymphatiques pulmonaires
 e nódulos linfáticos pulmonares
 d Lungenlymphknoten
 l lymphonodi pulmonales

3693 **pulmonary pleura; visceral pleura**
  f plèvre pulmonaire
  e pleura pulmonar; pleura visceral
  d Lungenfell
  l pleura pulmonalis

3694 **pulmonary plexus**
  f plexus pulmonaire
  e plexo pulmonar
  d Nervengeflecht der Lunge
  l plexus pulmonalis

3695 **pulmonary trunk**
  f tronc pulmonaire
  e tronco pulmonar
  d Truncus pulmonalis
  l truncus pulmonalis

  * **pulmonary valve → 4818**

3696 **pulmonary veins**
  f veines pulmonaires
  e venas pulmonares
  d Lungenvenen
  l venae pulmonales

3697 **pulp cavity of tooth**
  f cavité pulpaire de la dent
  e cavidad pulpar del diente
  d Zahnhöhle
  l cavum dentis

3698 **pulpy kidney disease; Clostridium perfringens type D infection**
  f entérotoxémie à Clostridium perfringens type D
  e enfermedad del riñón pulposo; basquilla; infección por Clostridium perfringens tipo D
  d Clostridium-perfringens-Typ-D-Enterotoxämie

3699 **pulse rate**
  f fréquence du pouls
  e frecuencia del pulso
  d Pulsfrequenz

3700 **pupil**
  f pupille
  e pupila
  d Pupille; Sehloch
  l pupilla

3701 **pupillary membrane**
  f membrane pupillaire
  e membrana pupilar
  d Membrana pupillaris
  l membrana pupillaris

3702 **purulent**
  f purulent
  e purulento
  d eitrig

3703 **pus**
  f pus
  e pus
  d Eiter

3704 **pustular**
  f pustuleux
  e pustuloso
  d pustulös

3705 **pustule**
  f pustule
  e pústula
  d Pustel
  l pustula

3706 **putamen**
  f putamen
  e putamen
  d Putamen
  l putamen

3707 **pyelitis**
  f pyélite
  e pielitis
  d Pyelitis

3708 **pygostyle**
  f pygostyle
  e pigostilo
  d Pygostyl
  l pygostylus

3709 **pyloric antrum**
  f antre pylorique
  e antro pilórico
  d Pylorusvorhof
  l antrum pyloricum

3710 **pyloric canal**
  f canal pylorique
  e canal pilórico
  d Canalis pyloricus
  l canalis pyloricus

3711 **pyloric glands**
  f glandes pyloriques
  e glándulas pilóricas
  d Pylorusdrüsen
  l glandulae pyloricae

3712 **pyloric opening**
  f orifice pylorique

e orificio pilórico
d Ostium pyloricum
l ostium pyloricum

3713 pyloric sphincter
f sphincter pylorique
e músculo esfínter del píloro
d Schließmuskel des Magenpförtners;
Sphincter pylori
l musculus sphincter pylori

3714 pylorus
f pylore
e píloro
d Pylorus; Magenpförtner
l pylorus

3715 pyoderma; pyogenic dermatitis
f pyodermite
e piodermatitis
d Pyodermie; Eiterausschlag

3716 pyogenic bacteria
f bactéries pyogènes
e bacterias piogénicas
d Eitererreger

* **pyogenic dermatitis** → 3715

3717 pyometritis
f pyométrite
e piometritis
d eitrige Gebärmutterentzündung

3718 pyramidal tract; corticospinal tract
f tractus pyramidal
e tracto piramidal
d Pyramidenbahn
l tractus pyramidalis; tractus corticospinalis

3719 pyramid of medulla oblongata
f pyramide de la moelle allongée
e pirámide de la médula oblongada
d Pyramis medullae oblongatae
l pyramis medullae oblongatae

3720 pyrantel
f pyrantel

e pirantel
d Pyrantel
l pyrantelum

3721 pyrethrin
f pyréthrine
e piretrina
d Pyrethrin

3722 pyrexia
f pyrexie
e pirexia
d Pyrexie

3723 pyridoxine; vitamin B6
f pyridoxine; vitamine B6
e piridoxina; vitamina B6
d Pyridoxin; Vitamin B6
l pyridoxinum

* **pyriform lobe** → 3480

3724 pyrimethamine
f pyriméthamine
e pirimetamina
d Pyrimethamin
l pyrimethaminum

* **pyrithidium bromide** → 3725

3725 pyritidium bromide; pyrithidium bromide
f bromure de pyritidium
e bromuro de piritidio
d Pyritidium-Bromid
l pyritidii bromidum

3726 pyrogen
f pyrogène
e pirógena
d Pyrogen

3727 pyrrolizidine alkaloid
f alcaloïde pyrrolizidinique
e alcaloide pirrolizidínico
d Pyrrolizidinalkaloid

Q

3728 Q fever; Coxiella burnetii infection
 f fièvre Q; infection à Coxiella burnetii
 e fiebre Q; infección por Coxiella burnetii
 d Q-Fieber; Coxiellose

3729 quadrate bone
 f os carré
 e hueso cuadrado
 d Quadratum
 l quadratum

3730 quadrate lobe of liver
 f lobe carré du foie
 e lóbulo cuadrado del hígado
 d Quadratlappen der Leber
 l lobus quadratus hepatis

3731 quadrate lumbar muscle
 f muscle carré des lombes
 e músculo cuadrado de los lomos
 d viereckiger Lendenmuskel
 l musculus quadratus lumborum

3732 quadrate muscle of thigh
 f muscle carré fémoral/crural
 e músculo cuadrado del muslo

 d viereckiger Oberschenkelmuskel
 l musculus quadratus femoris

3733 quadrate pronator muscle
 f muscle carré pronateur
 e músculo pronador cuadrado
 d viereckiger Einwärtsdreher
 l musculus pronator quadratus

3734 quadriceps muscle of thigh
 f muscle quadriceps fémoral
 e músculo cuádriceps del muslo
 d vierköpfiger Kniegelenkstrecker
 l musculus quadriceps femoris

* quadrigeminal body → 721, 3915

3735 quarantine
 f quarantaine
 e cuarentena
 d Quarantäne

3736 quarter of hoof
 f quartier du sabot
 e cuartas partes del casco
 d Trachtenwand

* quill → 624

3737 quittor; fistula of lateral cartilage of hoof
 f javart cartilagineux
 e gabarro cartilaginoso
 d Hufknorpelfistel

R

* rabbit myxomatosis → 2920

3738 **rabbit papillomavirus**
  f papillomavirus léporin
  e papillomavirus de los conejos
  d Kaninchen-Papillomavirus

3739 **rabbit spirochaetosis; Treponema cuniculi infection**
  f spirochétose des léporidés; infection à Treponema cuniculi
  e espiroquetosis del conejo; sífilis del conejo; infección por Treponema cuniculi
  d Kaninchenspirochätose; Treponema-cuniculi-Infektion

3740 **rabid**
  f rabique
  e rabioso
  d tollwütig

3741 **rabies**
  f rage
  e rabia
  d Tollwut
  l lyssa

3742 **rabies virus**
  f virus de la rage
  e virus de la rabia
  d Tollwut-Virus

3743 **rachis**
  f rachis; hampe creuse
  e raquis
  d Federschaft
  l rachis

3744 **radial artery**
  f artère radiale
  e arteria radial
  d Arteria radialis
  l arteria radialis

3745 **radial carpal bone; scaphoid**
  f os radial du carpe
  e hueso carporradial; hueso escafoideo
  d Os carpi radiale; Os scaphoideum
  l os carpi radiale; os scaphoideum

3746 **radial extensor of carpus**
  f muscle extenseur du carpe
  e músculo extensor carporradial
  d innerer Karpalstrecker
  l musculus extensor carpi radialis

3747 **radial flexor muscle of carpus**
  f muscle fléchisseur radial du carpe; muscle grand palmaire
  e músculo flexor carporradial
  d innerer Karpalbeuger
  l musculus flexor carpi radialis

3748 **radial incisure of ulna**
  f incisure radiale
  e escotadura radial
  d Incisura radialis ulnae
  l incisura radialis ulnae

3749 **radial nerve**
  f nerf radial
  e nervio radial
  d Nervus radialis; Speichennerv
  l nervus radialis

3750 **radial veins**
  f veines radiales
  e venas radiales
  d Venae radiales
  l venae radiales

3751 **radiate sternocostal ligaments**
  f ligaments sterno-costaux rayonnées
  e ligamentos esternocostales radiados
  d Ligamenta sternocostalia radiata
  l ligamenta sternocostalia radiata

3752 **radiation**
  f radiation; rayonnement
  e radiación
  d Strahlung

3753 **radiation injury; radiation sickness**
  f mal des rayons; trouble par irradiation
  e herida por radiación
  d Strahlenschaden

* **radiation sickness** → 3753

* **radiation therapy** → 3762

3754 **radioactive contamination**
  f contamination radioactive
  e radiocontaminación
  d radioaktive Kontamination

3755 **radioactive fallout**
  f retombée radioactive
  e precipitación radioactiva atmosférica
  d radioaktive Ausschüttung

* **radioactive isotope** → 3761

3756 radioactivity
  f radioactivité
  e radioactividad
  d Radioaktivität

3757 radiocarpal articulation
  f articulation radio-carpienne
  e articulación radiocarpiana
  d Articulatio radiocarpea
  l articulatio radiocarpea

* **radiocarpal joint** → 198

3758 radiocarpal ligament
  f ligament radio-carpien
  e ligamento radiocarpiano
  d Ligamentum radiocarpeum
  l ligamentum radiocarpeum

3759 radiography
  f radiographie
  e radiografía
  d Röntgenographie

3760 radioimmunoassay
  f dosage radio-immunologique
  e dosificación radioinmunológica
  d Radioimmunoassay

3761 radioisotope; radioactive isotope
  f isotope radioactif
  e isotopo radioactivo
  d Radioisotop

3762 radiotherapy; radiation therapy
  f radiothérapie
  e radioterapía
  d Radiotherapie; Strahlentherapie

3763 radioulnar articulation
  f articulation radio-ulnaire
  e articulación radiocubital
  d Radioulnargelenk
  l articulatio radioulnaris

3764 radioulnar ligament
  f ligament radio-ulnaire
  e ligamento radiocubital
  d Ligamentum radioulnare
  l ligamentum radioulnare

3765 radius
  f radius
  e radio
  d Radius; Speiche
  l radius

3766 rafoxanide
  f rafoxanide
  e rafoxanida
  d Rafoxanid
  l rafoxanidum

* **Raillietina infestation** → 3767

3767 raillietinosis; Raillietina infestation
  f railliétinose
  e raillietinosis
  d Raillietinose

3768 ramus of mandible
  f branche de la mandibule
  e rama de la mandíbula
  d Unterkieferast
  l ramus mandibulae

3769 raphe of pharynx
  f raphé du pharynx
  e rafe faríngeo
  d Rachennaht
  l raphe pharyngis

3770 reaction
  f réaction
  e reacción
  d Reaktion

3771 recovery (*from illness*); recuperation
  f guérison
  e recuperación
  d Genesung
  l restitutio

3772 rectal ampulla
  f ampoule rectale
  e ampolla rectal
  d Ampulle des Mastdarms
  l ampulla recti

3773 rectal arteries
  f artères rectales
  e arterias rectales
  d Arteriae rectales; Mastdarmarterien
  l arteriae rectales

3774 rectal atresia
  f atrésie rectale
  e atresia rectal
  d Rektalatresie
  l atresia recti

3775 rectal exploration; rectal palpation
  f exploration rectale
  e exploración rectal
  d rektale Untersuchung

* **rectal palpation** → 3775

3776 rectal plexuses
f plexus rectaux
e plexos rectales
d Plexus rectales
l plexus rectales

3777 rectal prolapse
f prolapsus rectal
e prolapso rectal
d Prolapsus ani; Rektumprolaps;
Mastdarmvorfall
l prolapsus ani

3778 rectal temperature
f température rectale
e temperatura rectal
d Rektaltemperatur

3779 rectal veins
f veines rectales
e venas rectales
d Mastdarmvenen
l venae rectales

3780 rectococcygeal muscle
f muscle recto-coccygien
e músculo rectocoxígeo
d Musculus rectococcygeus
l musculus rectococcygeus

3781 rectogenital pouch
f cul-de-sac recto-génital
e excavación rectogenital
d Excavatio rectogenitalis
l excavatio rectogenitalis

3782 rectourethral fistula
f fistule recto-urétrale
e fístula rectouretral
d Mastdarm-Harnröhrenfistel

3783 rectourethral muscle
f muscle recto-uréthral
e músculo rectouretral
d Musculus rectourethralis
l musculus rectourethralis

3784 rectovaginal fistula
f fistule recto-vaginale
e fístula rectovaginal
d Mastdarm-Scheidenfistel

3785 rectovaginal septum
f cloison recto-vaginale
e septo rectovaginal
d Septum rectovaginale
l septum rectovaginale

3786 rectricial bulb muscle
f muscle du bulbe des rectrices
e músculo del bulbo de las rectrices
d Musculus bulbi rectricium
l musculus bulbi rectricium

3787 rectum
f rectum
e recto
d Rektum; Mastdarm
l rectum

* **recuperation** → 3771

* **recurrence** → 3805

3788 recurrent arteries
f artères récurrentes
e arterias recurrentes
d rücklaufende Arterien
l arteriae recurrentes

**3789 recurrent iridocyclitis; periodic
ophthalmia**
f irido-cyclite du cheval; fluxion périodique
e iridociclitis periódica de los caballos
d periodische Augenentzündung
l iridocyclitis recidivans

3790 recurrent laryngeal nerve
f nerf laryngé récurrent
e nervio laríngeo recurrente
d Nervus laryngeus recurrens
l nervus laryngeus recurrens

* **red blood cell** → 1540

3791 red bone marrow
f moelle rouge
e médula ósea roja
d rotes Knochenmark
l medulla ossium rubrum

3792 redia
f rédie
e redia
d Redie

3793 red mite of poultry
f pou rouge; dermanysse des volailles
e ácaro rojo de las gallinas
d rote Vogelmilbe
l Dermanyssus gallinae

3794 red nucleus
f noyau rouge
e núcleo rojo
d roter Haubenkern
l nucleus ruber

3795 **red pulp of spleen**
f pulpe rouge de la rate
e pulpa roja del bazo
d rote Milzpulpa
l pulpa lienis rubra

* **reflected part of hoof wall** → 408

3796 **reflex**
f réflexe
e reflejo
d Reflex

3797 **region**
f région
e región
d Region; Gegend
l regio

3798 **region of mandibular joint**
f région de l'articulation temporo-
mandibulaire
e región de la articulación
temporomandibular
d Kiefergelenkgegend
l regio articulationis temporomandibularis

3799 **regions of the face**
f régions de la face
e regiones de la cara
d Gesichtsregionen
l regiones faciei

3800 **regions of the head**
f régions de la tête
e regiones de la cabeza
d Kopfregionen
l regiones capitis

3801 **regions of the neck**
f régions du cou
e regiones del cuello
d Halsregionen
l regiones colli

3802 **regurgitation**
f régurgitation
e regurgitación
d Regurgitieren

* **Reichert's recess** → 969

3803 **reinfection**
f réinfection
e reinfección
d Reinfektion

3804 **Reissner's membrane; vestibular wall of cochlear duct**
f membrane vestibulaire (de Reissner)
e pared vestibular del conducto coclear;
membrana vestibular (de Reissner)
d Reissnersche Membran
l membrana vestibularis; paries vestibularis
ductus cochlearis

3805 **relapse; recurrence**
f récidive
e recidiva
d Rezidiv

3806 **renal artery**
f artère rénale
e arteria renal
d Nierenarterie
l arteria renalis

3807 **renal calculus; nephrolith**
f calcul rénal
e cálculo renal
d Nierenstein; Nephrolith
l calculus renalis

3808 **renal calyx**
f calice rénal
e cáliz renal
d Nierenkelch
l calix renalis

3809 **renal clearance**
f clairance rénale
e aclaramiento renal
d Nieren-Clearance

3810 **renal corpuscle; malpighian corpuscle**
f corpuscule rénal (de Malpighi)
e corpúsculo renal
d Nierenkörperchen
l corpusculum renale

3811 **renal cortex; cortex of kidney**
f cortex rénal
e corteza del riñón
d Nierenrinde
l cortex renis

* **renal function** → 2378

3812 **renal ganglia**
f ganglions rénaux
e ganglios renales
d Nierenganglien
l ganglia renalia

3813 renal hilus; hilus of kidney
 f hile rénal
 e hilio renal
 d Nierenhilus
 l hilus renalis

3814 renal impression (*on the liver*)
 f empreinte rénale
 e impresión renal
 d Niereneindruck
 l impressio renalis

 * **renal insufficiency** → 2377

3815 renal medulla; medulla of kidney
 f medulla rénale
 e médula del riñón
 d Nierenmark
 l medulla renis

3816 renal papilla
 f papille rénale
 e papila renal
 d Nierenpapille
 l papilla renis

3817 renal pelvis
 f pelvis rénal; bassinet
 e pelvis renal
 d Nierenbecken
 l pelvis renalis

3818 renal plexus
 f plexus rénal
 e plexo renal
 d Nervengeflecht der Niere
 l plexus renalis

3819 renal pyramid
 f pyramide rénale (de Malpighi)
 e pirámide renal
 d Markpyramide
 l pyramis renalis

3820 renal tubules; uriniferous tubules
 f tubules rénaux
 e túbulos renales
 d Nierenkanälchen; Harnkanälchen
 l tubuli renales

3821 renal vein
 f veine rénale
 e vena renal
 d Nierenvene
 l vena renalis

 * **Renibacterium salmoninarum infection** →
   396

3822 renin
 f rénine
 e renina
 d Renin

3823 rennet
 f présure
 e cuajo
 d Lab

3824 reovirus
 f réovirus
 e reovirus
 d Reovirus

3825 reproductive disorder
 f trouble de la reproduction
 e trastorno del sistema reproductivo
 d Fortpflanzungsstörung

 * **RES** → 3839

3826 residue
 f résidu
 e residuo
 d Rückstand

 * **resistance to disease** → 1334

3827 resistant to antibiotics
 f antibiorésistant
 e resistente a los antibióticos
 d antibiotikaresistent

3828 resorantel
 f résorantel
 e resorantel
 d Resorantel
 l resorantelum

3829 respiration; breathing
 f respiration
 e respiración
 d Respiration; Atmung

3830 respiratory disease
 f maladie respiratoire
 e enfermedad respiratoria
 d Erkrankung des Atmungssystems

3831 respiratory disorder
 f trouble respiratoire
 e trastorno respiratorio
 d Atmungsstörung

3832 respiratory syncytial virus
 f virus syncytial respiratoire
 e virus sincitial respiratorio
 d respiratorisches Syncytial-Virus

3833 **respiratory system**
  f appareil respiratoire
  e aparato respiratorio
  d Atmungsapparat
  l apparatus respiratorius; systema respiratorium

* **respiratory tubules** → 3521

* **retained placenta** → 3489

3834 **rete mirabile**
  f réseau admirable
  e red admirable
  d Rete mirabile; Wundernetz
  l rete mirabile

3835 **reticular formation of mesencephalon**
  f formation réticulaire du mésencéphale
  e formación reticular del mesencéfalo
  d Formatio reticularis mesencephali
  l formatio reticularis mesencephali

3836 **reticular groove**
  f sillon réticulaire
  e surco del retículo
  d Haubenrinne
  l sulcus reticuli

3837 **reticular layer of corium**
  f stratum reticulare; derme profond
  e estrato reticular
  d Stratum reticulare; untere Schicht der Lederhaut
  l stratum reticulare corii

3838 **reticulitis; inflammation of the reticulum**
  f réticulite
  e reticulitis
  d Reticulitis; Haubenentzündung

3839 **reticulo-endothelial system; RES**
  f système réticulo-endothélial; SRE
  e sistema reticuloendotelial
  d retikuloendotheliales System; RES

3840 **reticuloendotheliosis**
  f réticuloendothéliose
  e reticuloendoteliosis
  d Retikuloendotheliose

3841 **reticulo-omasal opening**
  f ostium réticulo-omasal
  e orificio retículo-omásico
  d Haubenpsalteröffnung
  l ostium reticuloomasicum

3842 **reticuloperitonitis**
  f réticulopéritonite
  e reticuloperitonitis
  d Retikuloperitonitis

3843 **reticulospinal tract**
  f tractus réticulo-spinal
  e tracto retículoespinal
  d Tractus reticulospinalis
  l tractus reticulospinalis

3844 **reticulum**
  f réseau
  e retículo
  d Haube; Netzmagen
  l reticulum

3845 **retina**
  f rétine
  e retina
  d Retina; Netzhaut
  l retina

3846 **retinal atrophy**
  f atrophie rétinienne
  e atrofia de la retina
  d Netzhautatrophie

3847 **retinal detachment**
  f décollement rétinien
  e separación de la retina
  d Netzhautablösung
  l ablatio retinae

3848 **retinal dysplasia**
  f dysplasie rétinienne
  e displasia de la retina
  d Netzhautdysplasie

3849 **retinol; vitamin A**
  f rétinol; vitamine A
  e retinol; vitamina A
  d Retinol; Vitamin A
  l retinolum

3850 **retractor muscle of eyeball**
  f muscle rétracteur du bulbe de l'œil
  e músculo retractor del globo
  d Musculus retractor bulbi
  l musculus retractor bulbi

3851 **retractor muscle of penis**
  f muscle rétracteur du pénis
  e músculo retractor del pene
  d Musculus retractor penis
  l musculus retractor penis

3852 retractor muscle of the last rib
f muscle rétracteur de la dernière côte
e músculo retractor de la costilla
d Musculus retractor costae
l musculus retractor costae

3853 retromandibular vein
f veine rétromandibulaire
e vena retromandibular
d Vena retromandibularis
l vena retromandibularis

3854 retroperitoneal space
f espace rétropéritonéal
e espacio retroperitoneal
d Spatium retroperitoneale
l spatium retroperitoneale

3855 retropharyngeal lymphocentre
f lymphocentre rétropharyngien
e linfocentro retrofaríngeo
d Lymphocentrum retropharyngeum
l lymphocentrum retropharyngeum

3856 retrovirus
f rétrovirus
e retrovirus
d Retrovirus

3857 revaccination
f injection de rappel
e revacunación
d Revakzination; Wiederholungsimpfung

3858 rhabdovirus
f rhabdovirus
e rhabdovirus
d Rhabdovirus

3859 rhinencephalon
f rhinencéphale
e rinencéfalo
d Rhinenzephalon; Riechhirn
l rhinencephalon

3860 rhinitis
f rhinite
e rinitis
d Rhinitis; Nasenkatarrh

* rhinopharynx → 2954

3861 rhinosporidiosis; Rhinosporidium seeberi infection
f rhinosporidiose; infection à Rhinosporidium seeberi
e rinosporidiosis; infección por Rhinosporidium seeberi
d Rhinosporidiose; Rhinosporidium-seeberi-Infektion

* Rhinosporidium seeberi infection → 3861

3862 rhinotracheitis
f rhinotrachéite
e rinotraqueitis
d Rhinotracheitis

3863 rhinovirus
f rhinovirus
e rhinovirus
d Rhinovirus

* Rhizoctonia legumicola toxin → 4103

3864 rhombencephalon; hindbrain
f rhombencéphale; cerveau postérieur
e rombencéfalo
d Rhombenzephalon; Rautenhirn
l rhombencephalon

3865 rhomboid fossa
f fosse rhomboïde
e fosa romboidea
d Rautengrube
l fossa rhomboidea

3866 rhomboid muscle
f muscle rhomboïde
e músculo romboideo
d Rautenmuskel
l musculus rhomboideus

3867 rib
f côte
e costilla
d Rippe
l costa

3868 rib bone
f os costal
e hueso costal
d Rippenknochen
l os costale

3869 riboflavin; vitamin B2
f riboflavine; vitamine B2
e riboflavina; vitamina B2
d Riboflavin; Vitamin B2
l riboflavinum

3870 ribonucleic acid; RNA
f acide ribonucléique; ARN
e ácido ribonucléico; ARN
d Ribonukleinsäure; RNS

3871 ribosome
f ribosome
e ribosoma
d Ribosom

3872 rib region; costal region
f région costale
e región costal
d Rippenregion
l regio costalis

3873 rickets
f rachitisme
e raquitismo
d Rachitis; Knochenweiche

3874 rickettsia
f rickettsie
e rickettsia
d Rickettsie
l Rickettsia

* **rickettsial infection → 3875**

3875 rickettsiosis; rickettsial infection
f rickettsiose
e infección por Rickettsia
d Rickettsiose

3876 rictus
f rictus
e rictus
d Riktus
l rictus

3877 ridges of palate
f crêtes palatines; replis transversaux
e rugosidades palatinas
d Gaumenstaffeln
l rugae palatinae

3878 ridges of skin; dermal ridges
f crêtes dermiques; crêtes de la peau
e crestas de la piel
d Hautleisten
l cristae cutis

3879 rifampicin
f rifampicine
e rifampicina
d Rifampicin
l rifampicinum

3880 rifamycin
f rifamycine
e rifamicina
d Rifamycin
l rifamycinum

3881 Rift Valley fever
f fièvre de la Vallée du Rift
e fiebre del Valle del Rift
d Rifttalfiber

3882 right atrioventricular opening
f orifice auriculo-ventriculaire droit
e orificio atrioventricular derecho
d rechte Atrioventrikularöffnung
l ostium atrioventriculare dextrum

3883 right atrioventricular valve; tricuspid valve
f valvule atrio-ventriculaire droite; valvule tricuspide
e valva atrioventricular derecha; válvula tricúspide
d rechte Atrioventrikularklappe; Trikuspidalklappe
l valva atrioventricularis dextra; valva tricuspidalis

3884 right atrium
f oreillette droite du cœur
e atrio derecho
d rechter Vorhof
l atrium dextrum

3885 right azygous vein
f veine azygos droite
e vena ácigos derecha
d Vena azygos dextra
l vena azygos dextra

3886 right gastric artery
f artère gastrique droite
e arteria gástrica derecha
d Arteria gastrica dextra; rechte Magenarterie
l arteria gastrica dextra

3887 right gastro-epiploic artery
f artère gastro-épiploïque droite
e arteria gastroepiplóica derecha
d Arteria gastroepiploica dextra
l arteria gastroepiploica dextra

3888 right gastro-epiploic vein
f veine gastro-épiploïque droite
e vena gastroepiplóica derecha
d Vena gastroepiploica dextra
l vena gastroepiploica dextra

3889 right hepatic duct
f conduit hépatique droit; canal hépatique droit
e conducto hepático derecho
d rechter Gallengang
l ductus hepaticus dexter

3890 right lobe of liver
f lobe droit du foie; lobe hépatique droit
e lóbulo derecho del hígado
d rechter Leberlappen
l lobus hepatis dexter

3891 right lymphatic duct
 f conduit lymphatique droit
 e conducto linfático derecho
 d rechter Lymphgang
 l ductus lymphaticus dexter

3892 right ovarian vein
 f veine ovarique droite
 e vena ovárica derecha
 d rechte Eierstockvene
 l vena ovarica dextra

3893 right pulmonary artery
 f artère pulmonaire droite
 e arteria pulmonar derecha
 d rechte Lungenarterie
 l arteria pulmonalis dextra

3894 right ruminal artery
 f artère ruminale droite
 e arteria ruminal derecha
 d rechte Pansenarterie
 l arteria ruminalis dextra

3895 right testicular vein
 f veine testiculaire droite
 e vena testicular derecha
 d rechte Hodenvene
 l vena testicularis dextra

3896 right ventricle of heart
 f ventricule droit
 e ventrículo derecho
 d rechte Herzkammer
 l ventriculus dexter cordis

3897 rigor mortis; post mortem rigidity
 f rigidité cadavérique
 e rigidez cadavérica
 d Leichenstarre; Totenstarre

3898 rinderpest; cattle plague
 f peste bovine
 e peste bovina
 d Rinderpest
 l pestis bovina

3899 rinderpest virus
 f virus de la peste bovine
 e virus de la peste bovina
 d Rinderpest-Virus

3900 ringbone; phalangeal exostosis
 f formes phalangiennes; exostose
   phalangienne
 e exostosis falangea
 d Ringbein; Überbein; Krongelenkschale

3901 ring test
 f test de l'anneau
 e prueba del anillo
 d Ringprobe

 * **ringworm** → 1268

 * **RNA** → 3870

3902 roaring in horses; laryngeal hemiplegia
 f hémiplégie laryngienne; cornage
 e parálisis parcial de la laringe
 d Kehlkopflähmung des Pferdes
 l hemiplegia laryngis

3903 robenidine
 f robénidine
 e robenidina
 d Robenidin
 l robenidinum

3904 ronidazole
 f ronidazole
 e ronidazol
 d Ronidazol
 l ronidazolum

 * **ronnel** → 1667

3905 roof of fourth ventricle
 f toit du quatrième ventricule
 e techo del cuarto ventrículo
 d Dach der vierten Hirnkammer
 l tegmen ventriculi quarti

3906 roof of mesencephalon
 f toit du mésencéphale
 e tecto del mesencéfalo
 d Dach des Mittelhirns
 l tectum mesencephali

3907 root of infundibulum
 f racine de l'infundibulum
 e raíz del infundíbulo
 d Wurzel des Infundibulums
 l radix infundibuli

3908 root of lung
 f racine pulmonaire; pédicule du poumon
 e raíz del pulmón
 d Lungenwurzel
 l radix pulmonis

3909 root of mesentery
 f racine du mésentère
 e raíz del mesenterio
 d Gekrösewurzel
 l radix mesenterii

3910 root of nose
f racine du nez
e raíz de la nariz
d Nasenwurzel
l radix nasi

* **root of tail → 4433**

3911 root of tail region; base of tail region
f région de la racine de la queue
e región de la raíz de la cola
d Schwanzwurzelgegend
l regio radicis caudae

3912 root of tongue
f racine de la langue
e raíz de la lengua
d Zungenwurzel
l radix linguae

3913 root of tooth
f racine de la dent
e raíz del diente
d Zahnwurzel
l radix dentis

* **Rosenmüller's cavity → 3442**

3914 rostral bone; prenasal bone
f os du rostre
e hueso rostral
d Rüsselbein
l os rostrale

3915 rostral colliculus; quadrigeminal body
f colliculus rostral; tubercule quadrijumeau postérieur
e colículo rostral; tubérculo cuadrigémino
d Colliculus rostralis; vordere Vierhügel
l colliculus rostralis; corpora quadrigemina

3916 rostral communicating artery
f artère communicante rostrale
e arteria comunicante rostral
d vordere Verbindungsarterie
l arteria communicans rostralis

3917 rostral horn of lateral ventricle
f corne rostrale du ventricule latéral
e asta rostral
d Cornu rostralis
l cornu rostrale ventriculi lateralis

3918 rostral lobe of cerebellum
f lobe rostral du cervelet
e lóbulo rostral del cerebelo
d Lobus rostralis
l lobus rostralis

3919 rostral medullary velum
f voile médullaire rostral
e velo medular rostral
d Velum medullare rostrale; vorderes Marksegel
l velum medullare rostrale

3920 rostral nuclei of thalamus
f noyaux rostraux du thalamus
e núcleos rostrales del tálamo
d Nuclei rostrales thalami
l nuclei rostrales thalami

3921 rostral plane; disk of snout
f plan rostral; disque du groin
e plano rostral
d Rüsselscheibe
l planum rostrale

3922 rostral stylopharyngeal muscle
f muscle stylo-pharyngien rostral
e músculo estilofaríngeo rostral
d Musculus stylopharyngeus rostralis
l musculus stylopharyngeus rostralis

3923 rostrum; snout
f rostre; groin
e rostro; pico
d Rostrum; Schnauze; Rüssel
l rostrum

3924 rotator muscles
f muscles rotateurs
e músculos rotadores
d Drehmuskeln
l musculi rotatores

3925 rotavirus
f rotavirus
e rotavirus
d Rotavirus

3926 round foramen
f foramen rond
e agujero redondo
d Foramen rotundum
l foramen rotundum

3927 round ligament of bladder
f ligament rond de la vessie
e ligamento redondo de la vejiga
d rundes Harnblasenband
l ligamentum teres vesicae

* **round ligament of femur → 2515**

3928 round ligament of uterus
f ligament rond de l'utérus
e ligamento redondo del útero

d rundes Mutterband
l ligamentum teres uteri

3929 round pronator muscle
f muscle rond pronateur
e músculo pronador redondo
d runder Einwärtsdreher
l musculus pronator teres

3930 round window of middle ear; fenestra of cochlea
f fenêtre ronde; fenêtre de la cochlée
e ventana de la cóclea; ventana redondo
d Schneckenfenster
l fenestra cochleae

* roundworm → 2970

3931 Rous sarcoma
f sarcome de Rous
e sarcoma de Rous
d Rous-Sarkom

3932 Rous sarcoma oncovirus
f virus du sarcome de Rous
e virus del sarcoma de Rous
d Rous-Sarkom-Oncovirus

3933 roxarsone
f roxarsone
e roxarsona
d Roxarson
l roxarsonum

3934 rubrospinal tract
f tractus rubro-spinal
e tracto rubroespinal
d Tractus rubrospinalis
l tractus rubrospinalis

3935 rumen; paunch
f rumen; panse
e rumen; panza
d Rumen; Pansen
l rumen

3936 rumen fistula
f fistule du rumen
e fístula de la panza
d Pansenfistel

3937 rumen flora
f flore du rumen
e flora de la panza
d Pansenflora

* rumen fluke → 3287

3938 rumenitis
f ruménite
e rumenitis
d Ruminitis; Rumenitis

3939 rumenotomy
f ruménotomie
e rumenotomía
d Pansenschnitt; Rumenotomie

3940 rumen protozoa
f protozoaires du rumen
e protozoarios de la panza
d Pansenprotozoen

3941 ruminal atony
f atonie du rumen
e atonía del rumen
d Pansenatonie; Pansenstillstand

3942 ruminal contents
f contenu du rumen
e contenido de la panza
d Panseninhalt

3943 ruminal islet
f insula du rumen
e insula del rumen
d Panseninsel
l insula ruminis

3944 ruminal lymph nodes
f nœuds lymphatiques du rumen
e nódulos linfáticos ruminales
d Pansenlymphknoten
l lymphonodi ruminales

3945 ruminal papillae
f papilles du rumen
e papilas del rumen
d Pansenpapillen
l papillae ruminis

3946 ruminal pillars
f piliers du rumen
e pilares del rumen
d Pansenpfeiler
l pilae ruminis

3947 ruminal plexuses
f plexus du rumen
e plexos ruminales
d Pansengeflechte
l plexus ruminales

3948 ruminal recess
f récessus ruminal
e receso del rumen
d Recessus ruminis
l recessus ruminis

3949 ruminal veins
 f veines ruminales
 e venas ruminales
 d Pansenvenen
 l venae ruminales

3950 ruminant digestion
 f digestion des ruminants
 e digestión de los rumiantes
 d Verdauung bei Wiederkäuer

3951 ruminant stomach
 f estomac du ruminant
 e estómago de rumiante
 d Wiederkäuermagen

3952 rumination
 f rumination
 e rumia
 d Wiederkauen

3953 ruminoreticular fold
 f pli rumino-réticulaire
 e pliegue ruminorreticular
 d Haubenpansenfalte
 l plica ruminoreticularis

3954 ruminoreticular groove
 f sillon rumino-réticulaire

 e surco ruminorreticular
 d Haubenpansenfurche
 l sulcus ruminoreticularis

3955 ruminoreticular opening
 f ostium rumino-réticulaire
 e orificio ruminorreticular
 d Pansenhaubenöffnung
 l ostium ruminoreticulare

3956 ruminoreticulum
 f ruminoréticulum
 e ruminorretículo
 d Ruminoreticulum; Pansen und Haube
 l ruminoreticulum

3957 rump of birds
 f croupion
 e rabadilla
 d Bürzel; Pyga
 l pyga

3958 rupture
 f rupture
 e ruptura
 d Ruptur
 l ruptura

S

3959 saccule
f saccule
e sáculo
d Sacculus
l sacculus

3960 sacral crest
f crête sacrale
e cresta sacra
d Crista sacralis
l crista sacralis

3961 sacral foramina
f foramens sacraux
e agujeros sacros
d Foramina sacralia
l foramina sacralia

3962 sacral ganglia
f ganglions sacraux/sacrés
e ganglios sacros
d Kreuzbeinganglien
l ganglia sacralia

3963 sacral lymph nodes
f nœuds lymphatiques sacraux
e nódulos linfáticos sacros
d Sakrallymphknoten
l lymphonodi sacrales

3964 sacral nerves
f nerfs sacrés
e nervios sacros
d Sakralnerven
l nervi sacrales

3965 sacral plexus
f plexus sacral/sacré
e plexo sacro
d Kreuzgeflecht
l plexus sacralis

3966 sacral region; croup
f région sacrale/sacrée
e región sacra
d Kreuzbeingegend; Kruppe
l regio sacralis

3967 sacral splanchnic nerves
f nerfs splanchniques sacraux
e nervios esplácnicos sacros
d Nervi splanchnici sacrales
l nervi splanchnici sacrales

3968 sacral tuber; medial angle of ilium
f tubérosité sacrée; angle de la croupe
e tuberosidad sacra
d Kreuzhöcker; medialer Darmbeinwinkel
l tuber sacrale

3969 sacral vertebra
f vertèbre sacrale/sacrée
e vértebra sacra
d Kreuzwirbel
l vertebra sacralis

3970 sacrocaudal muscles
f muscles sacro-coccygiens
e músculos sacrocaudales/sacrococcígeos
d Kreuzbein-Schwanzmuskeln
l musculi sacrocaudales/sacrococcygis

3971 sacroiliac articulation
f articulation sacro-iliaque
e articulación sacroilíaca
d Kreuzbein-Darmbein-Gelenk
l articulatio sacroiliaca

3972 sacroiliac ligaments
f ligaments sacro-iliaques
e ligamentos sacroilíacas
d Gelenkbänder des Kreuzdarmbeingelenkes
l ligamenta sacroiliaca

3973 sacrotuberal ligament
f ligament sacrotubéreux
e ligamento sacrotuberoso
d Ligamentum sacrotuberale
l ligamentum sacrotuberale

3974 sacrum
f sacrum
e sacro
d Kreuzbein
l os sacrum

3975 saddle joint
f articulation en selle
e articulación sellar
d Sattelgelenk
l articulatio sellaris

3976 sagittal suture
f suture sagittale
e sutura sagital
d Sagittalnaht
l sutura sagittalis

3977 salinomycin
f salinomycine
e salinomicina
d Salinomycin
l salinomycinum

3978 saliva
f salive
e saliva
d Speichel

3979 salivary gland
f glande salivaire
e glándula salivar
d Speicheldrüse

3980 salivation
f salivation
e salivación
d Salivation; Speichelfluß

* **Salmonella gallinarum infection** → 1758

* **Salmonella infection** → 3981

3981 salmonellosis; Salmonella infection; paratyphoid
f salmonellose; infection à Salmonella
e salmonelosis; infección por Salmonella spp.
d Salmonellose; Salmonella-Infektion

3982 salmon poisoning of dogs; Neorickettsia helminthoeca infection
f néorickettsiose; infection à Neorickettsia helminthoeca
e enfermedad salmón de los perros; enfermedad de la intoxicación por salmón; infección por Neorickettsia helminthoeca
d Lachskrankheit der Hunde; Neorickettsia-helminthoeca-Infektion

3983 salpingitis
f salpingite
e salpingitis
d Salpingitis; Eileiterentzündung

* **sand flea** → 2357

3984 sand fly
f phlébotome
e flebótomo; mosca de arena; beatilla
d Engelsmücke; Schmetterlingsmücke; Phlebotome
l Phlebotomus

3985 saphenous artery
f artère saphène

e arteria safena
d Arteria saphena
l arteria saphena

3986 saphenous nerve
f nerf saphène
e nervio safeno
d Nervus saphenus
l nervus saphenus

3987 saphenous vein
f veine saphène
e vena safena
d Vena saphena
l vena saphena

* **Saprolegnia infection** → 3988

3988 saprolegniosis; Saprolegnia infection
f saprolegniose; infection à Saprolegnia
e saprolegniosis; infección por Saprolegnia spp.
d Saprolegniose

3989 sarcocystiosis; sarcosporidiosis
f sarcocystose; sarcosporidiose
e sarcocistosis; sarcosporidiosis
d Sarkozystiose; Sarkosporidiose; Sarkosporidienbefall

3990 sarcoma
f sarcome
e sarcoma
d Sarkom

3991 sarcoptic mange; sarcoptidosis
f gale sarcoptique
e sarna sarcóptica
d Sarkoptesräude

* **sarcoptidosis** → 3991

* **sarcoptiform mite** → 2678

* **sarcosporidiosis** → 3989

3992 sartorius muscle
f muscle couturier
e músculo sartorio
d Schneidermuskel; Darmbein-Schenkelmuskel
l musculus sartorius

3993 satratoxin; Stachybotrys atra toxin
f satratoxine
e satratoxina
d Satratoxin

* SBE → 4179

* scabies → 2677

3994 scale
 f scutum; écaille
 e escama
 d Schild
 l scutum

3995 scalene muscle
 f muscle scalène
 e músculo escaleno
 d Rippenhalter
 l musculus scalenus

3996 scaly leg mite
 f sarcopte changeant
 e ácaro de las patas
 d Kalkbeinmilbe
 l Knemidocoptes mutans

3997 scaly leg of fowls
 f gale des pattes
 e sarna aviar de las patas; patas escamosas
 d Kalkbein

3998 scapha; scaphoid fossa
 f fosse scaphoïde
 e escafa; fosa escafoidea
 d Tütenhöhlung der Ohrmuschel
 l scapha

* scaphoid → 3745

* scaphoid fossa → 3998

3999 scapula; shoulderblade
 f omoplate
 e omóplato; escápula
 d Skapula; Schulterblatt
 l scapula

4000 scapular cartilage
 f cartilage scapulaire; cartilage de
   l'omoplate
 e cartílago de la escápula
 d Schulterblattknorpel
 l cartilago scapulae

4001 scapular cartilage region
 f région du cartilage scapulaire
 e región del cartílago de la escápula
 d Schulterknorpelgegend
 l regio cartilaginis scapulae

* Scarpa's ganglion → 4895

4002 scent gland
 f glande odorifère
 e glándula odorífera
 d Hautduftdrüse
 l glandulae odoriferae

4003 schistosome
 f schistosome
 e esquistosoma
 d Schistosom; Pärchenegel
 l Schistosoma

* schistosomiasis → 4004

4004 schistosomosis; schistosomiasis
 f schistosomose
 e esquistosomosis
 d Schistosomose

4005 schizont
 f schizonte
 e esquizonte
 d Schizont

* Schlemm's canal → 4841

4006 sciatic foramen; ischiatic foramen
 f ouverture sciatique; ouverture ischiatique
 e agujero ciático
 d Foramen ischiadicum
 l foramen ischiadicum

4007 sciatic nerve; ischiatic nerve
 f nerf sciatique; nerf ischiatique
 e nervio ciático; nervio isquiático
 d Nervus ischiadicus; Hüftnerv
 l nervus ischiadicus

4008 sciatic notch; ischiatic incisure
 f échancrure sciatique
 e escotadura ciática
 d Incisura ischiadica; Sitzbeinausschnitt
 l incisura ischiadica

4009 sclera
 f sclérotique
 e esclerótica
 d Sklera; weiße Augenhaut
 l sclera

4010 sclerosis
 f sclérose
 e esclerosis
 d Sklerose

4011 scolex
 f scolex
 e escólex
 d Skolex

4012 scoliosis
f scoliose
e escoliosis
d Skoliose

4013 scrapie
f tremblante
e temblor enzoótico; prurigo lumbar
d Traberkrankheit
l paraplexia enzootica ovium

4014 scrapie agent
f agent de la tremblante du mouton
e agente del temblor enzoótico
d Scrapie-Erreger

4015 screw-worm fly
f Cochliomyia hominivorax; ver en vis;
    lucilie bouchère
e gusano borrenador; mosca gusanera
d Cochliomyia hominivorax;
    Schraubenwurmfliege
l Cochliomyia hominivorax

4016 scrotal hernia
f hernie scrotale
e hernia escrotal
d Skrotalhernie; Hodensackbruch
l hernia scrotalis

4017 scrotal raphe
f raphé des bourses
e rafe del escroto
d Hodensacknaht
l raphe scroti

4018 scrotal septum
f septum scrotal
e septo del escroto
d Scheidewand des Hodensacks
l septum scroti

4019 scrotal veins
f veines scrotales
e venas escrotales
d Hodensackvenen
l venae scrotales

4020 scrotum
f scrotum; bourse
e escroto; bolsa
d Hodensack; Skrotum
l scrotum

4021 scutellum; small scale
f scutelle
e escudillo
d Schuppe
l scutellum

4022 scutiform cartilage
f cartilage scutiforme
e cartílago escutiforme
d schildförmiger Knorpel
l cartilago scutiformis

4023 sebaceous cyst
f kyste sébacé
e quiste sebáceo
d Talgzyste

4024 sebaceous glands
f glandes sébacées
e glándulas sebáceas
d Haartalgdrüsen; Talgdrüsen
l glandulae sebaceae

* **seborrhea** → 4025

4025 seborrhoea; seborrhea
f séborrhée
e seborrea
d Seborrhoe

**4026 second and third carpal bones; carpal
    bones II & III**
f os carpal II et III; os capitato-trapézoïde
e hueso carpiano 2 y 3; hueso trapezoideo-
    grande
d zweiter und dritter Karpalknochen
l os carpale II et III

4027 secondary infection
f infection secondaire
e infección secundaria
d Sekundärinfektion

**4028 second carpal bone; trapezoid bone; carpal
    bone II**
f os carpal II; os trapézoïde
e hueso carpiano 2°; hueso trapezoideo
d Os carpale II
l os carpale II; os trapezoideum

* **second cervical vertebra** → 380

* **second phalanx** → 2835

4029 second tarsal bone; tarsal bone II
f os tarsal II; os cunéiforme intermédiaire
e hueso tarsiano 2°; hueso cuneiforme
    intermedio
d zweiter Tarsalknochen
l os tarsale II; os cuneiforme intermedium

* **sectorial tooth** → 679

4030 segmental bronchi
f bronches segmentaires
e bronquios segmentarios
d Segmentbronchen
l bronchi segmentales

4031 selenium deficiency
f carence en sélénium
e carencia en selenio
d Selenmangel

4032 selenium poisoning
f intoxication par le sélénium
e intoxicación por selenio
d Vergiftung durch Selen

4033 semen
f sperme; semence
e semen
d Samen; Sperma

4034 semicircular canals
f canaux semi-circulaires osseux
e canales semicirculares óseos
d Bogengänge
l canales semicirculares ossei

4035 semicircular ducts (of membranous labyrinth); membranous semicircular canals
f conduits semi-circulaires; canaux semi-circulaires
e conductos semicirculares
d häutige Bogengänge
l ductus semicirculares

* **semilunar bone** → 2251

* **semilunar ganglion** → 1799

4036 semilunar valves
f valvules semi-lunaires; valvules sigmoïdes
e válvulas semilunares; válvulas sigmoideas
d Semilunarklappen
l valvulae semilunares

4037 semimembranous muscle
f muscle demi-membraneux
e músculo semimembranoso
d halbhäutiger Muskel
l musculus semimembranosus

4038 seminal fluid
f liquide séminal
e líquido seminal
d Seminalplasma; Samenflüssigkeit
l liquor seminis

4039 semiplume
f semiplume
e semipluma
d Halbdune
l semipluma

4040 semispinal muscle
f muscle semi-spinalis/semi-épineux
e músculo semiespinal
d Musculus semispinalis
l musculus semispinalis

4041 semitendinous muscle
f muscle demi-tendineux
e músculo semitendinoso
d halbsehniger Muskel
l musculus semitendinosus

* **Senecio poisoning** → 4042

4042 seneciosis; Senecio poisoning
f sénéciose; hépatite enzootique du cheval
e intoxicación por Senecio
d Seneziose; Senecio-Vergiftung

4043 sensory disorder
f trouble sensoriel
e trastorno sensorio
d sensorische Störung

4044 sensory nerve
f nerf sensitif; nerf sensoriel
e nervio sensorio
d sensorischer Nerv; Sinnesnerv

* **sensory organs** → 3159

4045 septicaemia; septicemia
f septicémie
e septicemia
d Septikämie; Blutvergiftung

* **septicemia** → 4045

* **septum of tongue** → 2527

4046 seroconversion; serological conversion
f séroconversion; conversion sérologique
e conversión serológica
d Serokonversion

* **serodiagnosis** → 4048

4047 serological
f sérologique
e serológico
d serologisch

* serological conversion → 4046

4048 serological diagnosis; serodiagnosis
 f sérodiagnostic
 e serodiagnóstico
 d Serodiagnostik; serologische Diagnose

4049 serological survey
 f enquête sérologique
 e encuesta serológica
 d seroepidemiologische Untersuchung

4050 seromucous gland
 f glande séro-muqueuse
 e glándula seromucosa
 d seromuköse Drüse
 l glandula seromucosa

* serosa → 4054

4051 serositis
 f sérosite; inflammation d'une membrane séreuse
 e serositis
 d Serositis; Serosenentzündung

4052 serous
 f séreux
 e seroso
 d serös

4053 serous gland
 f glande séreuse
 e glándula serosa
 d seröse Drüse
 l glandula serosa

4054 serous membrane; serosa
 f tunique séreuse
 e túnica serosa
 d Serosa; Tunica serosa
 l tunica serosa

4055 serous pericardium
 f péricarde séreux
 e pericardio seroso
 d Serosa des Perikards
 l pericardium serosum

4056 serum
 f sérum
 e suero
 d Serum

4057 serum albumin
 f sérum-albumine
 e albúmina sérica
 d Serumalbumin

4058 serum gonadotrophin; PMS
 f gonadotrophine sérique
 e gonadotrofina sérica
 d Serumgonadotrophin
 l gonadotrophinum sericum

4059 serum neutralization
 f séroneutralisation
 e seroneutralización
 d Serumneutralisation

4060 serum protein
 f protéine sérique
 e proteína sérica
 d Serumprotein

4061 sesamoid ligaments
 f ligaments sésamoïdiens
 e ligamentos sesamoideos
 d Strahlbeinbänder
 l ligamenta sesamoidea

* Setaria infestation → 4062

4062 setariosis; Setaria infestation
 f sétariose
 e setariosis
 d Setariose

* sevin → 660

4063 shaft of femur
 f corps du fémur
 e cuerpo del hueso fémur
 d Mittelstück des Oberschenkelbeins
 l corpus ossis femoris

4064 shaft of rib
 f corps de la côte
 e cuerpo de la costilla
 d Mittelstück der Rippe
 l corpus costae

* sheath → 3604

4065 sheath of eyeball; bulbar fascia; Tenon's capsule
 f gaine du bulbe de l'œil; capsule de Tenon
 e vaina del globo; cápsula del globo (de Tenon)
 d Faszie des Augapfels; Tenonsche Kapsel
 l vagina bulbi; fascia bulbi

4066 sheep biting louse
 f mallophage du mouton
 e piojo de la oveja
 d Schafhaarling
 l Bovicola ovis

4067 sheep dip
f bain antiparasitaire
e baño parasiticido
d Schafbad; Badeverfahren

4068 sheep disease
f maladie des ovins
e enfermedad de las ovejas
d Schafkrankheit

4069 sheep head louse
f pou de la tête du mouton
e piojo de la cabeza de la oveja
d Linognathus ovillus
l Linognathus ovillus

4070 sheep ked
f mélophage du mouton
e melófago de la oveja
d Schaflausfliege
l Melophagus ovinus

4071 sheep nostril fly
f œstre du mouton
e rezno de la oveja
d Nasendasselfliege
l Oestrus ovis

4072 sheep pox
f clavelée
e viruela ovina
d Schafpocken
l variola ovina

4073 sheep pox virus; ovine poxvirus
f virus de la clavelée
e virus de la viruela ovina
d Schafpockenvirus

4074 shigellosis of foals; Actinobacillus equuli infection
f shigellose; infection à Actinobacillus equuli
e shigelosis de las yeguas; infección por Actinobacillus equuli
d Actinobacillus-equuli-Infektion

* **shin bone → 4569**

4075 short abductor of first digit
f muscle court abducteur du pouce (doigt I)
e músculo abductor corto del dedo 1°
d Musculus abductor digiti I brevis
l musculus abductor digiti I brevis

4076 short adductor muscle
f muscle adducteur court
e músculo aductor corto
d Musculus adductor brevis
l musculus adductor brevis

4077 short bone
f os court
e hueso corto
d kurzer Knochen
l os breve

4078 short digital extensor muscle
f muscle court extenseur des doigts
e músculo extensor digital corto
d kurzer Zehenstrecker
l musculus extensor digitorum brevis

4079 short digital flexor muscle
f muscle court fléchisseur des doigts
e músculo flexor digital corto
d kurzer Zehenbeuger
l musculus flexor digitorum brevis

4080 short-nosed cattle louse
f Haematopinus eurysternus
e piojo azul de cabeza corta
d kurznasige Rinderlaus
l Haematopinus eurysternus

4081 short palmar muscle
f muscle palmaire court
e músculo palmar corto
d Musculus palmaris brevis
l musculus palmaris brevis

4082 short peroneal muscle
f muscle court péronier
e músculo peroneo corto
d kurzer Wadenbeinmuskel
l musculus peroneus brevis

4083 shoulder
f épaule
e espalda
d Schulter

* **shoulderblade → 3999**

4084 shoulder girdle; thoracic girdle
f ceinture du membre thoracique
e cinturón del miembro torácico
d Schultergürtel
l cingulum membri thoracici

4085 shoulder joint
f articulation de l'épaule
e articulación del hombro
d Buggelenk; Schultergelenk
l articulatio humeri

4086 shoulder joint region
f région de l'articulation de l'épaule
e región de la articulación del hombro
d Schultergelenksgegend
l regio articulationis humeri

4087 shoulder region
f région scapulaire
e región escapular
d Schultergegend
l regio scapularis

4088 sigmoid arteries
f artères sigmoïdiennes
e arterias sigmoideas
d Arteriae sigmoideae
l arteriae sigmoideae

4089 sigmoid colon
f côlon sigmoïde
e colon sigmoideo
d S-förmiger Teil des Kolons
l colon sigmoideum

4090 sigmoid flexure of penis
f inflexion sigmoïde du pénis
e flexura sigmoidea del pene
d S-förmige Schleife des Penis
l flexura sigmoidea penis

4091 sigmoid veins
f veines sigmoïdiennes
e venas sigmoideas
d Venae sigmoideae
l venae sigmoideae

4092 simple joint
f articulation simple
e articulación simple
d einfaches Gelenk
l articulatio simplex

4093 sinciput
f sinciput
e sincipucio; bregma
d Vorderkopf
l sinciput

4094 sinus
f sinus
e seno
d Sinus; Hohlraum
l sinus

4095 sinusitis
f sinusite
e sinusitis
d Sinusitis; Nebenhöhlenerkrankung

* **siphonapterosis** → 1716

4096 skeletal musculature
f musculature du squelette
e musculatura del esqueleto
d Skelettmuskulatur
l musculi skeleti

4097 skeleton
f squelette
e esqueleto
d Skelett

4098 skin
f peau
e piel
d Kutis; Haut
l cutis

4099 skin disease; dermatosis
f maladie de la peau; dermatose
e dermatosis; dermatopatía
d Hautkrankheit; Dermopathie

4100 skin folds
f plis cutanés
e pliegues de la piel
d Hautfalten
l plicae cutis

4101 skin glands; cutaneous glands
f glandes cutanées
e glándulas de la piel
d Hautdrüse
l glandulae cutis

4102 skull
f crâne; boîte crânienne
e cráneo
d Schädel
l cranium

4103 slafranin; Rhizoctonia legumicola toxin
f slafranine
e eslafranina
d Slafranin

4104 slide agglutination test
f agglutination sur lame
e aglutinación en placa
d Objektträgeragglutination

* **slow virus** → 2492

4105 small colon (*of horse*)
f petit côlon (*du cheval*)
e colon delgado
d kleines Kolon
l colon tenue

4106 smaller psoas muscle
f muscle petit psoas
e músculo psóas menor
d kleiner Lendenmuskel
l musculus psoas minor

4107 small intestine
 f intestin grêle
 e intestino delgado
 d Dünndarm
 l intestinum tenue

* **small liver fluke** → **2418**

* **small pastern bone** → **2835**

4108 small round muscle
 f muscle petit rond
 e músculo redondo menor
 d kleiner runder Muskel
 l musculus teres minor

* **small scale** → **4021**

4109 small warble fly
 f hypoderme rayé
 e hipoderma rayado; barro
 d kleine Dasselfliege
 l Hypoderma lineatum

4110 smooth muscle
 f muscle lisse
 e músculo liso
 d glatte Muskulatur

* **snout** → **3923**

4111 soft palate
 f palais mou; voile du palais
 e paladar blando; velo del paladar
 d weicher Gaumen; Gaumensegel
 l palatum molle; velum palatinum

* **soft tick** → **265**

4112 sole
 f sole
 e suola
 d Sohle; Hufsohle
 l solea

4113 soleus muscle
 f muscle soléaire
 e músculo sóleo
 d Schollenmuskel
 l musculus soleus

4114 solitary lymphatic nodules
 f nodules lymphatiques solitaires
 e nodulillos linfáticos solitarios
 d Einzellymphknoten; Solitärlymphknoten
 l lymphonoduli solitarii

4115 solitary tract
 f tractus solitarius
 e tracto solitario
 d Tractus solitarius
 l tractus solitarius

4116 soluble antigen
 f antigène soluble
 e antígeno soluble
 d lösliches Antigen

4117 somatic cell
 f cellule somatique
 e célula somática
 d somatische Zelle

4118 somatomedin
 f somatomédine
 e somatomedina
 d Somatomedin

4119 somatostatin
 f somatostatine
 e somatostatina
 d Somatostatin

4120 somatotropin; growth hormone
 f somatotropine; hormone de croissance
 e somatotropina; hormona de crecimiento
 d Somatotropin; Wachstumshormon
 l somatotropinum

* **sparganosis** → **4162**

4121 spastic paresis
 f parésie spastique
 e paresis espástica
 d spastische Parese

* **spaying** → **3199**

4122 specimen
 f échantillon; prélèvement
 e muestra; espécimen
 d Exemplar; Probe

4123 spectinomycin
 f spectinomycine
 e espectinomicina
 d Spectinomycin
 l spectinomycinum

4124 spermatic cord
 f cordon spermatique
 e funículo espermático; cordón espermático
 d Samenstrang
 l funiculus spermaticus

4125 **spermatogenesis**
  f spermatogenèse
  e espermatogénesis
  d Spermatogenese

4126 **spermatozoon**
  f spermatozoïde
  e espermatozoide
  d Spermatozoon; Spermium

4127 **sphenoidal sinus**
  f sinus sphénoïdal
  e seno esfenoidal
  d Keilbeinhöhle
  l sinus sphenoidalis

4128 **sphenoid bone**
  f os sphénoïde
  e hueso esfenoides
  d Keilbein
  l os sphenoidalis

4129 **sphenopalatine artery**
  f artère sphéno-palatine
  e arteria esfenopalatina
  d Arteria sphenopalatina
  l arteria sphenopalatina

4130 **sphenopalatine foramen**
  f foramen sphéno-palatin
  e agujero esfenopalatino
  d Foramen sphenopalatinum
  l foramen sphenopalatinum

4131 **sphenopalatine sinus**
  f sinus sphéno-palatin
  e seno esfenopalatino
  d Sinus sphenopalatinus
  l sinus sphenopalatinus

4132 **spherical recess of vestibule**
  f récessus sphérique du vestibule
  e receso esférico del vestíbulo
  d Recessus sphaericus
  l recessus sphericus vestibuli

4133 **sphincter**
  f sphincter
  e músculo esfínter
  d Schließmuskel; Sphinkter
  l musculus sphincter

4134 **sphincter muscle of pupil**
  f muscle sphincter de la pupille
  e músculo esfínter de la pupila
  d Verengerer der Pupille
  l musculus sphincter pupillae

4135 **sphincter of common bile duct**
  f muscle sphincter du conduit cholédoque
  e músculo esfínter del conducto colédoco
  d Sphinkter des Lebergallengangs
  l musculus sphincter ductus choledochi

* **sphincter of hepatopancreatic ampulla** →
  3073

4136 **sphingolipidosis**
  f sphingolipidose
  e esfingolipidosis
  d Sphingolipidose

4137 **spina bifida**
  f spina bifida; myélodysraphie
  e espina bífida
  d Spaltwirbel
  l spina bifida

* **spinal arachnoid** → **258**

4138 **spinal cord**
  f moelle épinière
  e médula espinal
  d Rückenmark
  l medulla spinalis

4139 **spinal dura mater**
  f dure-mère rachidienne
  e duramadre espinal
  d harte Rückenmarkhaut
  l dura mater spinalis

4140 **spinal ganglion**
  f ganglion spinal/rachidien
  e ganglio espinal
  d Spinalganglion
  l ganglion spinale

4141 **spinal muscle**
  f muscle épineux/spinal
  e músculo espinoso
  d Dornmuskel
  l musculus spinalis

* **spinal nerve** → **30**

4142 **spinal nerves**
  f nerfs spinaux; nerfs rachidiens
  e nervios espinales; nervios raquídeos
  d Rückenmarksnerven
  l nervi spinales

4143 **spinal pia mater**
  f pie-mère spinale
  e piamadre espinal
  d weiche Rückenmarkhaut
  l pia mater spinalis

4144 spinal roots of accessory nerve
f racines spinales du nerf accessoire
e raíces espinales del nervio accesorio
d spinale Wurzeln des Nervus accessorius
l radices spinales nervi accessorii

4145 spinal veins
f veines spinales
e venas espinales
d Venae spinales
l venae spinales

* **spindle-shaped muscle → 1793**

* **spine → 4875**

4146 spine of frog; frog stay
f arête de la fourchette
e espina de la cuña
d Hahnenkamm des Hufes
l spina cunei

4147 spine of scapula
f épine scapulaire
e espina de la escápula
d Schulterblattgräte
l spina scapulae

4148 spinotectal tract
f tractus spino-tectal
e tracto espinotectal
d Tractus spinotectalis
l tractus spinotectalis

4149 spinothalamic tracts
f tractus spino-thalamiques
e tractos espinotalámicos
d Tractus spinothalamici
l tractus spinothalamici

4150 spinous foramen
f foramen épineux
e agujero espinoso
d Foramen spinosum
l foramen spinosum

4151 spinous layer of epidermis
f stratum spinosum de l'épiderme
e estrato espinoso
d Stratum spinosum; Stachelzellschicht
l stratum spinosum epidermidis

4152 spinous process
f processus épineux; apophyse épineuse
e apófisis espinosa
d Dornfortsatz
l processus spinosus

4153 spiral canal of cochlea
f canal spiral de la cochlée
e canal espiral de la cóclea
d Canalis spiralis cochleae
l canalis spiralis cochleae

4154 spiral ganglion; Corti's ganglion
f ganglion spiral; ganglion de Corti
e ganglio espiral; ganglio de Corti
d Spiralganglion
l ganglion spirale

4155 spiral loop of colon
f anse spirale du côlon
e asa espiral del colon
d Spiralschlinge des Colon ascendens
l ansa spiralis coli

4156 spiral membrane; tympanic wall of cochlear duct
f membrane spirale du conduit cochléaire
e pared timpánica del conducto coclear; membrana espiral
d Spiralmembran des Schneckengangs
l membrana spiralis; paries tympanicus ductus cochlearis

4157 spiral organ; organ of Corti
f organe spiral; organe de Corti
e órgano espiral
d Hörorgan; Cortisches Organ
l organum spirale

* **spiral plate of cochlea → 518**

4158 spiramycin
f spiramycine
e espiramicina
d Spiramycin
l spiramycinum

* **Spirocerca infestation → 4159**

4159 spirocercosis; Spirocerca infestation
f spirocercose
e espirocercosis; infestación por Spirocerca
d Spirozerkose

4160 spirochaete
f spirochète
e espiroqueta
d Spirochäte
l Spirochaetaceae

4161 spirochaetosis
f spirochétose
e espiroquetosis
d Spirochätose

* Spirometra infestation → 4162

4162 spirometrosis; sparganosis; Spirometra
      infestation
   f spirométrose; sparganose; infestation par
     Spirometra spp.
   e espirometrosis; esparganosis; infestación
     por Spirometra
   d Spirometrose; Sparganose

4163 spirurids
   f spirurides
   e espirúridos
   d Spiruriden
   l Spiruridae

4164 splanchnic ganglion
   f ganglion splanchnique
   e ganglio espláncnico
   d Ganglion splanchnicum
   l ganglion splanchnicum

4165 spleen
   f rate
   e bazo
   d Milz
   l lien

4166 splenectomy
   f splénectomie
   e esplenectomía
   d Splenektomie

4167 splenial bone
   f os splénial
   e hueso esplenial
   d Os spleniale
   l os spleniale

4168 splenic artery
   f artère liénale
   e arteria esplénica
   d Milzarterie
   l arteria lienalis

4169 splenic plexus
   f plexus splénique
   e plexo esplénico
   d Nervengeflecht der Milz
   l plexus lienalis

4170 splenic vein
   f veine splénique; veine liénale
   e vena esplénica
   d Milzvene
   l vena lienalis

4171 splenius muscle of head
   f muscle splénius de la tête
   e músculo esplenio de la cabeza
   d Riemenmuskel des Kopfes
   l musculus splenius capitis

4172 splenius muscle of neck
   f muscle splénius du cou
   e músculo esplenio del cuello
   d Riemenmuskel der Halswirbelsäule
   l musculus splenius cervicis

4173 splenomegaly; enlargement of spleen
   f splénomégalie
   e esplenomegalia
   d Splenomegalie; Milzvergrößerung

4174 splints in horses; metacarpal exostosis
   f suros; exostose métacarpienne
   e exostosis metacarpiana
   d Überbein; metakarpale Exostose

4175 spongiform
   f spongiforme
   e espongiforme
   d spongiform; spongiös; schwammig

4176 spongy body of penis
   f corps spongieux du pénis
   e cuerpo esponjoso del pene
   d Harnröhrenschwellkörper
   l corpus spongiosum penis

4177 spongy substance of bone
   f substance spongieuse de l'os
   e sustancia esponjosa
   d Spongiosa
   l substantia spongiosa

4178 sporadic
   f sporadique
   e esporádico
   d sporadisch

4179 sporadic bovine encephalomyelitis; SBE
   f encéphalomyélite sporadique bovine
   e encefalomielitis espóradica bovina
   d sporadische bovine Enzephalomyelitis

4180 sporidesmin; Pithomyces chartarum toxin
   f sporidesmine
   e esporidesmina
   d Sporidesmin

   * sporidesmin intoxication → 1613

4181 sporocyst
   f sporocyste
   e esporoquiste; esporocisto
   d Sporozyste

* Sporothrix schenkii infection → 4182

4182 sporotrichosis; Sporothrix schenkii
     infection
  f sporotrichose; infection à Sporothrix
    schenkii
  e esporotricosis; infección por Sporothrix
    schenkii
  d Sporotrichose; Sporothrix-schenkii-
    Infektion

4183 sporozoite
  f sporozoïte
  e esporozoito
  d Sporozoit

4184 spring viraemia of carp
  f virémie printanière de la carpe
  e viremia de primavera de la carpa
  d Frühlingsvirämie der Karpfen

4185 spring viraemia rhabdovirus
  f rhabdovirus de la virémie printanière
  e rhabdovirus de la viremia de primavera
  d Rhabdovirus der Frühlingsvirämie

4186 spumavirus; foamy virus
  f spumavirus
  e spumavirus
  d Spumavirus

4187 spur; metatarsal spur
  f ergot métatarsal
  e espolón
  d Sporn
  l calcar metatarsale

4188 squamous bone
  f os squameux
  e hueso escamoso
  d Os squamosum
  l os squamosum

4189 squamous part of temporal bone
  f partie squameuse de l'os temporal
  e porción escamosa
  d Schläfenbeinschuppe
  l pars squamosa ossis temporalis

4190 squamous suture
  f suture squameuse
  e sutura escamosa
  d Schuppennaht
  l sutura squamosa

* squint → 4229

4191 stable fly
  f stomox
  e mosca brava
  d gemeiner Wadenstecher
  l Stomoxys calcitrans

4192 stachybotryotoxicosis; Stachybotrys atra
     intoxication
  f stachybotryotoxicose
  e estaquibotriotoxicosis; intoxicación por
    Stachybotrys atra
  d Stachybotryotoxikose; Stachybotrys-atra-
    Vergiftung

* Stachybotrys atra intoxication → 4192

* Stachybotrys atra toxin → 3993

* stanolone → 178

4193 stapedial muscle
  f muscle stapédien/de l'étrier
  e músculo estapedio
  d Steigbügelmuskel
  l musculus stapedius

4194 stapedial nerve
  f nerf stapédien
  e nervio estapedio
  d Nervus stapedius; Steigbügelnerv
  l nervus stapedius

4195 stapes; stirrup
  f étrier
  e estribo
  d Steigbügel
  l stapes

4196 staphylococcal mastitis
  f mammite staphylococcique
  e mastitis estafilocócica
  d Staphylokokken-Mastitis

4197 staphylococcus
  f staphylocoque
  e estafilococo
  d Staphylokokk
  l Staphylococcus

4198 steatitis; yellow fat disease
  f inflammation du tissu adipeux
  e inflamación del panículo adiposo
  d Steatitis; Gelbfettkrankheit

* steatosis → 1628

* stellate ganglion → 839

4199 stenosis
f sténose
e estenosis
d Stenose; Verengung

* **Stephanofilaria infestation** → 4200

4200 stephanofilariosis; Stephanofilaria infestation
f stéphanofilariose
e estefanofilariosis; infestación por
Stephanofilaria
d Stephanofilariose; Sommerwunden

4201 stephanurosis; Stephanurus dentatus infestation
f stéphanurose
e estefanurosis; infestación por Stephanurus
d Stephanurose

* **Stephanurus dentatus infestation** → 4201

* **sterility** → 2186

4202 sternal bursa; keel bursa
f bourse sternale
e bolsa esternal
d Bursa sternalis
l bursa sternalis

* **sternal crest** → 2368

4203 sternal lymph nodes
f nœuds lymphatiques sternaux
e nódulos linfáticos esternales
d Brustbeinlymphknoten
l lymphonodi sternales

4204 sternal region
f région sternale
e región esternal
d Brustbeingegend
l regio sternalis

4205 sternal rib; true rib
f côte sternale
e costilla esternal; costilla verdadera
d wahre Rippe
l costa sternalis; costa vera

4206 sternal synchondrosis
f synchondrose sternale
e sincondrosis esternal
d Brustbeinfuge
l synchondrosis sternalis

4207 sternebra
f sternèbre

e estérnebra
d Sternebra; Brustbeinstück
l sternebra

4208 sternocephalic muscle
f muscle sterno-céphalique
e músculo esternocefálico
d Brustbein-Kopfmuskel
l musculus sternocephalicus

4209 sternocleidomastoid vein
f veine sterno-cléido-mastoïdienne
e vena esternocleidomastoidea
d Vena sternocleidomastoidea
l vena sternocleidomastoidea

4210 sternocoracoclavicular membrane; coracoclavicular membrane
f membrane sterno-coraco-claviculaire
e membrana esternocoracoclavicular
d Membrana sternocoracoclavicularis
l membrana sternocoracoclavicularis

4211 sternocoracoid articulation
f articulation sterno-coracoïdienne
e articulación esternocoracoidea
d Articulatio sternocoracoidea
l articulatio sternocoracoidea

4212 sternocoracoid muscle
f muscle sterno-coracoïdien
e músculo esternocoracoideo
d Musculus sternocoracoideus
l musculus sternocoracoideus

4213 sternocostal articulation
f articulation sterno-costale
e articulación esternocostal
d Brustbein-Rippen-Gelenk
l articulatio sternocostalis

4214 sternohyoid muscle
f muscle sterno-hyoïdien
e músculo esternohioideo
d Musculus sternohyoideus
l musculus sternohyoideus

4215 sternomandibular muscle
f muscle sterno-mandibulaire
e músculo esternomandibular
d Musculus sternomandibularis
l musculus sternomandibularis

4216 sternomastoid muscle
f muscle sterno-mastoïdien
e músculo esternomastoideo
d Musculus sternomastoideus
l musculus sternomastoideus

4217 sternopericardiac ligaments
  f ligaments sterno-péricardiques
  e ligamentos esternopericárdicos
  d Ligamenta sternopericardiaca
  l ligamenta sternopericardiaca

4218 sternothyroid muscle
  f muscle sterno-thyroïdien
  e músculo esternotiroideo
  d Musculus sternothyreoideus
  l musculus sternothyroideus

4219 sternotracheal muscle
  f muscle sterno-trachéal
  e músculo esternotraqueal
  d Musculus sternotrachealis
  l musculus sternotrachealis

4220 sternum; breastbone
  f sternum
  e esternón
  d Sternum; Brustbein
  l sternum

4221 steroid
  f stéroïde
  e esteroide
  d Steroid

  * **Sticker's sarcoma → 4629**

4222 stifle joint
  f articulation du grasset; articulation
     fémoro-tibio-patellaire
  e articulación de la rodilla
  d Kniegelenk
  l articulatio genus

  * **stifle region → 2381**

  * **stilboestrol → 1303**

  * **Stilesia infestation → 4223**

4223 stilesiosis; Stilesia infestation
  f stilésiose
  e stilesiosis; estilesiosis
  d Stilesiose

4224 stillbirth
  f mortinatalité
  e muerte al nacer
  d Totgeburt; Fruchttod

4225 stillborn
  f mort-né
  e nacido muerto
  d totgeboren

  * **stirrup → 4195**

4226 stomach
  f estomac
  e estómago
  d Magen
  l ventriculus; gaster

  * **stomach botfly → 2043**

  * **stomach tube → 2947**

4227 stomach ulcer; gastric ulcer
  f ulcère gastrique
  e úlcera gástrico
  d Magengeschwür
  l ulcus ventriculi

4228 stomatitis
  f stomatite
  e estomatitis
  d Stomatitis

4229 strabismus; squint
  f strabisme
  e estrabismo
  d Strabismus; Schielen

4230 straight muscle of abdomen
  f muscle droit de l'abdomen
  e músculo recto del abdomen
  d gerader Bauchmuskel
  l musculus rectus abdominis

4231 straight muscle of head
  f muscle droit de la tête
  e músculo recto de la cabeza
  d gerader Kopfmuskel
  l musculus rectus capitis

4232 straight muscle of thigh
  f muscle droit fémoral/de la cuisse
  e músculo recto del muslo
  d gerader Schenkelmuskel
  l musculus rectus femoris

4233 straight muscle of thorax
  f muscle droit du thorax
  e músculo recto del tórax
  d gerader Brustkorbmuskel
  l musculus rectus thoracis

4234 straight seminiferous tubules
  f tubes séminifères droits
  e túbulos seminíferos rectos
  d gerade Hodenkanälchen
  l tubuli seminiferi recti

4235 **strangles; Streptococcus equi infection**
  f gourme; infection à Streptococcus equi
  e papera equina; adenitis contagiosa;
    infección por Streptococcus equi
  d Druse; Streptococcus-equi-infektion
  l adenitis streptococcalis

4236 **strangulated hernia**
  f hernie étranglée
  e hernia estrangulada
  d eingeklemmte Hernie
  l hernia incarcerata

4237 **streptobacillosis; Streptobacillus**
    moniliformis infection
  f streptobacillose
  e infección por Streptobacillus moniliformis
  d Streptobazillose; Streptobacillus-
    moniliformis-Infektion

  * **Streptobacillus moniliformis infection →**
    4237

4238 **streptococcal mastitis**
  f mammite streptococcique
  e mastitis estreptocócica
  d Streptokokken-Mastitis

4239 **streptococcus**
  f streptocoque
  e estreptococo
  d Streptokokk
  l Streptococcus

  * **Streptococcus equi infection → 4235**

4240 **streptomycin**
  f streptomycine
  e estreptomicina
  d Streptomycin
  l streptomycinum

4241 **stress**
  f stress
  e stress
  d Streß; Belastung

4242 **striated body**
  f corps strié
  e cuerpo estriado
  d Streifenkörper
  l corpus striatum

4243 **striated muscle**
  f muscle strié
  e músculo estriado
  d quergestreifter Muskul

4244 **strongyles**
  f strongles
  e estróngilos
  d Palisadenwürmer
  l Strongylidae

  * **strongyliasis → 4246**

  * **Strongyloides infestation → 4245**

4245 **strongyloidosis; Strongyloides infestation**
  f strongyloïdose
  e estrongiloidosis
  d Strongyloidose; Befall mit
    Zwergfadenwürmern

4246 **strongylosis; strongyliasis; Strongylus**
    infestation
  f strongylose
  e estrongilosis
  d Strongylose; Befall mit Palisadenwürmern

  * **Strongylus infestation → 4246**

  * **struck → 3207**

  * **stunting → 4753**

4247 **styloglossus muscle**
  f muscle stylo-glosse
  e músculo estilogloso
  d Musculus styloglossus
  l musculus styloglossus

4248 **stylohyoid bone**
  f stylo-hyal
  e estilohioideo
  d Stylohyoid
  l stylohyoideum

4249 **stylohyoid muscle**
  f muscle stylo-hyoïdien
  e músculo estilohioideo
  d Musculus stylohyoideus
  l musculus stylohyoideus

4250 **styloid process of radius**
  f processus styloïde du radius
  e apófisis estiloidea del radio
  d Griffelfortsatz der Speiche
  l processus styloideus radii

4251 **styloid process of temporal bone**
  f processus styloïde de l'os temporal
  e apófisis estiloidea del hueso temporal
  d Griffelfortsatz des Schläfenbeins
  l processus styloideus ossis temporalis

4252 stylomastoid artery
f artère stylo-mastoïdienne
e arteria estilomastoidea
d Arteria stylomastoidea
l arteria stylomastoidea

4253 stylomastoid foramen
f foramen stylo-mastoïdien
e agujero estilomastoideo
d Foramen stylomastoideum
l foramen stylomastoideum

4254 stylomastoid vein
f veine stylo-mastoïdienne
e vena estilomastoidea
d Vena stylomastoidea
l vena stylomastoidea

4255 subarachnoidal cisterns
f citernes subarachnoïdiennes
e cisternas subaracnoideas
d Subarachnoidealzisternen
l cisternae subarachnoideales

* subarachnoidal space → 4256

**4256 subarachnoid cavity; subarachnoidal
       space**
f cavum sous-arachnoïdien; espace sous-
   arachnoïdien
e cavidad subaracnoidea
d Subarachnoidealraum
l cavum subarachnoideale

4257 subclavian artery
f artère sous-clavière
e arteria subclavia
d Arteria subclavia
l arteria subclavia

4258 subclavian muscle
f muscle sous-clavier
e músculo subclavio
d Unterschlüsselbeinmuskel
l musculus subclavius

4259 subclavian nerve
f nerf sous-clavier
e nervio subclavio
d Nervus subclavius
l nervus subclavius

4260 subclavian plexus
f plexus sous-clavier
e plexo subclavio
d Plexus subclavius
l plexus subclavius

4261 subclavian vein
f veine sous-clavière
e vena subclavia
d Vena subclavia
l vena subclavia

4262 subclinical
f subclinique
e subclínico
d subklinisch

4263 subclinical mastitis
f mammite subclinique
e mastitis subclínica
d subklinische Mastitis

4264 subcommissural organ
f organe subcommissural
e órgano subcomisural
d Subkommissuralorgan
l organum subcommissurale

4265 subcoracoscapular muscles
f muscles subcoraco-scapulaires
e músculos subcoracoescapulares
d Musculi subcoracoscapulares
l musculi subcoracoscapulares

4266 subcostal muscles
f muscles sous-costaux
e músculos subcostales
d Unterrippenmuskeln
l musculi subcostales

* subcutaneous abdominal vein → 4321

4267 subcutaneous adipose tissue
f pannicule adipeux
e panículo adiposo
d Unterhautfettgewebe
l panniculus adiposus

4268 subcutaneous bursa
f bourse sous-cutanée
e bolsa subcutánea
d Hautschleimbeutel
l bursa subcutanea

4269 subcutaneous injection
f injection sous-cutanée; injection
   hypodermique
e inyección subcutánea; inyección
   hipodérmica
d subkutane Injektion

4270 subcutaneous mite of poultry
f acarien sous-cutané des volailles
e ácaro subcutáneo de las gallinas
d Unterhautmilbe des Geflügels
l Laminosioptes cysticola

* subcutaneous tissue → 4271

4271 subcutis; subcutaneous tissue; hypodermis
 f tissu sous-cutané; hypoderme
 e tela subcutánea; hipodermis
 d Subkutis; Unterhaut
 l tela subcutanea

4272 subdural space
 f cavum subdural
 e cavidad subdural
 d Subduralraum
 l cavum subdurale

4273 subextensor recess
 f récessus subextensorius
 e receso subextensor
 d Recessus subextensorius
 l recessus subextensorius

4274 subfornical organ
 f organe subfornical
 e órgano subfornical
 d Subfornikalorgan
 l organum subfornicale

4275 subiliac lymph nodes; prefemoral lymph nodes
 f nœuds lymphatiques subiliaques
 e nódulos linfáticos subilíacos
 d Kniefaltenlymphknoten
 l lymphonodi subiliaci

4276 sublethal
 f sublétal
 e subletal
 d subletal

4277 sublingual artery
 f artère sublinguale
 e arteria sublingual
 d Unterzungenarterie
 l arteria sublingualis

4278 sublingual caruncle
 f caroncule sublinguale
 e carúncula sublingual
 d Hungerwarze
 l caruncula sublingualis

4279 sublingual ganglion
 f ganglion sublingual
 e ganglio sublingual
 d Unterzungenganglion
 l ganglion sublinguale

4280 sublingual gland
 f glande sublinguale

 e glándula sublingual
 d Unterzungendrüse
 l glandula sublingualis

* submandibular ganglion → 2670

4281 submental artery
 f artère sous-mentale
 e arteria submentoniana
 d Arteria submentalis
 l arteria submentalis

4282 submucosal layer
 f toile sous-muqueuse
 e tela submucosa
 d Submukosa
 l tela submucosa

4283 submucous plexus; Meissner's plexus
 f plexus sous-muqueux (*de l'intestin*); plexus de Meissner
 e plexo submucoso; plexo de Meissner
 d Plexus submucosus
 l plexus submucosus

4284 suboccipital nerve
 f nerf sous-occipital; petit nerf occipital
 e nervio suboccipital
 d Nervus suboccipitalis
 l nervus suboccipitalis

4285 subpopliteal recess
 f récessus subpoplité/sous-poplité
 e receso subpoplíteo
 d Recessus subpopliteus
 l recessus subpopliteus

4286 subscapular artery
 f artère subscapulaire
 e arteria subescapular
 d Arteria subscapularis
 l arteria subscapularis

4287 subscapular lymph nodes
 f nœuds lymphatiques subscapulaires
 e nódulos linfáticos subescapulares
 d Lymphonodi subscapulares
 l lymphonodi subscapulares

4288 subscapular muscle
 f muscle sous-scapulaire
 e músculo subescapular
 d Unterschulterblattmuskel
 l musculus subscapularis

4289 subscapular nerve
 f nerf subscapulaire
 e nervio subescapular
 d Nervus subscapularis
 l nervus subscapularis

4290 **subscapular vein**
 f veine sous-scapulaire
 e vena subescapular
 d Vena subscapularis
 l vena subscapularis

4291 **subserous plexus**
 f plexus sous-séreux
 e plexo subseroso
 d Plexus subserosus
 l plexus subserosus

4292 **subthalamic nucleus**
 f noyau subthalamique
 e núcleo subtálamico
 d Nucleus subthalamicus
 l nucleus subthalamicus

4293 **subthalamus; ventral thalamus**
 f subthalamus
 e subtálamo
 d Subthalamus
 l subthalamus

4294 **suburethral diverticulum**
 f diverticule sous-urétral
 e divertículo suburetral
 d Diverticulum suburethrale
 l diverticulum suburethrale

4295 **sucking louse**
 f pou piqueur
 e piojo chupador
 d Laus
 l Anoplura

4296 **sudden death**
 f mort subite
 e muerte repentina
 d plötzlicher Tod

4297 **sudoriferous duct**
 f conduit sudorifère
 e conducto sudorífero
 d Ausführgang der Schweißdrüsen
 l ductus sudoriferus

* **sudoriferous glands → 4381**

4298 **sulci of brain**
 f scissures cérébrales
 e surcos del cerebro
 d Gehirnfurchen
 l sulci cerebri

* **sulcus → 1904**

4299 **sulfachlorpyridazine;**
     sulphachlorpyridazine
 f sulfachlorpyridazine
 e sulfacloropiridazina
 d Sulfachlorpyridazin
 l sulfachlorpyridazinum

4300 **sulfadiazine; sulphadiazine**
 f sulfadiazine
 e sulfadiazina
 d Sulfadiazin
 l sulfadiazinum

4301 **sulfadimethoxine; sulphadimethoxine**
 f sulfadiméthoxine
 e sulfadimetoxina
 d Sulfadimethoxin
 l sulfadimethoxinum

4302 **sulfadimidine; sulphadimidine;**
     sulfamethazine
 f sulfadimidine
 e sulfadimidina
 d Sulfadimidin
 l sulfadimidinum

4303 **sulfadoxine; sulphadoxine**
 f sulfadoxine
 e sulfadoxina
 d Sulfadoxin
 l sulfadoxinum

4304 **sulfafurazole; sulphafurazole;**
     sulfisoxazole
 f sulfafurazole
 e sulfafurazol
 d Sulfafurazol
 l sulfafurazolum

4305 **sulfaguanidine; sulphaguanidine**
 f sulfaguanidine
 e sulfaguanidina
 d Sulfaguanidin
 l sulfaguanidinum

4306 **sulfamerazine; sulphamerazine**
 f sulfamérazine
 e sulfamerazina
 d Sulfamerazin
 l sulfamerazinum

* **sulfamethazine → 4302**

4307 **sulfamethoxazole; sulphamethoxazole**
 f sulfaméthoxazole
 e sulfametoxazol
 d Sulfamethoxazol
 l sulfamethoxazolum

4308 sulfamethoxypyridazine;
   sulphamethoxypyridazine
  f sulfaméthoxypyridazine
  e sulfametoxipiridazina
  d Sulfamethoxypyridazin
  l sulfamethoxypyridazinum

4309 sulfamonomethoxine
  f sulfamonométhoxine
  e sulfamonometoxina
  d Sulfamonomethoxin
  l sulfamonomethoxinum

4310 sulfanilamide; sulphanilamide
  f sulfanilamide
  e sulfanilamida
  d Sulfanilamid
  l sulfanilamidum

4311 sulfapyrazole; sulphapyrazole
  f sulfapyrazole
  e sulfapirazol
  d Sulfapyrazol
  l sulfapirazolum

4312 sulfapyridine; sulphapyridine
  f sulfapyridine
  e sulfapiridina
  d Sulfapiridin
  l sulfapyridinum

4313 sulfaquinoxaline; sulphaquinoxaline
  f sulfaquinoxaline
  e sulfaquinoxalina
  d Sulfaquinoxalin
  l sulfaquinoxalinum

4314 sulfathiazole; sulphathiazole
  f sulfathiazole
  e sulfatiazol
  d Sulfathiazol
  l sulfathiazolum

  * **sulfisoxazole** → **4304**

4315 sulfonamide
  f sulfamide
  e sulfonamida
  d Sulfonamid

  * **sulphachlorpyridazine** → **4299**

  * **sulphadiazine** → **4300**

  * **sulphadimethoxine** → **4301**

  * **sulphadimidine** → **4302**

  * **sulphadoxine** → **4303**

  * **sulphafurazole** → **4304**

  * **sulphaguanidine** → **4305**

  * **sulphamerazine** → **4306**

  * **sulphamethoxazole** → **4307**

  * **sulphamethoxypyridazine** → **4308**

  * **sulphanilamide** → **4310**

  * **sulphapyrazole** → **4311**

  * **sulphapyridine** → **4312**

  * **sulphaquinoxaline** → **4313**

  * **sulphathiazole** → **4314**

4316 superciliary arch
  f arcade sourcilière
  e arco superciliar
  d Arcus superciliaris
  l arcus superciliaris

4317 superficial branch
  f rameau superficiel
  e ramo superficial
  d oberflächlicher Ast
  l ramus superficialis

4318 superficial cervical lymph nodes
  f nœuds lymphatiques cervicaux superficiels
  e nódulos linfáticos cervicales superficiales
  d oberflächliche Halslymphknoten
  l lymphonodi cervicales superficiales

4319 superficial cervical lymphocentre
  f lymphocentre cervical superficiel
  e linfocentro cervical superficial
  d oberflächliches Lymphzentrum des Halses
  l lymphocentrum cervicale superficiale

4320 superficial cervical vein
  f veine cervicale superficielle
  e vena cervical superficial
  d Vena cervicalis superficialis
  l vena cervicalis superficialis

4321 superficial cranial epigastric vein;
   subcutaneous abdominal vein; milk vein
  f veine épigastrique crâniale superficielle;
    veine sous-cutanée abdominale
  e vena epigástrica craneal superficial; vena
    subcutánea del abdomen

d Bauchhautvene; Milchader
l vena epigastrica cranialis superficialis;
   vena subcutanea abdominis

4322 superficial digital flexor muscle
f muscle fléchisseur superficiel des doigts
e músculo flexor digital superficial
d oberflächlicher Zehenbeuger
l musculus flexor digitorum superficialis

4323 superficial gland (*of third eyelid*)
f glande superficielle
e glándula superficial
d Nickhautdrüse; oberflächliche Drüse (*der Nickhaut*)
l glandula superficialis

4324 superficial gluteal muscle
f fessier superficiel
e músculo glúteo superficial
d Glutäus; oberflächlicher Kruppenmuskel
l musculus gluteus superficialis

4325 superficial inguinal lymph nodes
f nœuds lymphatiques inguinaux superficiels
e nódulos linfáticos inguinales superficiales
d Leistenlymphknoten
l lymphonodi inguinales superficiales proprii

4326 superficial lamina; superficial layer
f lame superficielle
e lámina superficial
d Lamina superficialis
l lamina superficialis

* **superficial layer** → **4326**

4327 superficial parotid lymph nodes
f nœuds lymphatiques parotidiens superficiels
e nódulos linfáticos parotídeos superficiales
d oberflächliche Parotislymphknoten
l lymphonodi parotidei superficiales

4328 superficial peroneal nerve
f nerf péronier superficiel
e nervio peroneo superficial
d Nervus peroneus superficialis
l nervus peroneus superficialis

4329 superficial pseudotemporal muscle
f muscle pseudotemporal superficiel
e músculo pseudotemporal superficial
d Musculus pseudotemporalis superficialis
l musculus pseudotemporalis superficialis

4330 superficial thoracic vein
f veine thoracique superficielle
e vena torácica superficial
d oberflächliche Thorakalvene
l vena thoracica superficialis

4331 superinfection
f surinfection
e superinfección
d Superinfektion

4332 superior vestibular area; acoustic area
f aire vestibulaire supérieure; aire acoustique
e área vestibular superior
d obere Vestibularfläche; akustischer Cortex
l area vestibularis superior

4333 supernumerary
f surnuméraire
e supernumerario
d überzählig

* **supernumerary digits** → **3534**

* **supernumerary limbs** → **3537**

* **supernumerary mammary glands** → **3536**

* **supernumerary teats** → **3542**

4334 superovulation
f superovulation
e superovulación
d Superovulation; Mehrfachovulation

4335 supinator muscle
f muscle supinateur
e músculo supinador
d Auswärtsdreher
l musculus supinator

4336 suppuration
f suppuration
e supuración
d Eiterung; Eitern

4337 supra-angular bone
f os supra-angulaire
e hueso supraangular
d Os supraangulare
l os supraangulare

4338 suprachoroid lamina
f lame suprachoroïdienne
e lámina supracoroidea
d Lamina suprachoroidea
l lamina suprachoroidea

* **supraciliary bones** → 4348

4339 supraclavicular nerves
f nerfs supraclaviculaires
e nervios supraclaviculares
d Nervi supraclaviculares
l nervi supraclaviculares

4340 supracondylar foramen
f foramen supracondylaire
e agujero supracondilar
d Foramen supracondylare
l foramen supracondylare

4341 supracondylar fossa of femur
f fosse supracondylaire
e fosa supracondilar
d Fossa supracondylaris
l fossa supracondylaris

4342 supracoracoid muscle
f muscle supracoracoïdien
e músculo supracoracoideo
d Musculus supracoracoideus
l musculus supracoracoideus

4343 suprajugal bone
f os suprajugal
e hueso suprayugal
d Os suprajugale
l os suprajugale

4344 supramammary muscle
f muscle supramammaire
e músculo supramamario
d Musculus supramammarius
l musculus supramammarius

4345 supraoptic commissures
f commissures supra-optiques
e comisuras supraópticas
d supraoptische Kommissuren
l commissurae supraopticae

4346 supraoptic nucleus
f noyau supra-optique
e núcleo supraóptico
d Nucleus supraopticus
l nucleus supraopticus

4347 supraorbital artery
f artère supra-orbitaire
e arteria supraorbitaria
d Arteria supraorbitalis
l arteria supraorbitalis

4348 supraorbital bones; supraciliary bones
f os supra-orbitaires

e huesos supraorbitarios
d Os supraorbitalia
l os supraorbitalia

4349 supraorbital canal
f canal supra-orbitaire
e canal supraorbitario
d Canalis supraorbitalis
l canalis supraorbitalis

4350 supraorbital foramen
f foramen supra-orbitaire
e agujero supraorbitario
d Foramen supraorbitale
l foramen supraorbitale

4351 supraorbital incisure
f échancrure sus-orbitaire
e escotadura supraorbitaria
d Incisura supraorbitalis
l incisura supraorbitalis

4352 supraorbital margin of frontal bone; orbital arch
f bord supra-orbitaire de l'os frontal; arcade orbitaire
e borde supraorbitario; arcade orbital
d Margo supraorbitalis
l margo supraorbitalis ossis frontalis

4353 supraorbital nerve
f nerf supra-orbitaire
e nervio supraorbitario
d Nervus supraorbitalis
l nervus supraorbitalis

4354 supraorbital tactile hairs
f poils tactiles supra-orbitaires
e pelos supraorbitarios
d Supraorbitalhaare
l pili supraorbitales

4355 supraorbital vein
f veine supra-orbitaire
e vena supraorbitaria
d Vena supraorbitalis
·*l* vena supraorbitalis

4356 suprarenal arteries
f artères surrénales
e arterias suprarrenales
d Nebennierenarterien
l arteriae suprarenales

4357 suprarenal plexus
f plexus surrénal
e plexo suprarrenal
d Plexus suprarenalis
l plexus suprarenalis

4358 suprarenal veins
f veines surrénales
e venas suprarrenales
d Nebennierenvenen
l venae suprarenales

4359 suprascapular artery
f artère suprascapulaire
e arteria supraescapular
d Arteria suprascapularis
l arteria suprascapularis

4360 suprascapular nerve
f nerf suprascapulaire
e nervio supraescapular
d Nervus suprascapularis
l nervus suprascapularis

* suprascapular notch → 2162

4361 suprascapular vein
f veine suprascapulaire
e vena supraescapular
d Vena suprascapularis
l vena suprascapularis

4362 supraspinal ligament
f ligament supra-épineux
e ligamento supraespinal
d Ligamentum supraspinale
l ligamentum supraspinale

4363 supraspinal region
f région supraspinale
e región supraespinosa
d Supraspinalgegend
l regio supraspinata

4364 supraspinous muscle
f muscle supra-épineux
e músculo supraespinoso
d oberer Grätenmuskel
l musculus supraspinatus

4365 supratrochlear artery
f artère supratrochléaire
e arteria supratroclear
d Arteria supratrochlearis
l arteria supratrochlearis

4366 supratrochlear foramen
f foramen supratrochléaire
e agujero supratroclear
d Foramen supratrochleare
l foramen supratrochleare

4367 supratrochlear nerve
f nerf supratrochléaire

e nervio supratroclear
d Nervus supratrochlearis
l nervus supratrochlearis

4368 sural arteries
f artères surales
e arterias surales
d Arteriae surales; Wadenarterien
l arteriae surales

4369 suramin sodium
f suramine sodique
e suramina sódica
d Suramin
l suraminum natricum

4370 surgery
f chirurgie
e cirugía
d chirurgie

4371 surgical instrument
f instrument de chirurgie
e instrumento quirúrgico
d chirurgisches Instrument

4372 surgical operation
f opération chirurgicale
e operación quirúrgica; intervención quirúrgica
d operativer Eingriff

4373 surra; Trypanosoma evansi infection
f surra
e surra
d Surra; Trypanosoma-evansi-Infektion

4374 susceptibility
f susceptibilité; sensibilité
e susceptibilidad
d Empfänglichkeit; Anfälligkeit

4375 suspensory apparatus of udder
f appareil suspenseur des mamelles
e aparato suspensor de las mamas
d Aufhängapparat des Euters
l apparatus suspensorius mammarum

4376 suspensory ligament of ovary
f ligament suspenseur de l'ovaire
e ligamento suspensor del ovario
d Ligamentum suspensorium ovarii
l ligamentum suspensorium ovarii

4377 suspensory ligament of penis
f ligament suspenseur du pénis
e ligamento suspensor del pene
d Aufhängeband des Penis
l ligamentum suspensorium penis

4378 suture
f suture
e sutura
d Sutur; Knochennaht
l sutura

* **sutures of skull** → 1154

4379 suxamethonium
f suxaméthonium
e suxametonio
d Suxamethonium
l suxamethonium

4380 swayback; enzootic ataxia of lambs
f ataxie enzootique de l'agneau
e ataxia enzoótica de los corderos
d enzootische Ataxie der Lämmer

4381 sweat glands; sudoriferous glands
f glandes sudoripares
e glándulas sudoríferas
d Schweißdrüsen
l glandulae sudoriferae

4382 sweat pore; pore of sweat duct
f pore sudorifère
e poro sudorífero
d Schweißpore
l porus sudoriferus

* **swim bladder inflammation** → 84

* **swine disease** → 3468

4383 swine dysentery; Treponema hyodysenteriae infection
f entérite à Treponema hyodysenteriae
e disentería porcina; enteritis por Treponema hyodysenteriae
d Schweinedysenterie; Treponema-hyodysenteriae-Infektion

4384 swine erysipelas; Erysipelothrix rhusiopathiae infection
f rouget du porc; infection à Erysipelothrix rhusiopathiae
e mal rojo del cerdo; infección por Erysipelothrix rhusiopathiae
d Schweinerotlauf; Erysipelothrix-rhusiopathiae-Infektion
l erysipelas suis

4385 swine fever; classical swine fever; hog cholera
f peste porcine classique
e peste porcina clásica
d klassische Schweinepest; europäische Schweinepest
l pestis suum classica

4386 swine fever virus
f virus de la peste porcine classique
e virus de la peste porcina clásica
d Schweinepest-Virus

4387 swine influenza
f influenza porcin; grippe porcine
e influenza porcina
d Schweineinfluenza

4388 swine influenzavirus
f virus de l'influenza porcin
e virus de la influenza porcina
d Schweineinfluenza-Virus

4389 swine pox; pig pox
f variole du porc
e viruela porcina
d Schweinepocken

4390 swine poxvirus
f poxvirus porcin
e poxvirus porcino
d Schweinepockenvirus

4391 swine vesicular disease
f maladie vésiculeuse du porc
e enfermedad vesicular porcina
d vesikuläre Schweinekrankheit
l morbus vesicularis suum

4392 swine vesicular disease enterovirus
f virus de la maladie vésiculeuse du porc
e virus de la enfermedad vesicular porcina
d porzines Bläschenkrankheit-Enterovirus

4393 sylvian fissure; lateral fissure of brain
f fissure sylvienne; fissure latérale du cerveau
e cisura silviana
d Sylvius'sche Fissur; Fissura lateralis cerebri
l fissura sylvia; fissura lateralis cerebri

4394 symblepharon
f symblépharon
e simbléfaron
d Symblepharon

4395 sympathetic nervous system
f système orthosympathique; partie sympathique
e sistema nervioso simpático; porción simpática
d sympathisches Nervensystem
l pars sympathica systematis nervosi autonomici

4396 sympathetic trunk
 f tronc sympathique
 e tronco simpático
 d Grenzstrang
 l truncus sympathicus

4397 sympatholytic
 f sympatholytique
 e simpaticolítico
 d Sympatholytikum

4398 sympathomimetic
 f sympathomimétique
 e simpatomimético
 d Sympathomimetikum

4399 symphysis; fibrocartilaginous junction
 f symphyse; articulation fibro-cartilagineuse
 e sínfisis
 d Symphyse
 l symphysis; junctura fibrocartilaginea

4400 symptom
 f symptôme
 e síntoma
 d Symptom; Krankheitszeichen

4401 synarthrosis; fibrous joint
 f synarthrose
 e articulación fibrosa
 d Synarthrose
 l junctura fibrosa

4402 synchondrosis
 f synchondrose
 e sincondrosis
 d Synchondrose; Knorpelfuge
 l synchondrosis

4403 syndactyly
 f syndactylie
 e sindactilia
 d Syndaktylie

4404 syndesmosis
 f syndesmose
 e sindesmosis
 d Syndesmose
 l syndesmosis

4405 syndrome
 f syndrome
 e síndrome
 d Syndrom

4406 synechia
 f synéchie
 e sinequia
 d Synechie

4407 syngamosis; Syngamus infestation
 f syngamose
 e singamosis
 d Syngamose

* Syngamus infestation → 4407

4408 synostosis
 f synostose
 e sinostosis
 d Synostose
 l synostosis

* synovia → 4410

4409 synovial bursa
 f bourse synoviale
 e bolsa sinovial
 d Schleimbeutel
 l bursa synovialis

4410 synovial fluid; synovia
 f synovie; liquide synovial
 e sinovia; líquido sinovial
 d Synovia; Gelenkschmiere
 l synovia

4411 synovial joint
 f articulation synoviale
 e articulación sinovial
 d Spaltgelenk
 l junctura synovialis

4412 synovial membrane
 f membrane synoviale
 e membrana sinovial
 d Synovialmembran
 l membrana synovialis

4413 synovial sheath
 f gaine synoviale
 e vaina sinovial
 d Synovialscheide
 l vagina synovialis

4414 synovial sheath of tendon
 f gaine synoviale tendineuse
 e vaina sinovial tendinosa
 d Sehnenscheide
 l vagina synovialis tendinis

4415 synovial villi
 f villosités synoviales
 e villosidades sinoviales
 d Synovialzotten
 l villi synoviales

4416 synovitis
 f synovite
 e sinovitis
 d Synovitis

4417 synsacrum; pelvic vertebrae
 f synsacrum
 e sinsacro
 d Synsacrum
 l synsacrum

4418 syringe
 f seringue
 e jeringa
 d Spritze

4419 syringeal bulla
 f bulle de la syrinx
 e bulla de la siringe
 d Stimmkopfblase
 l bulla syringealis

4420 syringeal cartilages
 f cartilages de la syrinx
 e cartílagos de la siringe

 d Stimmkopfknorpeln
 l cartilagines syringeales

4421 syringeal muscles
 f muscles de la syrinx
 e músculos de la siringe
 d Stimmkopfmuskeln
 l musculi syringeales

4422 syrinx; vocal organ
 f syrinx; organe vocal
 e siringe
 d Stimmkopf
 l syrinx

4423 systemic disease
 f affection systémique
 e enfermedad sistémica
 d Systemerkrankung

4424 systole
 f systole
 e sistole
 d Systole

T

4425 T2 toxin; Fusarium T2 toxin
f toxine T2
e toxina T2
d T2-Toxin

* T3 → 4671

* T4 → 4566

4426 tachycardia
f tachycardie
e taquicardia
d Tachykardie

4427 tactile hairs
f poils tactiles
e pelos táctiles
d Sinneshaare; Tasthaare; Sinushaare
l pili tactiles

4428 tactile hairs of chin
f poils tactiles mentaux
e pelos mentonianos
d Kinntasthaare
l pili mentales

4429 taeniae coli
f ténias du côlon
e tenias del colon
d Taeniae coli
l taeniae coli

* Taenia infestation → 4431

4430 Taenia multiceps
f ténia à cénure
e tenia del cenuro
d Multiceps; Quesenbandwurm
l Taenia multiceps; Multiceps multiceps

* taeniasis → 4431

4431 taeniosis; taeniasis; Taenia infestation
f téniose; téniasis
e teniosis
d Täniose

4432 tail; coccyx
f queue; coccyx
e cola; cóccix
d Schwanz
l cauda; coccyx

* tail amputation → 1345

4433 tail base; root of tail
f racine de la queue; base de la queue
e raíz de la cola
d Schwanzwurzel
l radix caudae

* tail feather → 1721

* tail hairs → 4438

4434 taillessness; acaudia
f acaudie
e acaudia
d Schwanzlosigkeit

4435 tail muscles
f muscles de la queue
e músculos de la cola/coxígeos
d Schwanzmuskeln
l musculi caudae/coccygis

* tail nerves → 745

4436 tail of epididymis
f queue de l'épididyme
e cola del epidídimo
d Nebenhodenschwanz
l cauda epididymidis

4437 tail region
f région de la queue
e región de la cola; región coxígea
d Schwanzgegend
l regio caudalis; regio coccygea

4438 tail tuft; tail hairs
f crins de la queue; toupillon de la queue
e cerda de la cola
d Schweif; Schwanzhaare
l cirrus caudae

4439 tail veins; caudal veins; coccygeal veins
f veines caudales; veines de la queue
e venas caudales
d Schwanzvenen
l venae caudales

4440 talocalcaneal articulation
f articulation talo-calcanéenne
e articulación talocalcánea
d Articulatio talocalcanea
l articulatio talocalcanea

4441 talocalcaneal ligament
f ligament talo-calcanéen
e ligamento talocalcáneo
d Ligamentum talocalcaneum
l ligamentum talocalcaneum

* **talocrural joint** → 4450

4442 tapetum lucidum
 f tapetum lucidum; tapis clair
 e tapete lúcido
 d Tapetum lucidum
 l tapetum lucidum

* **tapeworm** → 841

* **tapeworm infestation** → 842

4443 tarsal arteries
 f artères tarsiennes
 e arterias tarsianas
 d Arteriae tarseae; Sprunggelenksarterien
 l arteriae tarseae

* **tarsal bone I** → 1706

* **tarsal bone II** → 4029

* **tarsal bone III** → 4530

* **tarsal bone IV** → 1752

4444 tarsal bones
 f os du tarse
 e huesos del tarso
 d Fußwurzelknochen
 l ossa tarsi

4445 tarsal glands; meibomian glands
 f glandes tarsales (de Meibomius)
 e glándulas tarsianas
 d Tarsaldrüsen; Meibomsche Drüsen
 l glandulae tarsales

* **tarsal hygroma** → 652

* **tarsal joint** → 2022

4446 tarsal ligaments
 f ligaments du tarse
 e ligamentos del tarso
 d Tarsalbänder
 l ligamenta tarsi

4447 tarsal muscle; eyelid muscle
 f muscle tarsal/tarsien
 e músculo tarsal
 d Augenlidmuskel
 l musculus tarsalis

* **tarsal pad** → 856

4448 tarsal region; hock region
 f région du tarse
 e región del tarso
 d Tarsalgegend; Fußwurzelgegend
 l regio tarsi

4449 tarsal veins
 f veines tarsiennes
 e venas tarsianas
 d Tarsalvenen
 l venae tarseae

4450 tarsocrural articulation; talocrural joint
 f articulation tibio-tarsienne
 e articulación tarsocrural
 d Rollgelenk; oberes Sprunggelenk
 l articulatio tarsocruralis; articulatio talocruralis

4451 tarsometatarsal articulations
 f articulations métatarsiennes
 e articulaciones tarsometatarsianas
 d Fußwurzel-Mittelfuß-Gelenke
 l articulationes tarsometatarseae

4452 tarsometatarsal ligaments
 f ligaments tarso-métatarsiens
 e ligamentos tarsometatarsianos
 d Ligamenta tarsometatarsea
 l ligamenta tarsometatarsea

4453 tarsometatarsus
 f tarso-métatarse
 e tarsometatarso
 d Tarsometatarsus
 l tarsometatarsus

* **tarsus** → 2021

4454 tarsus of eyelid
 f tarse des paupières
 e tarso palpebral
 d Lidplatte; Tarsus
 l tarsus inferior/superior

4455 taste bud
 f bourgeon du goût
 e calicillo gustativo; botón gustativo; papila del gusto
 d Geschmacksknospe
 l caliculus gustatorius

* **Taylorella equigenitalis infection** → 1046

4456 teat
 f mamelon; trayon; papille de la mamelle
 e papila de la mama; pezón; teta
 d Zitze; Strich
 l papilla mammae

4457 teat canal; papillary duct of teat
 f conduit papillaire; canal du trayon
 e conducto papilar del ubre
 d Zitzenkanal; Strichkanal
 l ductus papillaris uberis

4458 teat cannula
 f canule de trayon
 e cánula del pezón
 d Zitzenkanüle

4459 teat dipping; teat disinfection
 f trempage des trayons
 e enjuagado de los pezones
 d Zitzentauchen; Zitzendesinfektion

 * **teat disinfection → 4459**

4460 teat orifice
 f orifice papillaire de la tétine; ostium
    papillaire du trayon
 e orificio de la papila de la mama
 d Zitzenöffnung
 l ostium papillare

4461 teat sphincter
 f muscle sphincter du trayon
 e músculo esfínter de la papila
 d Zitzenschließmuskel
 l musculus sphincter papillae

4462 tegmental nuclei
 f noyaux du tegmentum
 e núcleos del tegmento
 d Nuclei tegmenti
 l nuclei tegmenti mesencephalica

4463 tela choroidea
 f toile choroïdienne
 e tela coroidea
 d Tela chorioidea
 l tela choroidea

4464 telangiectasis
 f télangiectasie
 e telangiectasia
 d Telangiektasie

4465 telencephalon; endbrain
 f télencéphale; cerveau terminal
 e telencéfalo
 d Telenzephalon; Endhirn
 l telencephalon

 * **temperature regulation → 4517**

4466 temple
 f tempe

 e sién
 d Schläfe
 l tempus

4467 temporal arteries
 f artères temporales
 e arterias temporales
 d Arteriae temporales; Schläfenarterien
 l arteriae temporales

4468 temporal bone
 f os temporal
 e hueso temporal
 d Schläfenbein
 l os temporal

4469 temporal crest
 f crête temporale
 e cresta temporal
 d Crista temporalis
 l crista temporalis

4470 temporal fossa
 f fosse temporale
 e fosa temporal
 d Schläfengrube
 l fossa temporalis

4471 temporal horn of lateral ventricle
 f corne temporale de la ventricule latérale
 e asta temporal
 d Cornu temporalis
 l cornu temporale ventriculi lateralis

4472 temporal line
 f ligne temporale
 e línea temporal
 d Linea temporalis
 l linea temporalis

4473 temporal lobe
 f lobe temporal
 e lóbulo temporal
 d Temporallappen
 l lobus temporalis

4474 temporal meatus
 f méat temporal
 e meato temporal
 d Schläfengang
 l meatus temporalis

4475 temporal muscle
 f muscle temporal
 e músculo temporal
 d Schläfenmuskel
 l musculus temporalis

4476 **temporal region**
f région temporale
e región temporal
d Schläfengegend
l regio temporalis

4477 **temporal veins**
f veines temporales
e venas temporales
d Schläfenvenen
l venae temporales

* **temporal wing of sphenoid bone** → 4971

* **temporary teeth** → 1220

* **temporomandibular articulation** → 2665

4478 **tendinitis**
f tendinite
e tendinitis
d Tendinitis; Sehnenentzündung

4479 **tendinous cords of heart**
f cordages tendineux du cœur
e cuerdas tendinosas
d Sehnenfäden der Herzklappen
l chordae tendinae

4480 **tendon**
f tendon
e tendón
d Sehne
l tendo

4481 **tendon sheath**
f gaine tendineuse
e vaina del tendón
d Sehnenfach
l vagina tendinis

* **Tenon's capsule** → 4065

4482 **tenosynovitis**
f ténosynovite; tendovaginalite
e tenosinovitis
d Tenosynovitis; Tendovaginitis

4483 **tensor muscle of antebrachial fascia**
f muscle tenseur du fascia antébrachial
e músculo tensor de la fascia del antebrazo
d Musculus tensor fasciae antebrachii
l musculus tensor fasciae antebrachii

4484 **tensor muscle of fascia lata**
f tenseur du fascia lata
e músculo tensor de la fascia lata
d Spanner der Schenkelfaszie
l musculus tensor fasciae latae

4485 **teratogen**
f tératogène
e teratógeno
d Teratogen

4486 **teratogenicity**
f tératogénicité
e teratogenicidad
d Teratogenität

4487 **teratology**
f tératologie
e teratología
d Teratologie

4488 **teratoma**
f tératome
e teratoma
d Teratom

4489 **terminal arch**
f arc terminal
e arco terminal
d Endbogen
l arcus terminalis

* **terminal cone** → 2748

4490 **terminal filament of spinal cord**
f filum terminale
e filum terminal
d Endfaden des Rückenmarks
l filum terminale

4491 **terminal ganglion**
f ganglion terminal
e ganglio terminal
d Ganglion terminale
l ganglion terminale

4492 **terminal nerve**
f nerf terminal
e nervio terminal
d Nervus terminalis
l nervus terminalis

4493 **terminal nerve corpuscles; encapsulated nerve endings**
f corpuscules nerveux terminaux
e corpúsculos nervios terminales
d Nervenendkörperchen
l corpuscula nervosa terminalia

4494 **terminal recesses** (*of equine kidney*)
f récessus terminaux
e recesos terminales
d Recessus terminales
l recessus terminales

4495 Teschen disease; porcine viral encephalomyelitis
f maladie de Teschen
e enfermedad de Teschen; encefalomielitis vírica porcina
d Teschener Krankheit; ansteckende Schweinelähmung

* **Teschen virus** → 3553

* **testicle** → 4500

4496 testicular appendix
f appendice du testicule; hydatide sessile de Morgagni
e apéndice del testículo
d Hodenanhängsel
l appendix testis

4497 testicular artery
f artère testiculaire
e arteria testicular
d Hodenarterie
l arteria testicularis

4498 testicular hypoplasia
f hypoplasie testiculaire
e hipoplasia testicular
d Hodenhypoplasie

* **testicular mesentery** → 2780

4499 testicular plexus
f plexus testiculaire
e plexo testicular
d Hodennervengeflecht
l plexus testicularis

4500 testis; testicle
f testicule
e testículo
d Hoden
l testis

4501 test kit; test strip
f coffret de diagnostic; trousse de diagnostic
e equipo de prueba
d Teststreifen

4502 testosterone
f testostérone
e testosterona
d Testosteron
l testosteronum

* **test strip** → 4501

4503 tetanus; Clostridium tetani intoxication
f tétanos; intoxication à Clostridium tetani
e tétanos; intoxicación por Clostridium tetani
d Tetanus; Starrkrampf; Clostridium-tetani-Intoxikation

4504 tetany
f tétanie
e tetania
d Tetanie

4505 tetracaine
f tétracaïne
e tetracaína
d Tetracain
l tetracainum

4506 tetracycline
f tétracycline
e tetraciclina
d Tetracyclin
l tetracyclinum

4507 tetramisole
f tétramisole
e tetramisol
d Tetramisol
l tetramisolum

* **TGE** → 4628

* **TGE virus** → 4627

4508 thalamencephalon; interbrain
f thalamencéphale
e talamoencéfalo
d Thalamushirn
l thalamencephalon

4509 thalamic nuclei; nuclei of thalamus
f noyaux du thalamus
e núcleos talámicos
d Thalamuskerne
l nuclei thalami

4510 thalamus
f thalamus
e tálamo
d Thalamus; Seehügel
l thalamus

4511 theca of ovarian follicle
f thèque des follicules ovariques
e teca del folículo ovárico
d Theca folliculi
l theca folliculi

* **Theileria infection** → 4512

* **Theileria lawrencei infection** → 1091

* **Theileria parva infection** → 1422

* **theileriasis** → 4512

4512 theileriosis; theileriasis; Theileria infection
 f theilériose; infection par Theileria spp.
 e teileriosis
 d Theileriose; Theileria-Infektion

4513 thelaziosis; eyeworm infestation
 f thélaziose; infestation par Thelazia spp.
 e filariosis ocular; infestación por Thelazia
 d Thelaziose

* **thelitis** → 2657

4514 therapeutic dose
 f dose thérapeutique
 e dosis terapéutica
 d therapeutische Dosis

4515 therapy; treatment
 f thérapeutique; traitement
 e terapia; tratamiento
 d Therapie; Behandlung

4516 theriogenology
 f thériogénologie
 e teriogenología
 d Theriogenologie

4517 thermoregulation; temperature regulation
 f thermorégulation
 e termorregulación
 d Temperaturregulation

* **thiabendazole** → 4567

4518 thiacetarsamide sodium
 f thiacétarsamide sodique
 e tiacetarsamida sódica
 d Thiacetarsamid-Natrium
 l thiacetarsamidum natricum

4519 thiamine; vitamin B1
 f thiamine; vitamine B1
 e tiamina; vitamina B1
 d Thiamin; Vitamin B1
 l thiaminum

4520 thick muscle of gizzard
 f muscle principal du gésier
 e músculo principal de la molleja
 d Hauptmuskel des Muskelmagens
 l musculus crassus

4521 thigh
 f fémur
 e fémur
 d Femur; Oberschenkel
 l femur

* **thighbone** → 1665

4522 thigh region
 f région de la cuisse
 e región del muslo
 d Oberschenkelgegend
 l regio femoris

4523 thin muscle of gizzard
 f muscle accessoire du gésier
 e músculo accesorio de la molleja
 d Zwischenmuskel des Muskelmagens
 l musculus tenuis

4524 thiopental sodium
 f thiopental sodique
 e tiopental sódico
 d Thiopental-Natrium
 l thiopentalum natricum

4525 third carpal bone; carpal bone III
 f os carpal III; os capitatum
 e hueso carpiano 3°; hueso grande
 d Os carpale III
 l os carpale III; os capitatum

4526 third eyelid; nictitating membrane
 f repli semi-lunaire; membrane nictitante
 e pliegue semilunar; tercer párpado;
   membrana nictitante
 d drittes Augenlid; Nickhaut
 l plica semilunaris conjunctivae; palpebra
   III

4527 third metacarpal bone; cannon bone
 f os métacarpien III
 e hueso metacarpiano 3°
 d Hauptmittelfußknochen; Röhrbein
 l os metacarpale III

4528 third metatarsal bone; cannon bone
 f os métatarsien III
 e hueso metatarsiano 3°
 d dritter Hauptmittelfußknochen; Röhrbein
 l os metatarsale III

4529 third peroneal muscle
 f muscle troisième péronier
 e músculo terco peroneo
 d dritter Wadenbeinmuskel
 l musculus peroneus tertius

* third phalanx → 1340

4530 third tarsal bone; tarsal bone III
 f os tarsal III; os cunéiforme latéral
 e hueso tarsiano 3°; hueso cuneiforme lateral
 d dritter Tarsalknochen
 l os tarsale III; os cuneiforme laterale

4531 third ventricle
 f troisième ventricule; ventricule moyen
 e tercer ventrículo
 d dritte Hirnventrikel
 l ventriculus tertius

4532 thoracic aorta
 f aorte thoracique
 e aorta torácica
 d Brustaorta
 l aorta thoracica

4533 thoracic aortic plexus
 f plexus aortique thoracique
 e plexo aórtico torácico
 d Plexus aorticus thoracicus
 l plexus aorticus thoracicus

4534 thoracic cardiac nerves
 f nerfs cardiaques thoraciques
 e nervios cardíacos torácicos
 d Nervi cardiaci thoracici
 l nervi cardiaci thoracici

4535 thoracic cavity
 f cavité thoracique
 e cavidad torácica
 d Brustkorbhöhle
 l cavum thoracis

4536 thoracic duct; left lymphatic duct
 f conduit thoracique; canal thoracique
 e conducto torácico; conducto linfático
   izquierdo
 d Brustlymphgang; Milchbrustgang
 l ductus thoracicus

4537 thoracic ganglia
 f ganglions thoraciques
 e ganglios torácicos
 d Brustkorbganglien
 l ganglia thoracica

* thoracic girdle → 4084

4538 thoracic limb; foreleg
 f membre thoracique; membre antérieur
 e miembro torácico
 d Schultergliedmaße; Vorderextremität
 l membrum thoracicum

4539 thoracic mammary gland
 f mamelle thoracique
 e mama torácica
 d thorakale Milchdrüse
 l mamma thoracica

4540 thoracic muscles; chest muscles
 f muscles du thorax
 e músculos del tórax
 d Brustmuskeln
 l musculi thoracis

4541 thoracic nerves
 f nerfs thoraciques
 e nervios torácicos
 d Brustkorbnerven
 l nervi thoracici

4542 thoracic vertebra
 f vertèbre thoracique
 e vértebra torácica
 d Brustwirbel
 l vertebra thoracica

4543 thoracodorsal artery
 f artère thoraco-dorsale
 e arteria toracodorsal
 d Arteria thoracodorsalis
 l arteria thoracodorsalis

4544 thoracodorsal nerve
 f nerf thoraco-dorsal
 e nervio toracodorsal
 d Nervus thoracodorsalis
 l nervus thoracodorsalis

4545 thoracodorsal vein
 f veine thoraco-dorsale
 e vena toracodorsal
 d Vena thoracodorsalis
 l vena thoracodorsalis

4546 thoracolumbar fascia
 f fascia thoraco-lombaire
 e fascia toracolumbar
 d Fascia thoracolumbalis
 l fascia thoracolumbalis

4547 thorax; chest
 f thorax
 e tórax
 d Thorax; Brustkorb
 l thorax

* thorny-headed worms → 20

* throat → 1630

* **thrombocyte** → 479

4548 thrombocytic
 f thrombocytaire; plaquettaire
 e trombocitario
 d Thrombozyten-

4549 thrombophlebitis
 f thrombophlébite
 e tromboflebitis
 d Thrombophlebitis

4550 thrombosis
 f thrombose
 e trombosis
 d Thrombose

4551 thumb; first digit of fore limb
 f pouce; doigt I de la main
 e dedo I° de la mano
 d Daumen; Pollex
 l digitus I

4552 thymic veins
 f veines thymiques
 e venas tímicas
 d Thymusvenen
 l venae thymicae

4553 thymus
 f thymus
 e timo
 d Thymus
 l thymus

4554 thyroarytenoid muscle
 f muscle thyro-aryténoïdien
 e músculo tiroaritenoideo
 d Musculus thyreoarytaenoideus
 l musculus thyroarytenoideus

4555 thyrocervical trunk
 f tronc thyro-cervical
 e tronco tirocervical (de Theile)
 d Truncus thyreocervicalis
 l truncus thyrocervicalis

4556 thyroglossal duct (of His)
 f conduit thyréo-glosse; canal de His
 e conducto tirogloso (de His)
 d Ductus thyreoglossus
 l ductus thyroglossus

4557 thyrohyoid articulation
 f articulation thyro-hyoïdienne
 e articulación tirohiodiea
 d Articulatio thyreohyoidea
 l articulatio thyrohyoidea

4558 thyrohyoid muscle
 f muscle thyro-hyoïdien
 e músculo tirohioideo
 d Musculus thyreohyoideus
 l musculus thyrohyoideus

4559 thyroid cartilage
 f cartilage thyroïdien
 e cartílago tiroideo
 d Schildknorpel
 l cartilago thyr(e)oidea

4560 thyroid function
 f fonction thyroïdienne
 e función de la glándula tiroidea
 d Schilddrüsenfunktion

4561 thyroid gland
 f glande thyroïde
 e glándula tiroidea
 d Thyreoidea; Schilddrüse
 l glandula thyr(e)oidea

4562 thyroid hormone
 f hormone thyroïdienne
 e hormona tiroidea
 d Schilddrüsenhormon

* **thyroid stimulating hormone** → 4565

4563 thyroid veins
 f veines thyroïdiennes
 e venas tiroideas
 d Schilddrüsenvenen
 l venae thyr(e)oideae

4564 thyropharyngeal muscle
 f muscle thyro-pharyngien
 e músculo tirofaríngeo
 d Musculus thyreopharyngeus
 l musculus thyropharyngeus

4565 thyrotrophin; thyrotropin; thyroid
 stimulating hormone; TSH
 f thyrotrophine; hormone thyréotrope
 e tirotrofina; tirotropina; hormona
 tireotrópica
 d Thyreotrophin
 l thyrotrophinum

* **thyrotropin** → 4565

4566 thyroxine; T4
 f thyroxine
 e tiroxina
 d Thyroxin; T4

4567 tiabendazole; thiabendazole
f tiabendazole; thiabendazole
e tiabendazol
d Tiabendazol; Thiabendazol
l tiabendazolum

4568 tiaprost
f tiaprost
e tiaprost
d Tiaprost
l tiaprostum

4569 tibia; shin bone
f tibia
e tibia
d Tibia; Schienbein
l tibia

4570 tibial cochlea
f cochlée tibiale
e cóclea de la tibia
d Schraubenkamm
l cochlea tibiae

4571 tibial crest; cranial margin of tibia
f crête tibiale; bord cranial du tibia
e borde craneal de la tibia
d Schienbeingräte
l margo cranialis tibiae

4572 tibial dyschondroplasia
f dyschondroplasie tibiale
e discondroplasia tibial
d Schienbeindyschondroplasie

4573 tibial nerve
f nerf tibial
e nervio tibial
d Nervus tibialis
l nervus tibialis

4574 tibial tarsal bone; ankle bone; astragalus
f talus; astragale
e talo; astrágalo
d Talus; Rollbein
l talus

4575 tibial tuberosity
f tubérosité tibiale
e tuberosidad de la tibia
d Schienbeinbeule
l tuberositas tibiae

4576 tibiofibular articulation
f articulation tibio-fibulaire
e articulación tibioperonea
d Schienbein-Wadenbein-Gelenk
l articulatio tibiofibularis

4577 tibiofibular ligament
f ligament tibio-fibulaire
e ligamento tibioperoneo
d Ligamentum tibiofibulare
l ligamentum tibiofibulare

4578 tibiotarsus
f tibiotarse
e tibiotarso
d Tibiotarsus
l tibiotarsus

4579 tick
f tique
e garrapata
d Zecke
l Ixodoidea

4580 tickborne fever; Ehrlichia phagocytophilia infection
f fièvre à tiques; infection à Ehrlichia phagocytophilia
e fiebre transmitida por garrapatas; infección por Ehrlichia phagocytophilia
d Zeckenfieber; Ehrlichia-phagocytophilia-Infektion

4581 tick infestation; argasidosis; ixodidosis
f infestation par les tiques
e infestación por garrapatas
d Zeckenbefall

4582 tick paralysis
f paralysie à tiques
e parálisis por garrapatas
d Zeckenparalyse

* tip of nose → 245

* tip of tongue → 4593

4583 tissue
f tissu
e tejido
d Gewebe
l tela

4584 tissue culture
f culture tissulaire
e cultivo de tejidos
d Gewebekultur

* titer → 4585

4585 titre; titer
f titre
e título
d Titer

4586 T lymphocyte
f lymphocyte T
e linfocito T
d T-Lymphozyt

* **toe bone** → 3431

4587 toe of hoof
f pince de l'onglon
e porción dorsal de la pared
d Rückenteil des Hufes
l pars dorsalis ungulae

* **toe pad** → 1313

4588 toes; digits of hind limb
f orteils; doigts du pied
e dedos del pie
d Hinterzehen
l digiti pedis

4589 togavirus
f togavirus
e togavirus
d Togavirus

4590 toltrazuril
f toltrazuril
e toltrazurilo
d Toltrazuril
l toltrazurilum

4591 tongue
f langue
e lengua
d Zunge
l lingua

4592 tongue muscles; lingual muscles
f muscles de la langue
e músculos de la lengua
d Zungenmuskeln
l musculi linguae

4593 tongue tip; tip of tongue
f apex de la langue; pointe de la langue
e vértice de la lengua; punta de la lengua
d Zungenspitze
l apex linguae

* **tongueworm infestation** → 2530

4594 tonsil
f amygdale; tonsille
e amígdala; tonsila
d Mandel
l tonsilla

4595 tonsillar fossa
f fosse tonsillaire; fosse amygdalienne
e fosa tonsilar
d Mandelgrube
l fossa tonsillaris

4596 tonsillar pits
f cryptes amygdaliennes
e fositas tonsilares
d Mandelgrübchen
l fossulae tonsillares

4597 tonsillitis
f amygdalite
e amigdalitis
d Tonsillitis; Mandelentzündung

4598 tooth
f dent
e diente
d Zahn
l dens

4599 tooth of axis; odontoid process
f dent de l'axis; processus odontoïde
e diente del eje
d Zahn des zweiten Halswirbels
l dens axis

4600 tooth pulp
f pulpe dentaire
e pulpa del diente
d Zahnmark
l pulpa dentis

4601 tooth socket
f alvéole dentaire
e alveolo dental
d Zahnfach
l alveolus dentalis

4602 torticollis
f torticolis
e torticolis
d Tortikollis; Halsverkrümmung; Schiefhals

4603 toxaemia; toxemia
f toxémie
e toxemia
d Toxämie

4604 toxaemic; toxemic
f toxémique
e toxemico
d toxämisch

4605 toxascariosis; Toxascaris infestation
f toxascariose; infestation par Toxascaris spp.
e toxascariosis; infestación por Toxascaris
d Toxaskaridose; Spulwurmbefall

* Toxascaris infestation → 4605

* toxemia → 4603

* toxemic → 4604

4606 toxic
  f toxique
  e tóxico
  d toxisch; giftig

4607 toxicity
  f toxicité
  e toxicidad
  d Toxizität; Giftigkeit

* toxicosis → 3531

4609 toxin
  f toxine
  e toxina
  d Toxin

* Toxocara infestation → 4610

4610 toxocarosis; Toxocara infestation
  f toxocarose; infestation par Toxocara
  e toxocarosis; infestación por Toxocara
  d Toxokarose; Spulwurmbefall

4611 toxoid
  f anatoxine
  e anatoxina; toxoide
  d Toxoid

* Toxoplasma gondii infection → 4612

4612 toxoplasmosis; Toxoplasma gondii infection
  f toxoplasmose; infection à Toxoplasma gondii
  e toxoplasmosis
  d Toxoplasmose; Toxoplasma-Infektion

4613 trabeculae of spleen; Billroth's cords
  f trabécules de la rate
  e trabéculas esplénicas; cordones de Billroth
  d Milztrabekeln
  l trabeculae lienis

4614 trace element deficiency
  f carence en oligoéléments
  e carencia de oligoelementos
  d Spurenelementmangel

4615 trachea; windpipe
  f trachée
  e tráquea
  d Luftröhre
  l trachea

4616 tracheal bifurcation
  f bifurcation de la trachée
  e bifurcación de la tráquea
  d Luftröhrengabelung
  l bifurcatio tracheae

4617 tracheal cartilages; tracheal rings
  f cartilages trachéaux; anneaux de la trachée
  e cartílagos traqueales
  d Knorpelspangen der Luftröhre
  l cartilagines tracheales

4618 tracheal coils
  f anses trachéales
  e asas traqueales
  d Luftröhrenwindungen
  l ansae tracheales

4619 tracheal glands
  f glandes trachéales
  e glándulas traqueales
  d Luftröhrendrüsen
  l glandulae tracheales

4620 tracheal muscle
  f muscle trachéal
  e músculo traqueal
  d Luftröhrenmuskel
  l musculus trachealis

* tracheal rings → 4617

4621 tracheitis
  f trachéite
  e traqueitis
  d Tracheitis; Luftröhrenentzündung

4622 tracheobronchial lymph nodes
  f nœuds lymphatiques trachéo-bronchiques
  e nódulos linfáticos traqueobronquiales
  d Lymphonodi tracheobronchales
  l lymphonodi tracheobronchales

4623 tracheobronchitis
  f trachéobronchite
  e traqueobronquitis
  d Tracheobronchitis

4624 tracheolateral muscle
  f muscle trachéo-latéral
  e músculo traqueolateral
  d Musculus tracheolateralis
  l musculus tracheolateralis

4625 tracheotomy
f trachéotomie
e traqueotomía
d Tracheotomie

4626 tragus
f tragus
e trago
d Tragus; Ohrecke
l tragus

* transmissible disease → 1043

4627 transmissible gastroenteritis coronavirus; TGE virus
f virus de la gastro-entérite transmissible du porc
e virus de la gastroenteritis transmisible del cerdo
d porcines Gastroenteritis-Coronavirus

4628 transmissible gastroenteritis of swine; TGE
f gastro-entérite transmissible du porc
e gastroenteritis transmisible del cerdo; TGE
d übertragbare Gastroenteritis des Schweines

4629 transmissible venereal sarcoma; Sticker's sarcoma
f sarcome de Sticker
e tumor venéreo; sarcoma de Sticker
d venerische Lymphosarkomatose; Stickersarkom

4630 transverse abdominal muscle
f muscle transverse de l'abdomen
e músculo transverso del abdomen
d querer Bauchmuskel
l musculus transversus abdominis

4631 transverse acetabular ligament
f ligament transverse de l'acétabulum
e ligamento transverso del acetábulo
d Ligamentum transversum acetabuli
l ligamentum transversum acetabuli

4632 transverse arteries
f artères transverses
e arterias transversas
d querverlaufende Arterien
l arteriae transversae

4633 transverse artery of face
f artère transverse de la face
e arteria transversa de la cara
d Arteria transversa faciei
l arteria transversa faciei

4634 transverse arytenoid muscle
f muscle aryténoïdien transverse
e músculo aritenoideo transverso
d Musculus arytenoideus transversus
l musculus arytenoideus transversus

4635 transverse colon
f côlon transverse
e colon transverso
d Querkolon
l colon transversum

4636 transverse fibres of pons
f fibres pontiques transverses
e fibras transversas del puente
d Fibrae pontis transversae
l fibrae pontis transversae

4637 transverse fissure of cerebrum
f fissure transverse du cerveau
e cisura transversa del cerebro
d Querspalt des Gehirns
l fissura transversa cerebri

4638 transverse foramen; foramen of transverse process
f foramen transversaire
e agujero transverso
d Foramen transversarium
l foramen transversarium

4639 transverse hyoid muscle
f muscle transverse de l'os hyoïde
e músculo hioideo transverso
d Musculus hyoides transversus
l musculus hyoides transversus

4640 transverse ligament
f ligament transverse
e ligamento transverso
d Ligamentum transversum
l ligamentum transversum

4641 transverse ligament of stifle; meniscal ligament
f ligament transverse du genou
e ligamento transverso de la rodilla
d Ligamentum transversum genus
l ligamentum transversum genus

4642 transverse nerve of neck
f nerf transverse du cou
e nervio transverso del cuello
d Nervus transversus colli
l nervus transversus colli

4643 transverse pectoral muscle
f muscle pectoral transverse
e músculo pectoral transverso
d querer Brustmuskel
l musculus pectoralis transversus

4644 transverse process
f processus transverse; apophyse transverse
e apófisis transversa
d Querfortsatz
l processus transversus

* **transverse process of atlas → 4970**

4645 transverse sinus of dura mater
f sinus transverse de la dure-mère
e seno transverso de la duramadre
d Sinus transversus
l sinus transversus durae matris

4646 transverse thoracic muscle
f muscle transverse du thorax
e músculo transverso del tórax
d querer Brustkorbmuskel
l musculus transversus thoracis

4647 transversospinal muscle
f muscle transverse épineux
e músculo transversoespinal
d Musculus transversospinalis
l musculus transversospinalis

4648 trapezius muscle
f muscle trapèze
e músculo trapecio
d Trapezmuskel
l musculus trapezius

* **trapezoid bone → 4028**

4649 trauma
f trauma
e trauma; traumatismo
d Trauma

4650 traumatic
f traumatique
e traumático
d traumatisch

* **treatment → 4515**

4651 trematode; fluke
f trématode; douve
e trematodo
d Trematode; Saugwurm
l Trematoda

4652 trematodosis
f trématodose
e trematodosis
d Trematodose; Befall mit Trematoden

4653 trenbolone
f trenbolone
e trenbolona
d Trenbolon
l trenbolonum

* **Treponema cuniculi infection → 3739**

* **Treponema hyodysenteriae infection →
  4383**

* **Treponema infection → 4654**

4654 treponematosis; Treponema infection
f tréponématose
e treponematosis
d Treponematose; Treponema-Infektion

4655 triangular ligament of liver
f ligament triangulaire du foie
e ligamento triangular del hígado
d Ligamentum triangulare hepatis
l ligamentum triangulare hepatis

4656 triceps muscle of arm
f muscle triceps brachial
e músculo tríceps braquial
d dreiköpfiger Oberarmmuskel
l musculus triceps brachii

4657 triceps muscle of calf of leg
f muscle triceps sural
e músculo tríceps sural
d dreiköpfiger Wadenmuskel
l musculus triceps surae

4658 trichinella
f trichine
e triquinela; triquina
d Trichinelle
l Trichinella

4659 trichinellosis; trichinosis
f trichinellose; trichinose
e triquinelosis; triquinosis
d Trichinellose

* **trichinosis → 4659**

* **trichlorfon → 2811**

4660 trichomonosis
f trichomonose
e tricomonosis
d Trichomonose

* **Trichonema infestation → 4661**

4661 trichonemosis; Trichonema infestation
f trichonémose
e triconemosis
d Trichonemose

* **Trichophyton gallinae infection** → 1631

4662 trichostrongyle
f trichostrongle
e tricostróngilo
d Trichostrongylus
l Trichostrongylus

4663 trichostrongylidosis
f trichostrongylidose
e tricostrongilidosis
d Trichostrongylidose

4664 trichostrongylids
f trichostrongylidés
e tricostrongílidos
d Trichostrongyliden
l Trichostrongylidae

4665 trichostrongylosis; Trichostrongylus infestation
f trichostrongylose
e tricostrongilosis
d Trichostrongylose

* **Trichostrongylus infestation** → 4665

4666 trichothecene toxin
f toxine trichothécène
e toxina tricoteceno
d Trichothecine-Toxin

4667 trichuriosis; Trichuris infestation
f trichuriose
e tricuriosis; infestación por Trichuris
d Trichuriose

* **Trichuris infestation** → 4667

4668 triclabendazole
f triclabendazole
e triclabendazol
d Triclabendazol
l triclabendazolum

* **tricuspid valve** → 3883

4669 trigeminal nerve
f nerf trijumeau
e nervio trigémino
d Nervus trigeminus
l nervus trigeminus

4670 trigone of bladder
f trigone vésical
e trígono vesical
d Harnblasendreieck
l trigonum vesicae

4671 triiodothyronine; T3
f triiodothyronine; T3
e triyodotironina
d Trijodthyronin

4672 trimethoprim
f triméthoprime
e trimetoprima
d Trimethoprim
l trimethoprimum

* **tritrichomonosis** → 549

4673 trocar
f trocart
e trocar
d Trokar

4674 trochanter
f trochanter
e trocánter
d Trochanter; Rollhügel
l trochanter

4675 trochanteric fossa
f fosse trochantérique
e fosa trocantérica
d Fossa trochanterica
l fossa trochanterica

4676 trochlea
f trochlée
e tróclea
d Trochlea; Rolle
l trochlea

4677 trochlea of femur
f trochlée du fémur
e tróclea del hueso fémur
d Kniescheibenrolle
l trochlea ossis femoris

4678 trochlea of humerus
f trochlée de l'humérus
e tróclea del húmero
d Gelenkrolle des Oberarmbeins
l trochlea humeri

4679 trochlea of radius
f trochlée du radius
e tróclea del radio
d Speichenwalze
l trochlea radii

4680 trochlea of talus
 f trochlée du talus; poulie astragalienne
 e tróclea del talo; tróclea del astrágalo
 d Talusgelenkrolle
 l trochlea tali

4681 trochlear nerve
 f nerf pathétique
 e nervio troclear
 d Nervus trochlearis
 l nervus trochlearis

4682 trochlear notch of ulna
 f incisure trochléaire
 e escotadura troclear
 d Incisura trochlearis
 l incisura trochlearis

* **trochoid articulation** → 3485

4683 trombiculid mites; velvet mites
 f trombiculidés
 e trombicúlidos; trombidios
 d Herbstgrasmilben
 l Trombiculidae

4684 trophoblast
 f trophoblaste
 e trofoblasto
 d Trophoblast

4685 tropical disease
 f maladie tropicale
 e enfermedad tropical
 d Tropenkrankheit

4686 tropical warble fly
 f berne; ver macaque
 e tórsalo; gusano moyocuil; mosca del berne
 d Dermatobia hominis
 l Dermatobia hominis

* **true rib** → 4205

4687 trunk
 f tronc
 e tronco
 d Rumpf; Stamm
 l truncus

4688 trypanocidal drug
 f trypanocide
 e medicamento tripanocido
 d Mittel gegen Trypanosomen

* **Trypanosoma equiperdum infection** → 1389

* **Trypanosoma evansi infection** → 4373

4689 trypanosome
 f trypanosome
 e tripanosoma
 d Trypanosom
 l Trypanosoma

* **trypanosomiasis** → 4690

4690 trypanosomosis; trypanosomiasis
 f trypanosomose
 e tripanosomosis
 d Trypanosomose; Trypanosomeninfektion

4691 trypsin
 f trypsine
 e tripsina
 d Trypsin

4692 tsetse fly
 f glossines
 e mosca tsé-tsé
 d Tsetsefliege
 l Glossina

* **TSH** → 4565

4693 tube agglutination test
 f agglutination en tube
 e aglutinación en tubos; aglutinación lenta
 d Blutserum-Langsamagglutination

4694 tuber cinereum
 f tuber cinereum
 e túber cinéreo
 d Tuber cinereum
 l tuber cinereum

4695 tubercle
 f tubercule
 e tubérculo
 d Tuberkel
 l tuberculum

4696 tubercle bacillus
 f bacille tuberculeux
 e bacilo tuberculoso
 d Tuberkelbazillus
 l Mycobacterium tuberculosis

4697 tubercle of rib
 f tubercule costal/de la côte
 e tubérculo de la costilla
 d Rippenhöckerchen
 l tuberculum costae

4698 tuberculin
f tuberculine
e tuberculina
d Tuberkulin

4699 tuberculin test
f tuberculination
e prueba de la tuberculina
d Tuberkulinprobe

4700 tuberosity of maxilla
f tubérosité maxillaire
e tuberosidad maxilar
d Oberkieferhöcker
l tuber maxillae

4701 tuberosity of scapular spine
f tubérosité de l'épine scapulaire
e tuberosidad del la espina de la escápula
d Höcker der Schulterblattgräte
l tuber spinae scapulae

4702 tuber vermis
f tuber vermis
e tuberosidad del vermis
d Klappenwulst des Kleinhirnwurms
l tuber vermis

4703 tularaemia; Francisella tularensis infection
f tularémie
e tularemia
d Tularämie

* tumor → 2975

* tumour → 2975

4704 tunica propria; proper tunic; proper coat
f tunique propre
e túnica propia
d Tunica propria; Eigenschicht
l tunica propria

* turbinate → 2935

* turbinate bones → 1554

4705 turkey bluecomb disease coronavirus
f coronavirus de la maladie de la crête bleue du dindon
e coronavirus de la enfermedad de la cresta azul del pavo
d Blaukammkrankheit-Coronavirus

4706 turkey herpesvirus
f herpèsvirus du dindon
e herpesvirus del pavo
d Puten-Herpesvirus

* tying-up syndrome of racehorses → 1563

4707 tylosin
f tylosine
e tilosina
d Tylosin
l tylosinum

4708 tympanic box; tympanic chamber
f chambre tympanique
e tímpano
d Trommel
l tympanum

4709 tympanic bulla
f bulle tympanique
e bulla timpánica
d Paukenblase
l bulla tympanica

4710 tympanic canal of cochlea
f rampe tympanique
e rampa del tímpano
d Paukentreppe
l scala tympani

4711 tympanic cavity
f cavum tympanique
e cavidad timpánica
d Paukenhöhle
l cavum tympani

* tympanic chamber → 4708

4712 tympanic membrane
f membrane du tympan
e membrana del tímpano
d Trommelfell
l membrana tympani

4713 tympanic nerve
f nerf tympanique
e nervio timpánico
d Paukenhöhlennerv
l nervus tympanicus

4714 tympanic opening of auditory tube
f ostium tympanique de la trompe auditive
e orificio timpánico de la trompa auditiva
d Paukenhöhlenmündung der Ohrtrompete
l ostium tympanicum tubae auditivae

4715 tympanic part of temporal bone
f partie tympanique de l'os temporal
e porción timpánica del temporal
d Paukenteil des Schläfenbeins
l pars tympanica ossis temporalis

4716 tympanic plexus
  f plexus tympanique
  e plexo timpánico
  d Plexus tympanicus
  l plexus tympanicus

4717 tympanic ring
  f anneau tympanique
  e anillo timpánico
  d Paukenring
  l anulus tympanicus

  * **tympanic wall of cochlear duct → 4156**

  * **tympanites → 4718**

4718 tympany; tympanites; bloat
  f tympanisme

  e timpanitis
  d Tympanie

4719 typhlitis
  f typhlite
  e tiflitis
  d Typhlitis; Blinddarmentzündung

4720 Tyzzer's disease; Bacillus piliformis infection
  f maladie de Tyzzer; infection à Bacillus piliformis
  e enfermedad de Tyzzer; infección por Bacillus piliformis
  d Bacillus-piliformis-Infektion

U

4721 udder
f pis
e ubre
d Euter
l uber

* **udder oedema** → 2662

4722 udder quarter
f quartier de la mamelle
e cuarto de la ubre
d Euterviertel

4723 udder region
f région du pis
e región de la ubre
d Euterregion
l regio uberis

4724 ulcer
f ulcère
e úlcera
d Geschwür
l ulcus

4725 ulcerative dermatosis of sheep
f dermatose ulcérative ovine
e dermatosis ulcerativa ovina
d ulzerative Dermatose des Schafes

4726 ulcerative lymphangitis; Corynebacterium pseudotuberculosis infection
f lymphangite ulcéreuse bactérienne; infection à Corynebacterium pseudotuberculosis
e linfangitis ulcerosa bacteriana; infección por Corynebacterium pseudotuberculosis
d geschwürige Lymphgefäßentzündung
l lymphangitis ulcerosa

4727 ulna
f ulna; cubitus
e cúbito
d Elle
l ulna

4728 ulnar artery
f artère ulnaire
e arteria cubital
d Arteria ulnaris
l arteria ulnaris

4729 ulnar carpal bone
f os ulnaire du carpe
e hueso carpocubital; hueso piramidal
d Os carpi ulnare
l os carpi ulnare; os triquetrum

4730 ulnar extensor of carpus; lateral ulnar muscle
f muscle extenseur ulnaire du carpe; muscle ulnaire latéral
e músculo extensor carpocubital; músculo cubital lateral
d äußerer Karpalstrecker
l musculus extensor carpi ulnaris; musculus ulnaris lateralis

4731 ulnar flexor muscle of carpus
f muscle fléchisseur ulnaire du carpe; muscle ulnaire médial
e músculo flexor carpocubital
d äußerer Karpalbeuger
l musculus flexor carpi ulnaris

4732 ulnar nerve
f nerf ulnaire
e nervio cubital
d Nervus ulnaris
l nervus ulnaris

4733 ulnar notch of radius
f incisure ulnaire
e escotadura cubital
d Incisura ulnaris radii
l incisura ulnaris radii

4734 ulnar vein
f veine ulnaire
e vena cubital
d Vena ulnaris
l vena ulnaris

4735 ultimobranchial body
f corps ultimobranchial
e cuerpo ultimobranquial
d Ultimobranchialkörper

4736 ultrasonic diagnosis
f échodiagnostic
e diagnóstico ultrasónico
d Ultraschalluntersuchung; Sonographie

4737 umbilical artery
f artère ombilicale
e arteria umbilical
d Nabelarterie
l arteria umbilicalis

4738 umbilical cord
f cordon ombilical
e cordón umbilical

d Nabelstrang; Nabelschnur
l funiculus umbilicalis

4739 umbilical hernia
f hernie ombilicale
e hernia umbilical
d Nabelbruch
l hernia umbilicalis

4740 umbilical region
f région ombilicale
e región umbilical
d Nabelgegend
l regio umbilicalis

4741 umbilical ring
f anneau ombilical
e anillo umbilical
d Nabelring
l anulus umbilicalis

4742 umbilical vein
f veine ombilicale
e vena umbilical
d Nabelvene
l vena umbilicalis

* **umbilicus** → 2957

* **unarmed tapeworm** → 423

* **Uncinaria infestation** → 4743

4743 uncinariosis; Uncinaria infestation
f uncinariose
e uncinariosis
d Uncinariose; Hakenwurmbefall

4744 uncinate fascicle
f faisceau unciné
e fascículo uncinado
d Fasciculus uncinatus
l fasciculus uncinatus

4745 uncinate process of ethmoid bone
f processus unciné de l'os ethmoïde
e apófisis uncinada del hueso etmoides
d hakenförmiger Fortsatz des Siebbeins
l processus uncinatus ossis ethmoidalis

4746 uncinate process of pancreas
f processus unciné du pancréas
e apófisis uncinada del páncreas
d hakenförmiger Fortsatz des Pankreas
l processus uncinatus pancreaticus

* **underfed** → 4747

4747 undernourished; underfed
f sous-alimenté
e subalimentado
d unterernährt

4748 unfeathered areas
f aptéries
e apteria
d Federraine
l apteria

4749 unfit for breeding
f inapte à la reproduction
e no apto para la reproducción
d zuchtuntauglich

4750 unhealthy
f malsain
e enfermizo
d ungesund

4751 uninfected
f non infecté
e no infecto
d nicht infiziert

4752 unpaired ganglion
f ganglion impair
e ganglio impar
d unpaariges Ganglion
l ganglion impar

4753 unthriftiness; impaired growth; stunting
f retard de développement
e retraso del crecimiento
d Kümmern; Kleinwuchs;
  Wachstumshemmung

4754 upper arm
f bras
e brazo
d Oberarm
l brachium

4755 upper beak
f rostre maxillaire; bec supérieur
e pico superior
d Oberschnabel
l rostrum maxillare

4756 upper eyelid
f paupière supérieure
e párpado superior
d Oberlid
l palpebra superior

* **upper jawbone** → 2696

4757 upper lip
f lèvre supérieure
e labio superior; labio maxilar
d Oberlippe
l labium superius; labium maxillare

4758 upper surface of foot
f dos du pied
e dorso del pie
d Fußrücken
l dorsum pedis

4759 urachus
f ouraque
e uraco
d Urachus
l urachus

4760 uraemia; uremia
f urémie
e uremia
d Uremie

* **uremia → 4760**

4761 ureter
f uretère
e uréter
d Harnleiter
l ureter

4762 ureteral glands
f glandes urétériques
e glándulas ureléricas
d Harnleiterdrüsen
l glandulae uretericae

4763 urethra
f urètre; urèthre
e uretra
d Harnröhre
l urethra

4764 urethral glands
f glandes urétrales
e glándulas uretrales
d Harnröhrendrüsen
l glandulae urethrales

4765 urethral process
f processus urétral
e proceso uretral
d Harnröhrenfortsatz
l processus urethrae

4766 urethral sphincter
f sphincter externe de l'urètre
e músculo esfínter externo de la uretra
d Schließmuskel der Harnröhre
l musculus sphincter urethrae

4767 urinary
f urinaire
e urinario
d Harn-

* **urinary bladder → 463**

4768 urinary calculus
f calcul urinaire
e cálculo urinario
d Harnstein

4769 urinary incontinence
f incontinence urinaire
e incontinencia urinaria
d Harninkontinenz
l incontinentia urinae

4770 urinary organs
f organes uropoïétiques
e órganos uropoyéticos
d Harnorgane
l organa uropoetica

4771 urinary tract disease
f maladie des voies urinaires
e enfermedad de las vías urinarias
d Uropathie

4772 urination disorder
f trouble urinaire
e trastorno urinario
d Harnabsatzstörung

4773 urine
f urine
e orina
d Harn
l urina

* **uriniferous tubules → 3820**

4774 urodeum
f urodéum
e urodeo
d Urodeum; Harnraum
l urodeum

4775 urogenital sinus
f sinus urogénital
e seno urogenital
d Sinus urogenitalis
l sinus urogenitalis

4776 urogenital system
f appareil urogénital
e aparato urogenital
d Harn-Geschlechts-Apparat
l apparatus urogenitalis; systema
  urogenitale

4777 urolithiasis
f urolithiase; lithiase urinaire
e urolitiasis
d Urolithiasis; Harnsteinkrankheit

4778 uropygial eminence
f proéminence uropygienne
e eminencia uropigial
d Eminentia uropygialis
l eminentia uropygialis

4779 uropygial gland; preen gland
f glande uropygienne
e glándula uropígea
d Bürzeldrüse
l glandula uropygialis

4780 urticaria; nettlerash
f urticaire
e urticaria
d Urtikaria

4781 uterine adnexa; uterine appendages
f annexes de l'utérus
e anexos del útero
d Anhangsgebilde der Gebärmutter
l adnexa uteri

* **uterine appendages** → 4781

4782 uterine artery
f artère utérine
e arteria uterina
d Uterusarterie
l arteria uterina

4783 uterine cavity
f cavité utérine
e cavidad del útero
d Gebärmutterhöhle
l cavum uteri

* **uterine cervix** → 840

4784 uterine crypts
f cryptes utérines
e criptas uterinas
d Uterinkrypten
l cryptae uterinae

4785 uterine disease
f maladie de l'utérus
e enfermedad del útero
d Uteruserkrankung

4786 uterine glands
f glandes utérines
e glándulas uterinas
d Gebärmutterdrüsen
l glandulae uterinae

* **uterine horns** → 2038

4787 uterine inertia
f inertie utérine
e inercia uterina
d Wehenschwäche
l inertia uteri

4788 uterine involution
f involution utérine
e involución del útero
d Uterusinvolution
l involutio uteri

* **uterine mesentery** → 2778

* **uterine mucosa** → 1468

* **uterine muscle** → 2917

4789 uterine prolapse; prolapse of uterus
f prolapsus utérin
e prolapso uterino
d Gebärmuttervorfall
l prolapsus uteri

4790 uterine torsion
f torsion utérine
e torsión del útero
d Gebärmutterverdrehung
l torsio uteri

* **uterine tube** → 3204

4791 uterine vein
f veine utérine
e vena uterina
d Gebärmuttervene
l vena uterina

4792 utero-ovarian ligament; ovarian ligament
f ligament utéro-ovarien
e ligamento propio del ovario; ligamento uteroovárico
d Ligamentum ovarii proprium
l ligamentum ovarii proprium

4793 uterovaginal plexus
f plexus utéro-vaginal
e plexo uterovaginal
d Plexus uterovaginalis
l plexus uterovaginalis

4794 uterus
f utérus
e útero; matriz
d Uterus; Gebärmutter
l uterus

4795 utricle
f utricule
e utrículo
d Utrikulus
l utriculus

* uveal tract → 4827

4796 uvula
f uvule
e uvula
d Zäpfchen
l uvula (palatina)

valve

V

4797 vaccinated
f vacciné
e vacunado
d vakziniert; geimpft

4798 vaccination
f vaccination
e vacunación
d Vakzination; Schutzimpfung

4799 vaccine
f vaccin
e vacuna
d Vakzine; Impfstoff

4800 vaccinia orthopoxvirus
f orthopoxvirus de la vaccine
e orthopoxvirus vacunal
d Vacciniavirus

4801 vagina
f vagin
e vagina
d Scheide
l vagina

4802 vaginal artery
f artère vaginale
e arteria vaginal
d Scheidenarterie
l arteria vaginalis

4803 vaginal discharge
f écoulement vaginal
e descargo vaginal
d Scheidenausfluß

4804 vaginal disease
f maladie du vagin
e enfermedad de la vagina
d Vaginalerkrankung

4805 vaginal fornix
f fornix du vagin; cul-de-sac vaginal
e fórnix de la vagina
d Scheidengewölbe
l fornix vaginae

4806 vaginal mucus
f mucus vaginal
e muco vaginal
d Vaginalschleim

4807 vaginal nerves
f nerfs vaginaux
e nervios vaginales
d Nervi vaginales; Scheidennerven
l nervi vaginales

4808 vaginal process of peritoneum
f processus vaginal du péritoine
e proceso vaginal del peritoneo
d Scheidenhautfortsatz
l processus vaginalis periton(a)ei

4809 vaginal prolapse; prolapse of vagina
f prolapsus vaginal
e prolapso vaginal
d Scheidenvorfall
l prolapsus vaginae

4810 vaginal tunic of testis
f tunique vaginale
e túnica vaginal
d Serosa des Hodens
l tunica vaginalis testis

* vaginal vestibule → 4903

4811 vaginitis
f vaginite
e vaginitis
d Vaginitis; Scheidenentzündung

4812 vagosympathetic trunk
f tronc vago-sympathique
e tronco vagosimpático
d Truncus vagosympathicus
l truncus vagosympathicus

4813 vagus nerve
f nerf vague; nerf pneumogastrique
e nervio vago; nervio neumogástrico
d Nervus vagus
l nervus vagus

4814 vallate papillae; circumvallate papillae
f papilles circumvallées; papilles calciformes
e papilas valladas
d umwallte Papillen
l papillae vallatae

4815 vallecula cerebelli
f vallécule du cervelet
e valécula del cerebelo
d Vallecula cerebelli
l vallecula cerebelli

4816 valve of aorta; aortic valve
f valve de l'aorte
e valva aórtica
d Aortenklappe
l valva aortae

4817 **valve of a vein; venous valve**
 f valvule veineuse
 e válvula venosa
 d Venenklappe
 l valvula venosa

4818 **valve of pulmonary trunk; pulmonary valve**
 f valve du tronc pulmonaire
 e valva del tronco pulmonar; válvula
   pulmonar
 d Pulmonalklappe
 l valva trunci pulmonalis

4819 **vane of feather**
 f vexille
 e vexilo; estandarte
 d Federfahne
 l vexillum

4820 **vanule**
 f vexille de la barbe
 e vexilo de la barba
 d Vexilla barbae
 l vexilla barbae

 * **Varroa jacobsoni infestation** → 4821

 * **varroasis** → 4821

4821 **varroosis; varroasis; Varroa jacobsoni
   infestation**
 f varroase; varroatose
 e varroasis
 d Varroatose; Befall mit Varroa jacobsoni

4822 **vasa vasorum; vessels of blood vessels**
 f vasa vasorum
 e vasa vasorum; vasos de los vasos
 d Vasa vasorum
 l vasa vasorum

4823 **vascular disease**
 f maladie vasculaire
 e enfermedad vascular
 d Angiopathie

4824 **vascular insufficiency**
 f insuffisance vasculaire
 e insuficiencia vascular
 d Gefäßinsuffizienz

4825 **vascular nerve**
 f nerf vasculaire
 e nervio vascular
 d Gefäßnerv
 l nervus vascularis

4826 **vascular system**
 f système vasculaire
 e sistema vascular
 d Gefäßsystem
 l systema vasorum

4827 **vascular tunic of eye; uveal tract**
 f tunique vasculaire du bulbe; uvée
 e túnica vasculosa del globo; úvea; tracto
   uveal
 d mittlere Augenhaut; Uvea
 l tunica vasculosa bulbi

 * **vas deferens** → 1238

4828 **vasectomized**
 f vasectomisé
 e vasectomizado
 d vasektomisiert

4829 **vasoconstriction**
 f vasoconstriction
 e vasoconstricción
 d Vasokonstriktion; Gefäßverengung

4830 **vasodilation**
 f vasodilatation
 e vasodilatación
 d Vasodilatation

 * **vasopressin** → 215

 * **Vater-Pacini corpuscle** → 2409

4831 **vector; disease vector**
 f vecteur; transmetteur
 e vector
 d Vektor; Überträger

4832 **vein**
 f veine
 e vena
 d Vene; Blutader
 l vena

4833 **veins of labyrinth**
 f veines labyrinthiques
 e venas del laberinto
 d Labyrinthvenen
 l venae labyrinthi

 * **velvet mites** → 4683

4834 **vena caval foramen**
 f foramen de la veine cave
 e agujero de la vena cava
 d Foramen venae cavae
 l foramen venae cavae

4835 venous arch
f arcade veineuse
e arco venoso
d Venenring
l arcus venosus

4836 venous circulation
f circulation veineuse
e circulación venosa
d venöser Kreislauf

4837 venous duct (*of Arantius*)
f conduit veineux (*d'Arantius*)
e conducto venoso
d Ductus venosus; Ductus Arantii
l ductus venosus

4838 venous network
f réseau veineux
e red venosa
d venöses Gefäßnetz
l rete venosum

4839 venous plexus
f plexus veineux
e plexo venoso
d Venengeflecht
l plexus venosus

4840 venous sinus
f sinus veineux
e seno venoso
d Sinus venosus
l sinus venosus

* venous sinuses of dura mater → 1153

4841 venous sinus of sclera; Schlemm's canal
f sinus veineux de la sclère; canal de Schlemm
e seno venoso de la esclerótica
d Schlemmscher Kanal
l sinus venosus sclerae

* venous valve → 4817

4842 vent; external opening of cloaca
f orifice externe du cloaque
e orificio externo de la cloaca
d Kloakenöffnung
l ventus

4843 vent gland; cloacal gland
f glande cloacale
e glándula cloacal
d Kloakendrüse
l glandula ventis

4844 ventral branch
f rameau ventral
e ramo ventral
d ventraler Ast
l ramus ventralis

4845 ventral cerebellar veins
f veines cérébelleuses ventrales
e venas ventrales del cerebelo
d Venae cerebelli ventrales; ventrale Kleinhirnvenen
l venae cerebelli ventrales

4846 ventral colon
f côlon ventral
e colon ventral
d ventrales Kolon
l colon ventrale

4847 ventral column of spinal cord
f cordon ventral de la moelle épinière
e cordón ventral del médula espinal
d Ventralstrang des Rückenmarks
l funiculus ventralis

4848 ventral external ophthalmic vein
f veine ophtalmique externe ventrale
e vena oftálmica externa ventral
d Vena ophthalmica externa ventralis
l vena ophthalmica externa ventralis

4849 ventral horn of spinal cord
f corne ventrale de la moelle épinière
e asta ventral del médula espinal
d Ventralhorn der grauen Substanz
l cornu ventrale

4850 ventral nasal concha
f cornet nasal ventral
e concha nasal ventral
d ventrale Nasenmuschel
l concha nasalis ventralis

4851 ventral nuclei of thalamus
f noyaux ventraux du thalamus
e núcleos ventrales del tálamo
d ventrale Thalamuskerne
l nuclei ventrales thalami

4852 ventral oblique muscle (*of eyeball*)
f muscle oblique ventral (*du bulbe de l'œil*)
e músculo oblicuo ventral
d Musculus obliquus ventralis
l musculus obliquus ventralis

4853 ventral perineal artery
f artère périnéale ventrale
e arteria perineal ventral
d Arteria perinealis ventralis
l arteria perinealis ventralis

4854 **ventral petrosal sinus**
 f sinus pétreux ventral
 e seno petroso ventral
 d Sinus petrosus ventralis
 l sinus petrosus ventralis

4855 **ventral rectus muscle** (*of eyeball*)
 f muscle droit ventral (*du bulbe de l'œil*)
 e músculo recto ventral
 d Musculus rectus ventralis
 l musculus rectus ventralis

4856 **ventral region of neck**
 f région ventrale du cou
 e región ventral del cuello
 d ventrale Halsregion
 l regio colli ventralis

4857 **ventral sac of rumen**
 f sac ventral du rumen
 e saco ventral del rumen
 d ventraler Pansensack
 l saccus ventralis ruminis

4858 **ventral sagittal sinus**
 f sinus sagittal ventral
 e seno sagital ventral
 d Sinus sagittalis ventralis
 l sinus sagittalis ventralis

4859 **ventral serrated muscle of neck**
 f muscle dentelé ventral du cou
 e músculo serrato ventral del cuello
 d Musculus serratus ventralis cervicis
 l musculus serratus ventralis cervicis

4860 **ventral serrated muscle of thorax**
 f muscle dentelé ventral du thorax
 e músculo serrato ventral del tórax
 d Musculus serratus ventralis thoracis
 l musculus serratus ventralis thoracis

4861 **ventral spinal artery**
 f artère spinale ventrale
 e arteria espinal ventral
 d Arteria spinalis ventralis
 l arteria spinalis ventralis

4862 **ventral spinocerebellar tract; Flechsig's tract**
 f tractus spino-cérébelleux ventral; faisceau de Flechsig
 e tracto espinocerebeloso ventral
 d Tractus spinocerebellaris ventralis; Flechsigsches Bündel
 l tractus spinocerebellaris ventralis

 * **ventral thalamus** → 4293

4863 **ventral thoracic lymphocentre**
 f lymphocentre thoracique ventral
 e linfocentro torácico ventral
 d Lymphocentrum thoracicum ventrale
 l lymphocentrum thoracicum ventrale

4864 **ventral vagal trunk**
 f tronc vagal ventral
 e tronco vago ventral
 d ventraler Vagusstamm
 l truncus vagalis ventralis

4865 **ventricle of heart**
 f ventricule du cœur
 e ventrículo del corazón
 d Herzkammer
 l ventriculus cordis

 * **ventricle of larynx** → 2431

4866 **ventricular**
 f ventriculaire
 e ventricular
 d ventrikulär

4867 **ventromedial nucleus of hypothalamus**
 f noyau ventro-médial de l'hypothalamus
 e núcleo hipotalámico ventromedial
 d Nucleus hypothalamicus ventromedialis
 l nucleus hypothalamicus ventromedialis

4868 **venule**
 f veinule
 e vénula
 d Venula; kleinste Vene
 l venula

 * **vermifuge** → 201

4869 **vermis of cerebellum**
 f vermis
 e vermis del cerebelo
 d Vermis cerebelli; Kleinhirnwurm
 l vermis cerebelli

4870 **vertebra**
 f vertèbre
 e vértebra
 d Vertebra; Wirbel
 l vertebra

4871 **vertebral arch**
 f arc vertébral
 e arco de la vértebra
 d Wirbelbogen
 l arcus vertebrae

4872 vertebral artery
  f artère vertébrale
  e arteria vertebral
  d Arteria vertebralis; Wirbelarterie
  l arteria vertebralis

4873 vertebral body
  f corps vertébral
  e cuerpo de la vértebra
  d Wirbelkörper
  l corpus vertebrae

4874 vertebral canal
  f canal rachidien
  e canal vertebral
  d Wirbelkanal
  l canalis vertebralis

4875 vertebral column; spine; backbone
  f colonne vertébrale; rachis
  e columna vertebral; raquis
  d Wirbelsäule
  l columna vertebralis

4876 vertebral foramen
  f trou vertébral
  e agujero vertebral
  d Wirbelloch
  l foramen vertebrale

* **vertebral fossa → 725**

* **vertebral head → 1131**

4877 vertebral nerve
  f nerf vertébral
  e nervio vertebral
  d Wirbelnerv
  l nervus vertebralis

4878 vertebral notch
  f incisure vertébrale; échancrure vertébrale
  e escotadura vertebral
  d Wirbeleinschnitt
  l incisura vertebralis

4879 vertebral vein
  f veine vertébrale
  e vena vertebral
  d Vertebralvene
  l vena vertebralis

4880 vertebral venous plexuses
  f plexus veineux vertébraux
  e plexos vertebrales
  d venöse Wirbelgeflechte
  l plexus vertebrales

4881 vesical arteries
  f artères vésicales
  e arterias vesicales
  d Harnblasenarterien
  l arteriae vesicales

4882 vesical plexuses; plexuses of bladder
  f plexus vésicaux
  e plexos vesicales
  d Harnblasengeflechte
  l plexus vesicales

4883 vesical veins
  f veines vésicales
  e venas vesicales
  d Harnblasenvenen
  l venae vesicales

4884 vesicle
  f vésicule
  e vesícula
  d Vesikel; Bläschen
  l vesicula

4885 vesicogenital pouch
  f cul-de-sac vésico-génital
  e excavación vesicogenital
  d Excavatio vesicogenitalis
  l excavatio vesicogenitalis

4886 vesicular exanthema calicivirus
  f calicivirus de l'exanthème vésiculaire
  e virus del exantema vesicular
  d vesikuläres Exanthem-Virus

4887 vesicular exanthema of swine
  f exanthème vésiculaire porcin
  e exantema vesicular porcino
  d vesikuläres Exanthem des Schweines

4888 vesicular gland
  f glande vésiculaire
  e glándula vesicular
  d Samenblasendrüse
  l glandula vesicularis

* **vesicular ovarian follicles → 1881**

4889 vesicular stomatitis
  f stomatite vésiculeuse contagieuse
  e estomatitis vesiculosa
  d Stomatitis vesicularis
  l stomatitis vesicularis

4890 vesicular stomatitis virus
  f virus de la stomatite vésiculeuse
  e virus de la estomatitis vesicular
  d vesikuläres Stomatitis-Virus

4891 vesiculovirus
f vésiculovirus
e vesiculovirus
d Vesiculovirus

* **vessels of blood vessels → 4822**

4892 vestibular artery
f artère vestibulaire
e arteria vestibular
d Arteria vestibularis
l arteria vestibularis

4893 vestibular canal of cochlea
f rampe vestibulaire
e rampa del vestíbulo
d Vorhofstreppe
l scala vestibuli

* **vestibular fenestra → 3190**

4894 vestibular fold of larynx
f pli vestibulaire du larynx
e pliegue vestibular
d Taschenfalte
l plica vestibularis

4895 vestibular ganglion; Scarpa's ganglion
f ganglion vestibulaire; ganglion de Scarpa
e ganglio vestibular
d Vorhofsganglion
l ganglion vestibulare

4896 vestibular ligament of larynx
f ligament vestibulaire du larynx
e ligamento vestibular
d Taschenband
l ligamentum vestibulare

4897 vestibular nuclei
f noyaux vestibulaires
e núcleos vestibulares
d Vestibularkerne
l nuclei vestibulares

4898 vestibular root of vestibulocochlear nerve
f racine vestibulaire du nerf vestibulo-cochléaire
e raíz vestibular del nervio vestíbulococlear
d Vorhofswurzel des Gleichgewichts-Hörnervs
l radix vestibularis nervi vestibulocochlearis

4899 vestibular veins
f veines vestibulaires
e venas vestibulares
d Vestibularvenen
l venae vestibulares

* **vestibular wall of cochlear duct → 3804**

4900 vestibule
f vestibule
e vestíbulo
d Vorhof
l vestibulum

4901 vestibule of mouth
f vestibule de la bouche
e vestíbulo de la boca
d Vorhof der Mundhöhle
l vestibulum oris

4902 vestibule of nose
f vestibule nasal
e vestíbulo de la nariz
d Nasenvorhof
l vestibulum nasi

4903 vestibule of vagina; vaginal vestibule
f vestibule du vagin
e vestíbulo de la vagina
d Scheidenvorhof
l vestibulum vaginae

4904 vestibulocochlear nerve (8th cranial)
f nerf vestibulo-cochléaire; nerf auditif
e nervio vestibulococlear; nervio auditivo; octavo nervio craneal
d Gleichgewichts-Hörnerv
l nervus vestibulocochlearis

4905 vestibulocochlear nuclei
f noyaux du nerf vestibulo-cochléaire
e núcleos del nervio vestibulococlear
d Nuclei nervi vestibulocochlearis
l nuclei nervi vestibulocochlearis

4906 vestibulocochlear organ
f organe vestibulo-cochléaire
e órgano vestíbulococlear
d Gleichgewichts- und Gehörorgan
l organum vestibulocochleare

4907 vestibulospinal tract; Deiters' tract
f tractus vestibulo-spinal; faisceau de Deiters
e tracto vestíbuloespinal
d Tractus vestibulospinalis
l tractus vestibulospinalis

4908 veterinarian; veterinary surgeon
f vétérinaire
e veterinario; médico veterinario
d Tierarzt

4909 veterinary education
f enseignement vétérinaire
e educación veterinaria
d tierärztliche Ausbildung

* **veterinary faculty** → 4916

4910 veterinary history
f histoire vétérinaire
e historia veterinaria
d Veterinärgeschichte

4911 veterinary hygiene
f hygiène vétérinaire
e higiene veterinaria
d Veterinärhygiene

4912 veterinary jurisprudence
f jurisprudence vétérinaire
e jurisprudencia veterinaria
d Berufs- und Standesrecht

4913 veterinary medicine
f médecine vétérinaire
e medicina veterinaria
d Veterinärmedizin; Tierheilkunde

4914 veterinary practice
f pratique vétérinaire
e práctica veterinaria
d tierärztliche Praxis

4915 veterinary profession
f profession vétérinaire
e profesión veterinaria
d tierärztlicher Beruf; Tierärzteschaft

4916 veterinary school; veterinary faculty
f école vétérinaire
e escuela veterinaria; facultad de veterinaria
d tierärztliche Hochschule; tierärztliche Fakultät

4917 veterinary science
f science vétérinaire
e ciencia veterinaria
d Veterinärwissenschaft

4918 veterinary service
f service vétérinaire
e servicio veterinario
d Veterinärdienst; Veterinärwesen

* **veterinary surgeon** → 4908

* **VHS** → 4926

* **VHS virus** → 1940

* **Vibrio anguillarum infection** → 4919

4919 vibriosis of fish; Vibrio anguillarum infection
f infection à Vibrio anguillarum
e vibriosis de los peces; infección por Vibrio anguillarum
d Vibriose; Vibrio-anguillarum-Infektion

4920 vibrissa
f vibrisse
e vibrisa
d Gitterhaar (am Naseneingang)
l vibrissa

4921 villus
f villosité
e villosidad
d Villus; Zotte
l villus

4922 viraemia; viremia
f virémie
e viremia
d Virämie

4923 viral antigen
f antigène viral
e antígeno viral
d Virusantigen

4924 viral disease
f maladie virale; virose
e enfermedad vírica; virosis
d virale Krankheit

4925 viral haemorrhagic disease of rabbits
f maladie hémorragique virale du lapin
e enfermedad vírica hemorrágica de los conejos
d virale hämorrhagische Krankheit der Kaninchen

4926 viral haemorrhagic septicaemia of fish; VHS
f septicémie hémorragique virale
e septicemia vírica hemorrágica; VHS
d virale hämorrhagische Septikämie der Fische

* **viremia** → 4922

4927 virginiamycin
f virginiamycine
e virginiamicina
d Virginiamycin
l virginiamycinum

4928 virulence
f virulence
e virulencia
d Virulenz

4929 virulent
f virulent
e virulento
d virulent

4930 virus
f virus
e virus
d Virus

4931 viscera
f viscères
e vísceras
d Eingeweide
l viscera

4932 visceral layer of pericardium; epicardium
f lame viscérale du péricarde
e lámina visceral del pericardio
d Lamina visceralis pericardii; Epikard
l lamina visceralis pericardii

4933 visceral peritoneum
f péritoine viscéral
e peritoneo visceral
d viszerales Blatt des Bauchfells
l peritoneum viscerale

* **visceral pleura** → 3693

4934 visna; ovine lentiviral meningoencephalitis
f visna
e visna
d Visna

4935 visna-maedi lentivirus
f virus visna/maedi
e virus visna-maedi
d Visna-Maedi-Virus

* **vitamin A** → 3849

* **vitamin B1** → 4519

* **vitamin B2** → 3869

* **vitamin B6** → 3723

* **vitamin B12** → 1200

4936 vitamin C; ascorbic acid
f vitamine C; acide ascorbique
e vitamina C; acido ascórbico
d Vitamin C; Askorbinsäure

* **vitamin D2** → 1533

* **vitamin D3** → 875

* **vitamin deficiency** → 367

* **vitelline duct** → 3114

4937 vitellus; yolk
f vitellus; jaune d'œuf
e vitelo
d Vitellus; Dotter
l vitellus

4938 vitiligo
f vitiligo
e vitíligo
d Vitiligo

4939 vitreous body
f corps vitré
e cuerpo vítreo
d Glaskörper
l corpus vitreum

4940 vitreous chamber of eye
f chambre vitrée du bulbe
e cámara vítrea del globo
d Camera vitrea bulbi
l camera vitrea bulbi

4941 vitreous humour
f humeur vitrée
e humor vítreo
d Glaskörperflüssigkeit
l humor vitreus

4942 vitreous membrane; hyaloid membrane
f membrane vitrée; membrane hyaloïde
e membrana vítrea
d Membrana vitrea
l membrana vitrea

4943 vocal cord
f pli vocal; corde vocale
e pliegue vocal; cuerdo vocal
d Stimmfalte
l plica vocalis

4944 vocal ligament of larynx
f ligament vocal
e ligamento vocal
d Stimmband
l ligamentum vocale

* **vocal organ** → 4422

4945 vocal process of arytenoid cartilage
  f processus vocal
  e apófisis vocal
  d Stimmbandfortsatz
  l processus vocal

4946 voluntary muscle
  f muscle volontaire
  e músculo voluntario
  d willkürlicher Muskul

4947 vomer
  f vomer
  e vómer
  d Vomer; Pflugscharbein
  l vomer

4948 vomeronasal duct
  f conduit voméro-nasal
  e conducto vomeronasal
  d Ductus vomeronasalis
  l ductus vomeronasalis

4949 vomeronasal nerve
  f nerf voméro-nasal
  e nervio vomeronasal
  d Nervus vomeronasalis
  l nervus vomeronasalis

4950 vomeronasal organ
  f organe voméro-nasal; organe de Jacobson
  e órgano vomeronasal; órgano de Jacobson

  d Nasenbodenorgan; Jacobsonsches Organ
  l organum vomeronasale

4951 vomiting
  f vomissement
  e vómito
  d Erbrechen

* vomiting and wasting disease → 3556

4952 vomitoxin; deoxynivalenol; Fusarium roseum toxin
  f vomitoxine; désoxynivalénol
  e vomitoxina
  d Vomitoxin; Deoxynivalenol

4953 vorticose veins
  f veines vorticineuses
  e venas vorticosas
  d Venae vorticosae
  l venae vorticosae

4954 vulva
  f vulve
  e vulva
  d Vulva; Scham
  l vulva

4955 vulvovaginitis
  f vulvovaginite
  e vulvovaginitis
  d Vulvovaginitis

W

4956 wall of hoof; horn plate
 f paroi de l'onglon; paroi cornée
 e pared córnea
 d Hufwand; Hornplatte
 l paries corneus

* **warble fly infestation** → 2083

* **wasting** → 1444

4957 water metabolism disorder
 f trouble du métabolisme hydrique
 e trastorno del metabolismo del agua
 d Störung im Wasserhaushalt

4958 wattles
 f barbillons
 e barbillas
 d Kehllappen
 l palea

4959 weakness
 f faiblesse
 e debilidad
 d Schwäche

4960 Wesselsbron flavivirus
 f flavivirus de Wesselsbron
 e flavivirus de Wesselsbron
 d Wesselsbron-Virus

* **Wharton's duct** → 2667

4961 whipworm
 f trichocéphale
 e verme látigo
 d Peitschenwurm
 l Trichuris

* **whirling disease of trout** → 2922

4962 whiskers; labial tactile hairs
 f moustache; poils tactiles labiaux
 e pelos labiales
 d Schnurhaare; Barthaare; Lippenhaare
 l pili labiales

* **white blood cell** → 2503

4963 white commissure
 f commissure blanche
 e comisura blanca
 d weiße Kommissur
 l commissura alba

4964 white heifer disease
 f maladie des génisses blanches
 e enfermedad de las novillas blancas
 d Weißfärsenkrankheit

4965 white line of abdomen
 f ligne blanche
 e línea blanca
 d weiße Linie
 l linea alba abdominis

* **white line of hoof** → 4968

4966 white matter
 f substance blanche
 e sustancia blanca
 d weiße Substanz
 l substantia alba

4967 white pulp of spleen
 f pulpe blanche de la rate
 e pulpa blanca del bazo
 d weiße Milzpulpa
 l pulpa lienis alba

4968 white zone; white line of hoof
 f ligne blanche
 e zona blanca
 d weiße Linie
 l zona alba

* **windpipe** → 4615

* **wind sucking** → 86

4969 wing
 f aile
 e ala
 d Flügel
 l ala

* **wing feather** → 1722

4970 wing of atlas; transverse process of atlas
 f aile de l'atlas
 e ala del atlas; apófisis transversa del atlas
 d Atlasflügel
 l ala atlantis; processus transversus atlantis

4971 wing of basisphenoid bone; temporal wing of sphenoid bone
 f aile du basisphénoïde

e ala del hueso basiesfenoides
d Temporalflügel des Keilbeins
l ala ossis basisphenoidalis

4972 wing of ilium
f aile iliaque
e ala del hueso ilion
d Darmbeinflügel
l ala ossis ilii

4973 wing of nostril
f aile du nez
e ala de la nariz
d Nasenflügel
l ala nasi

4974 wing of presphenoid bone; orbital wing of sphenoid bone
f aile du présphénoïde
e ala del hueso presfenoides
d Orbitalflügel des Keilbeins
l ala ossis presphenoidalis

4975 wing of sacrum
f aile du sacrum
e ala del hueso sacro
d Kreuzbeinflügel
l ala ossis sacri

* **wish bone** → **928**

4976 withdrawal period; withholding time
f temps d'attente; délai d'attente
e tiempo de espera; pazo de retirado
d Karenzzeit; Wartezeit

* **withholding time** → **4976**

* **Wohlfahrtia myiasis** → **4977**

4977 wohlfahrtiosis; Wohlfahrtia myiasis
f wohlfahrtiose

e wohlfahrtiosis
d Wohlfahrtiose

4978 wolffian duct; mesonephric duct
f canal mésonéphrique; canal de Wolff
e conducto mesonéfrico (de Wolff)
d Wolffscher Gang; Urnierengang
l ductus mesonephricus

4979 wolf tooth; first premolar
f dent de loup; première prémolaire
e diente del lobo
d Wolfszahn; erster Prämolar
l dens lupinus

* **wooden tongue of cattle** → **51**

4980 wool fibre
f fibre de laine
e fibra de la lana
d Wollfaser

4981 wool hair
f jarre
e pelo de la lana
d Wollhaar
l pili lanei

4982 worm burden
f nombre d'helminthes
e madeja de vermes
d Wurmbürde

* **worm infestation** → **1985**

4983 wound
f plaie
e herida
d Wunde

* **Wrisberg's tubercle** → **1193**

* **wrist** → **692**

X

4984 xenobiotic; foreign substance
f xénobiotique
e xenobiótico
d Xenobiotikum; Fremdstoff

4985 xenodiagnosis
f xénodiagnose
e xenodiagnóstico
d Xenodiagnose

4986 xiphisternum; xiphoid process
f xiphisternum; processus xiphoïde

e apófisis xifoidea
d Processus xiphoideus; Schwertfortsatz
l processus xiphoideus

* xiphoid process → 4986

4987 xiphoid region
f région xiphoïde
e región xifoidea
d Schwertfortsatzgegend
l regio xiphoidea

4988 xylazine
f xylazine
e xilazina
d Xylazin
l xylazinum

Y

4989 yeasts
f levures
e levaduras
d Hefepilze; Sproßpilze

4990 yellow bone marrow
f moelle jaune
e médula ósea amarilla
d gelbes Knochenmark
l medulla ossium flava

* **yellow fat disease** → **4198**

4991 yellow ligaments; interarcuate ligaments
f ligaments jaunes
e ligamentos amarillos
d Ligamenta flava
l ligamenta flava

4992 yellow spot of retina
f tache jaune de la rétine
e mácula lútea
d gelber Fleck der Netzhaut
l macula lutea retinae

4993 yellow tunic of abdomen
f tunique jaune de l'abdomen
e túnica amarilla abdominal
d gelbe Bauchhaut
l tunica flava abdominis

* **Yersinia infection** → **4994**

* **Yersinia pseudotuberculosis infection** →
  3662

* **Yersinia ruckeri infection** → **1475**

4994 yersiniosis; Yersinia infection
f yersiniose; infection à Yersinia
e yersiniosis
d Yersiniose; Yersinia-Infektion

* **yolk** → **4937**

Z

4995 zearalenone; Fusarium roseum toxin
 f zéaralénone
 e zearalenona
 d Zearalenon

4996 zinc deficiency
 f carence en zinc
 e carencia de cinc
 d Zinkmangel

4997 zinc poisoning
 f intoxication par le zinc; zincisme
 e intoxicación por cinc
 d Zinkvergiftung

 * **Zinn's membrane → 917**

4998 zonary placenta
 f placenta zonaire
 e placenta zonal
 d Gürtelplazenta
 l placenta zonaria

4999 zoonosis
 f zoonose
 e zoonosis
 d Zoonose

5000 zygomatic arch
 f arcade zygomatique
 e arco cigomático
 d Jochbeinbogen
 l arcus zygomaticus

5001 zygomatic bone
 f os zygomatique
 e hueso cigomático
 d Jochbein
 l os zygomaticum

5002 zygomatic muscle
 f muscle zygomatique
 e músculo cigomático
 d Jochbeinmuskel
 l musculus zygomaticus

5003 zygomatic nerve
 f nerf zygomatique
 e nervio cigomático
 d Nervus zygomaticus; Jochbeinnerv
 l nervus zygomaticus

5004 zygomatic process
 f processus zygomatique
 e apófisis cigomática
 d Processus zygomaticus
 l processus zygomaticus

5005 zygomatic region
 f région zygomatique
 e región cigomática
 d Jochgegend
 l regio zygomatica

5006 zygomatic salivary gland; orbital gland
 f glande zygomatique
 e glándula cigomática; glándula orbitaria
 d maxillare Backendrüse
 l glandula zygomatica; glandula orbitalis

5007 zygomatic tactile hairs
 f poils tactiles zygomatiques
 e pelos zigomáticos
 d Jochbogentasthaare
 l pili zygomatici

5008 zymogen
 f zymogène
 e cimógeno
 d Zymogen

5009 zymosan
 f zymosan
 e cimosana
 d Zymosan

Français

abaisseur de la lèvre inférieure 1265
abattage d'urgence 705
abcès 19
abdomen 1
aberration du goût 3466
abomasite 16
abomasum 17
absence congénitale du cœur 24
absence de développement 100
absence de granulocytes 105
absence de poils 321
absence des membres 138
absorption intestinale 2308
acanthocéphales 20
acanthocéphalose 21
acanthose 22
acardie 24
acaricide 25
acarien 2846
acarien auriculaire du chien 1416
acarien responsable de la gale déplumante 1264
acarien sous-cutané des volailles 4270
acariose trachéale des abeilles 23
acaudie 4434
acétabulum 39
acétonémie 40
acétonurie 41
achalasie 42
acheilie 43
achondroplasie 44
acide ascorbique 4936
acide nalidixique 2927
acide nicotinique 3009
acide oxolinique 3213
acide ribonucléique 3870
acidose 47
acrodermatite 49
acromion 50
ACTH 1096
actinobacillose 51
actinomycètes 52
actinomycose 53
acuariose 56
acupuncture 57
adénite 66
adénocarcinome 67
adénohypophyse 68
adénomatose 70
adénomatose pulmonaire ovine 3689
adénome 69
adénovirose 72
adénovirus 71
adénovirus aviaire 346

adénovirus bovin 525
adénovirus canin 638
adhérence 73
administration orale 3136
administration parentérale 3317
adrénaline 1502
adventice 79
ægyptianellose aviaire 82
aérobie 83
aérocystite 84
aéromonose par Aeromonas salmonicida 85
aérophagie 86
aérosacculite 107
afébrile 88
affection articulaire 283
affection systémique 4423
aflatoxicose 89
aflatoxine 90
agalactie 98
agalaxie 98
agalaxie contagieuse 1040
agammaglobulinémie 99
agénésie 100
agent antibactérien 204
agent antifongique 218
agent antimétéorisant 208
agent antiviral 232
agent de la tremblante du mouton 4014
agent pathogène 3345
agglutination 101
agglutination en tube 4693
agglutination sur lame 4104
agnathie 104
agranulocytose 105
aigu 58
aiguille hypodermique 2217
aile 4969
aile de l'atlas 4970
aile du basisphénoïde 4971
aile du nez 4973
aile du présphénoïde 4974
aile du sacrum 4975
aile iliaque 4972
aile postacétabulaire de l'ilium 3563
aile préacétabulaire de l'ilium 3587
aine 1903
aire acoustique 2184, 4332
aire vestibulaire inférieure 2184
aire vestibulaire supérieure 4332
aisselle 369
albendazole 111
albinisme 112
albuminurie 113

alcaloïde 119
alcaloïde de l'ergot de seigle 1536
alcaloïde pyrrolizidinique 3727
alcalose 120
aldostérone 114
alfaprostol 116
alfaxalone 117
aliment médicamenteux 2742
allanto-chorion 884, 885
allantoïde 121
allergène 122
allergie 125
allergisant 123
allocortex 126
allotriophagie 3466
alopécie 127
alphavirus 128
altération pathologique 3349
altération post-mortem 3568
altrénogest 130
alvéole dentaire 4601
alvéoles pulmonaires 3690
alvéus de l'hippocampe 135
amaigrissement 1444
amantadine 136
amélie 138
amicarbalide 140
amidostomose 141
amikacine 142
amitraz 143
amnios 144
amoxicilline 145
amphiarthrose 146
amphotéricine B 147
ampicilline 148
ampoule de la trompe utérine 151
ampoule du bréchet 570
ampoule du canal déférent 150
ampoule rectale 3772
amprolium 149
amputation de la queue 1345
amygdale 4594
amygdale linguale 2528
amygdale palatine 3232
amygdale pharyngienne 3443
amygdalite 4597
amyloïdose 154
amylose 154
amyotrophie 155
anaérobie 158
analeptique 163
analgésique 164
analyse biologique 449
anaphylaxie 169
anaplasmose 170
anastomose 171
anastomose artério-veineuse 276

bartonellose 410
base de la corne 414
base de la queue 4433
base du cœur 413
basihyal 415
basipode 419
bassin 3373
bassinet 3817
bêta-bloquant 435
bêta-hémolytique 436
bêtaméthasone 437
bec 421
bec de l'olécrâne 174
bec inférieur 2579
bec supérieur 4755
benzathine benzylpénicilline
    428
benzimidazole 429
benzocaïne 430
benzylpénicilline 431
berne 4686
besnoïtiose 433
bézoard 438
bifurcation de la trachée 4616
bile 445
bilirubine 448
biocénose 450
biologie moléculaire 2851
biopsie 452
biotechnologie 453
birnavirus 455
bithionol 457
blastomycose nord-américaine
    3025
blépharite 464
blessure 2218
blocage 3055
boîte crânienne 4102
boiterie 2412
boldénone 509
bord coronal de la paroi du
    sabot 1074
bord cranial du tibia 4571
bord soléaire du sabot 1905
bord supra-orbitaire de l'os
    frontal 4352
borréliose 523
bosselures du cæcum 618
botulisme 524
bouche 2868
boulet 2805
boulet des ongulés 2791
bourgeon du goût 4455
bourrelet 1085
bourrelet acétabulaire 38
bourrelet glénoïdal 1850
bourse 4020
bourse cloacale (de Fabricius)
    950
bourse omentale 3109

bourse ovarienne 3192
bourse podotrochléaire 3528
bourse sous-cutanée 4268
bourse sternale 4202
bourse synoviale 4409
Bovicola bovis 713
brachygnathie 564
branche de la mandibule 3768
branche latérale de la
    fourchette 2460
branche médiale de la
    fourchette 2725
branchiomycose 567
bras 4754
bréchet 2368
bromociclène 576
bromocriptine 577
bromophos 578
bromure de pyritidium 3725
bromure d'homidium 2024
bronche 590
bronches lobaires 2547
bronches segmentaires 4030
bronchestasie 584
bronchite 585
bronchite infectieuse aviaire
    352
broncho-pneumonie 588
bronchospasme 581
brotianide 591
brucelline 593
brucellose 594
brucellose bovine 527
brucellose ovine 3206
brucellose porcine 3552
buiatrie 602
bulbe oculaire 1603
bulbe olfactif 3094
bulbe pileux 1943
bulbe rachidien 2746
bulle de la syrinx 4419
bulle lacrymale 2395
bulle tympanique 4709
bunamidine 607
bunostomose 608
bunyavirus 609
buparvaquone 610
bupivacaïne 611
buquinolate 612
bursectomie 613
bursite 614
bursite infectieuse aviaire 2174
bursite sternale 570
buséréline 615

cæcum 621
caillette 17
caillot 472
calamus 624
calcaneum 1692

calcinose 627
calcul 628
calcul biliaire 1795
calcul rénal 3807
calcul urinaire 4768
calice rénal 3808
calicivirus 633
calicivirus de l'exanthème
    vésiculaire 4886
calicivirus félin 1640
calliphore 493
cambendazole 634
campylobactériose 635
canal 1398
canal anal 162
canal artériel (de Botal) 272
canal central de la moelle
    épinière 792
canal cervical 829
canal cholédoque 1003
canal déférent 1238
canal de Gärtner 1804
canal de Havers 3046
canal de His 4556
canal de l'épendyme 792
canal de Santorini 32
canal de Schlemm 4841
canal de Sténon 3326
canal de Wharton 2667
canal de Wirsung 2641
canal de Wolff 4978
canal du nerf hypoglosse 2086
canal éjaculateur 1436
canal fémoral 1656
canal galactophore 2404
canal hépatique commun 1007
canal hépatique droit 3889
canal hépatique gauche 2479
canalicule 636, 1401
canalicules aberrants 13
canalicules alvéolaires 131
canalicules biliaires 447
canalicules efférents du
    testicule 1432
canalicules excréteurs de la
    glande lacrymale 1562
canalicules interlobulaires 2245
canalicules prostatiques 1402
canal infra-orbitaire 2192
canal inguinal 2210
canal mésonéphrique 4978
canal naso-lacrymal 2950
canal nourricier de l'os 3046
canal omasique 3103
canal omphalo-mésentérique
    3114
canal paramésonéphrotique
    (de Mueller) 3286
canal pylorique 3710

canal rachidien 4874
canal spiral de la cochlée 4153
canal supra-orbitaire 4349
canal thoracique 4536
canal vitellin 3114
canaux semi-circulaires 4035
canaux semi-circulaires osseux 4034
candidose 637
cannibalisme 647
canule de trayon 4458
capelet 652
capillaires aériens 3521
Capillaria 1953
capillariose 648
capripoxvirus 655
capsule adipeuse du rein 74
capsule articulaire 2358
capsule de l'onglon 2026
capsule de Tenon 4065
capsule du cristallin 2489
capsule du glomérule 1853
capsule fibreuse 1685
capture des animaux 656
capuchon céphalique du spermatozoaire 1962
carazolol 657
carbachol 658
carbadox 659
carbaryl 660
carcinome 666
cardia 667
cardiomégalie 675
cardiopathie 1970
cardiovirus 678
carence 1240, 3048
carence en cuivre 1056
carence en iode 2334
carence en magnésium 2640
carence en oligoéléments 4614
carence en sélénium 4031
carence en vitamines 367
carence en zinc 4996
carence minérale 2838
carène sternale 2368
carie dentaire 1253
caroncule 698
caroncule cutanée 1196
caroncule lacrymale 2396
caroncule sublinguale 4278
carpe 692
carpo-métacarpe 691
cartilage articulaire 286
cartilage aryténoïde 299
cartilage auriculaire 1021
cartilage costal 1101
cartilage cricoïde 1170
cartilage de la cloison nasale 2945
cartilage de la conque 1021

cartilage de la troisième phalange 696
cartilage de l'omoplate 4000
cartilage du cœur 695
cartilage du manubrium sternal 2681
cartilage du nez 2933
cartilage élastique 1437
cartilage épiphysaire 1503
cartilage fibreux 1682
cartilage hyalin 2056
cartilage scapulaire 4000
cartilage scutiforme 4022
cartilages de la syrinx 4420
cartilages du larynx 2425
cartilages trachéaux 4617
cartilage thyroïdien 4559
cartilage unguéal 696
casque 515
castration 704
castré 703
castrer 702
cataracte 706
catarrhe 707
catécholamine 710
cathéter 712
cautère 770
cautérisation 770
cavité abdominale 5
cavité articulaire 2359
cavité buccale 3137
cavité crânienne 1127
cavité de la tige pituitaire 774
cavité de l'hypophyse 2090
cavité glénoïde 1849
cavité laryngée 2426
cavité médullaire 2747
cavité nasale 2934
cavité orale 3137
cavité orbitaire 3145
cavité pelvienne 3359
cavité péricardique 3389
cavité péritonéale 3411
cavité pharyngée 3437
cavité pleurale 3510
cavité pulpaire de la dent 3697
cavité thoracique 4535
cavité utérine 4783
cavum épidural 1498
cavum sous-arachnoïdien 4256
cavum subdural 4272
cavum tympanique 4711
cécité 465
céfaléxine 775
céfaloridine 776
céfalotine 777
céfotaxime 778
ceinture du membre pelvien 3363

ceinture du membre thoracique 4084
cellule 779
cellule hépatique 1991
cellule interstitielle 2295
cellule lymphoïde 2628
cellule osseuse 3173
cellule plasmatique 3504
cellule sanguine 470
cellules ethmoïdales osseuses 1547
cellule somatique 4117
cellules réticulaires 787
cément de la dent 1254
centre tendineux du périnée 798
cénure 976
céphalosporine 803
cératohyal 804
cerveau 823
cerveau antérieur 3637
cerveau moyen 2771
cerveau postérieur 3864
cerveau terminal 4465
cervelet 811
cervicite 838
cestode 841
cestodose 842
cétonémie 40
cétonurie 41
cétose 2374
cétrimide 843
châtré 703
châtrer 702
chambre antérieure de l'œil 199
chambre postérieure du bulbe de l'œil 3564
chambre tympanique 4708
chambre vitrée du bulbe 4940
champignon 1787
chanfrein 571
chaponnage 650
charbon bactéridien 202
charbon symptomatique 462
charnière imparfaite 1023
chélateur 850
chiasma optique 3130
chimioprophylaxie 851
chimiotactisme 852
chimiothérapeutique 853
chimiothérapie 855
chique 2357
chirurgie 4370
chlamydie 861
chlamydiose 862
chlamydiose ovine 1484
chloralose 864
chloramphénicol 865
chlorfenvinphos 954

Eomenacanthus stramineus 3576
éosinophilie 1487
épaule 4083
épendyme 1488
éperon 1535
épérythrozoonose 1489
épicondyle 1491
épi convergent 1948
épidémie 1493
épidémiologie 1494
épidémique 1492
épiderme 1495
épididyme 1496
épididymite 1497
épi divergent 1947
épiglotte 1501
épine iliaque 2116
épinéphrine 1502
épine scapulaire 4147
épiphyse 1505
épiphyse cérébrale 3476
épiploon 1507
épistaxis 1508
épithalamus 1509
épithélium 1510
épithélium germinal 1833
épithélium pigmentaire 3473
épizootie 1513
épizootique 1512
époophoron 1515
épreuve de fixation du complément 1018
épreuve d'hémagglutination 1920
épreuve diagnostique 1280
épreuve d'inhibition de l'hémagglutination 1919
épreuve intradermique 2324
épreuve virulente 846
équilibre acido-basique 45
éradication 1531
ergocalciférol 1533
ergométrine 1534
ergot 1276, 1535
ergotisme 1537
ergot métatarsal 4187
érosion du gésier 1839
érythroblastose 1539
érythroblastose aviaire 350
érythrocyte 1540
érythrocytose 1542
érythromycine 1543
érythropoïèse 1544
espace épidural 1498
espace intercostal 2233
espace interdigital 2240
espace rétropéritonéal 3854
espace sous-arachnoïdien 4256
essai contrôlé 1051

estomac 4226
estomac du ruminant 3951
estradiol 3083
estrone 3086
état fébril 1676
ethmoturbinaux 1554
étiologie 87
étisazole 1555
étomidate 1556
étrier 4195
étude contrôlée 1051
euthanasie 1558
évolution bénigne 426
examen clinique 943
exanthème 1559
exanthème coïtal équin 1518
exanthème vésiculaire porcin 4887
excreta 1560
excroissance osseuse 3179
exophtalmie 1566
exostose 1567
exostose métacarpienne 4174
exostose phalangienne 3900
exploration rectale 3775
extrémité 1600
extrémité caudale de la vertèbre 725
extrémité crâniale de la vertèbre 1131
extrémité du nez 245

face 1608
face occlusale de la dent 3069
facette articulaire 295
facteur causal 769
facteur favorisant 3591
facteur létal 2501
facteur prédisposant 3591
faiblesse 1219, 4959
faisceau atrio-ventriculaire de His 322
faisceau cunéiforme (de Burdach) 1622
faisceau de Deiters 4907
faisceau de Flechsig 4862
faisceau de Gowers 1379
faisceau gracile (de Goll) 1623
faisceau longitudinal médial 2719
faisceau mamillo-hypothalamique 2658
faisceau médial du télencéphale 2717
faisceau rétroflexe de Meynert 2814
faisceau unciné 4744
faisceaux propres 1621
fanon 1277
fanon métacarpien 1674

fanon métatarsien 1674
fascia 1620
fascia antébrachial 197
fascia axillaire 371
fascia brachial 555
fascia cervical 832
fascia iliaque 2113
fascia nuchal 3034
fascia orbitaire 3146
fascia pectoral 3354
fascia pelvien 3361
fascia thoraco-lombaire 4546
fascioloïdose 1624
fasciolose 1625
fatigue de la pondeuse en cage 623
faux du cerveau 1619
favus 1631
fébantel 1638
fébrile 1639
fèces 1617
fémur 1665, 4521
fenbendazole 1666
fenclophos 1667
fenêtre de la cochlée 3930
fenêtre du vestibule 3190
fenêtre ovale 3190
fenêtre ronde 3930
fenprostalène 1668
fentanyl 1669
fente de la glotte 1859
fente orbitaire 3147
fente palpébrale 3248
fesse 616
fessier moyen 2830
fessier profond 1228
fessier superficiel 4324
feuillet 3108
fibre 1677
fibre de laine 4980
fibre musculaire 2880
fibre nerveuse 2983
fibres arquées du cerveau 261
fibres cortico-réticulaires 1094
fibres du cristallin 2490
fibres périventriculaires 3415
fibres pontiques transverses 4636
fibres tecto-spinales latérales 2464
fibrille 1678
fibrine 1679
fibrinogène 1680
fibroblaste 1681
fibrocartilage 1682
fibrome 1684
fibula 1690
fiel 445
fièvre 1676
fièvre aphteuse 1740

hormone thyroïdienne 4562
humérus 2054
humeur aqueuse 255
humeur vitrée 4941
huppe 3518
hyaluronidase 2058
hydatide sessile de Morgagni 4496
hydatidose 1423
hydrargyrisme 2767
hydrate de chloral 863
hydrocéphalie 2059
hydrocortisone 2060
hydrothorax 2061
hydroxynaphtoate de béphénium 432
hygiène 2062
hygiène alimentaire 1736
hygiène de la viande 2704
hygiène du lait 2837
hygiène vétérinaire 4911
hygroma 2063
hygroma du coude 651
hygroma du tarse 652
hyménolépiose 2064
hyostrongylose 2069
hypercalcémie 2070
hyperfonctionnement 2071
hyperglycémie 2072
hyperimmunisation 2074
hyperlipémie 2075
hyperplasie 2076
hypersensibilité 2077
hyperthermie 2078
hyperthermie maligne du porc 2651
hypertrophie 2079
hypertrophie cardiaque 669
hyphomycose 2080
hypocalcémie 2081
hypoderme 4271
hypoderme du bœuf 1012
hypoderme rayé 4109
hypodermose 2083
hypogammaglobulinémie 2084
hypoglycémie 2088
hypomagnésémie 2089
hypopenne 97
hypophyse 2092
hypoplasie 2093
hypoplasie myofibrillaire 2914
hypoplasie testiculaire 4498
hypoprotéinémie 2094
hypostase 2095
hypothalamus 2096
hypothermie 2097
hypotrichose 2098
hypoxie 2099
hystérectomie 2100

îlots pancréatiques (de Langerhans) 3254
ictère 2353
idoxuridine 2103
Ig 2147
iléon 2112
ilion 2133
imidocarbe 2135
immun 2136
immuncomplexe 2137
immunisation 2141
immunité 2140
immunité active 54
immunité à médiation cellulaire 784
immunité colostrale 996
immunité croisée 1180
immunité naturelle 2956
immunité passive 3339
immunodéficience 2142
immunodiffusion 2143
immunofluorescence 2144
immunogène 2145
immunogénicité 2146
immunoglobuline 2147
immunologie 2150
immunologique 2148
immunopathologie 2149
immunostimulation 2151
immunostimulation non spécifique 3021
immunosuppresseur 2153
immunosuppression 2152
immunsérum 2139
inappétence 2155
inapte à la reproduction 4749
incisure 2161
incisure cardiaque du poumon 672
incisure de l'acétabulum 37
incisure fibulaire 1691
incisure glénoïdale 1851
incisure jugulaire 2364
incisure mandibulaire 2675
incisure radiale 3748
incisure scapulaire 2162
incisures costales du sternum 1103
incisure trochléaire 4682
incisure ulnaire 4733
incisure vertébrale 4878
inclinaison pelvienne 2163
incontinence urinaire 4769
incoordination 2165
index chimiothérapeutique 854
indigestion ingluviale 1178
inertie utérine 4787
infarctus 2168
infecté 2169
infectieux 2171

infection 2170
infection à Actinobacillus equuli 4074
infection à Actinobacillus pleuropneumoniae 3559
infection à Actinomyces bovis 53
infection à Arizona 266
infection à Bacillus larvae 139
infection à Bacillus piliformis 4720
infection à Bacteroides nodosus 2177
infection à Blastomyces dermatitidis 3025
infection à Borrelia 523
infection à Borrelia burgdorferi 2611
infection à Borrelia gallinarum 362
infection à Branchiomyces 567
infection à Brucella suis 3552
infection à Campylobacter 635
infection à Candida 637
infection à Chlamydia 862
infection à Chlamydia psittaci 3162
infection à Chrysosporium 1958
infection à Clostridium chauvoei 462
infection à Clostridium haemolyticum 385
infection à Clostridium novyi 460, 2652
infection à Clostridium perfringens 959
infection à Clostridium perfringens type A 1805
infection à Clostridium septicum 462
infection à Coccidioides immitis 964
infection à Corynebacterium ovis 700
infection à Corynebacterium pseudotuberculosis 4726
infection à Cowdria ruminantium 1975
infection à Coxiella burnetii 3728
infection à Cryptococcus neoformans 1186
infection à Cytauxzoon 1211
infection à Cytoecetes ondiri 3121
infection à Cytophaga columnaris 999
infection à Cytophaga psychrophila 1702

mélatonine 2755
mélioïdose 2756
mélophage du mouton 4070
membrane 2757
membrane alaire caudale 3574
membrane alaire crânienne 3632
membrane basilaire du conduit cochléaire 417
membrane cellulaire 785
membrane chorio-allantoïdienne 884
membrane de Bowman 200
membrane de Descemet 3565
membrane du tympan 4712
membrane hyaloïde 4942
membrane interosseuse 2286
membrane nictitante 4526
membrane pupillaire 3701
membranes fœtales 1671
membrane spirale du conduit cochléaire 4156
membrane sterno-coraco-claviculaire 4210
membrane synoviale 4412
membrane vestibulaire (de Reissner) 3804
membrane vitrée 4942
membre 2517
membre antérieur 4538
membre pelvique 3365
membre postérieur 3365
membre thoracique 4538
méninges 2760
méningite 2761
méningo-encéphalite 2762
ménisque articulaire 289
ménopon 3580
menton 858
mérozoïte 2769
mésencéphale 2771
mésenchyme 2772
mésentère 2774
mésocolon 2776
mésoduodénum 2777
mésomètre 2778
mésonéphros 2779
mésorchium 2780
mésorectum 2781
mésosalpinx 2782
mésovarium 2783
métacarpe 2792
métacercaire 2793
métallibure 2794
métapatagium 2795
métaphyse 2796
métastrongylose 2797
métatarse 2806
métencéphale 2807
météorifuge 208

méthadone 2808
méthémoglobine 2809
métomidate 2810
métrifonate 2811
métrite 2812
métrite contagieuse équine 1046
métronidazole 2813
microbe 2821
microbien 2815
microbiologie 2816
microcéphalie 2817
microcoque 2818
microgamète 2819
microgamonte 2820
micro-organisme 2821
microphtalmie 2822
microscopie 2823
microsporidies 2824
milieu de culture 1191
minéralocorticoïde 2840
miracidium 2845
mise bas 3334
mitochondries 2847
modificateur cardiovasculaire 676
modiolus 2848
moelle allongé 2746
moelle épinière 4138
moelle jaune 4990
moelle osseuse 513
moelle rouge 3791
mollet 631
molybdénose 2853
monensin 2854
moniéziose 2855
morantel 2858
morbillivirus 2859
morphologie 2861
morphologique 2860
morsure 456
mort 1218
mortalité 2862
mortalité embryonnaire 1449
mort cardiaque 1971
mort du fœtus 1670
mortinatalité 4224
mort-né 4225
mort subite 4296
morve 1841
mouche 1728
mouche d'automne 1609
mouche des cornes 2035
mouche domestique 2050
moucheron 1868
mouche verte 1899
moustache 4962
moustique 2864
MSH 2752
mucormycose 2870

mucosité 2874
mucus vaginal 4806
muelleriose 2875
mufle 2949
muqueuse 2871
muqueuse nasale 2940
muqueuse orale 3140
muscle 2879
muscle abaisseur de la mandibule 2666
muscle abducteur caudal de la jambe 718
muscle abducteur crânial de la jambe 1125
muscle abducteur de l'index (doigt II) 12
muscle accessoire du gésier 4523
muscle adducteur 61
muscle adducteur caudal de la mandibule 737
muscle adducteur court 4076
muscle adducteur des rectrices 64
muscle adducteur du doigt I 63
muscle adducteur du doigt II 65
muscle adducteur du doigt V 62
muscle adducteur externe de la mandibule 1584
muscle adducteur long 2559
muscle anconé 175
muscle articulaire 290
muscle articulaire de la hanche 291
muscle articulaire de l'épaule 292
muscle articulaire du genou 293
muscle aryténoïdien transverse 4634
muscle biceps du bras 441
muscle biceps fémoral 442
muscle bipenne 454
muscle brachial 556
muscle brachio-céphalique 560
muscle brachio-radial 563
muscle broncho-œsophagien 587
muscle buccinateur 600
muscle bulbo-caverneux 604
muscle bulbo-spongieux 604
muscle canin 642
muscle cardiaque 2912
muscle carré des lombes 3731
muscle carré fémoral/crural 3732
muscle carré pronateur 3733
muscle caudo-fémoral 766

partie tympanique de l'os
   temporal 4715
parturition 3334
parvovirose canine 644
parvovirus 3336
parvovirus de la panleucopénie
   féline 1649
parvovirus porcin 3558
passalurose 3338
pasteurellose 3340
pasteurellose aviaire 1754
pasteurellose bovine 1939
pathogène 3347
pathogenèse 3346
paume de la main 3244
paupière 1606
paupière inférieure 2580
paupière supérieure 4756
pavillon de l'oreille 3478
peau 4098
pecten du pubis 3352
pédicule du poumon 3908
pédoncule 3357
pédoncule cérébral 816
pédoncules cérébelleux 809
pelade 127
pelvis rénal 3817
pénicillaminc 3374
pénicillines 3375
pénicilline V 3449
pénis 3379
pennes de contour 1049
pentamidine 3381
pentastome 3382
pentobarbital 3383
percutané 3385
péricarde 3391
péricarde fibreux 1686
péricarde séreux 4055
péricardite 3390
périlymphe 3392
périmétrium 3394
périnatal 3395
périnée 3401
période post-natale 3572
période puerpérale 3688
périople 3403
périorbite 3406
périoste 3407
périostite 3408
péripneumonie 3514
péripneumonie contagieuse
   bovine 1041
péritoine 3412
péritoine pariétal 3321
péritoine viscéral 4933
péritonite 3413
péritonite virale du chat 1645
périvasculaire 3414
perméthrine 3417

péroné 1690
pérose 3418
pessulus 3420
peste aviaire 355
peste bovine 3898
peste des petits ruminants 3492
peste du canard 1395
peste équine 91
peste porcine africaine 93
peste porcine classique 4385
pesticide 3421
pestivirose ovine 520
pestivirus 3422
pestivirus de la diarrhée virale
   bovine 528
pétéchial 3424
pétéchie 3423
petit colôn 4105
petite circonférence de l'iris
   2498
petite douve du foie 2418
petit nerf occipital 4284
petit omentum 2497
phagocytaire 3429
phagocyte 3428
phagocytose 3430
phalange 3431
phalange distale 1340
phalange moyenne 2835
phalange proximale 1705
phallus 3432
pharmacie 3436
pharmacocinétique 3434
pharmacodynamie 3433
pharmacologie 3435
pharmacothérapie 1392
pharyngite 3445
pharynx 3446
phénobarbital 3447
phénothiazine 3448
phénoxyméthylpénicilline 3449
phénylbutazone 3450
phimosis 528
phlébite 3452
phlébotome 3984
phosphatase 3453
phospholipide 3454
phosphoprotéine 3455
photosensibilisation 3457
phtiriase 2578
phycomycose 3461
physiologie 3463
physiologique 3462
physocéphalose 3464
phytohémagglutinine 3465
pica 3466
picornavirus 3467
pied 1739
pie-mère de l'encéphale 817
pie-mère spinale 4143

piétin 2177
pigment biliaire 446
pilier 3475
pilier accessoire du rumen 33
pilier caudal du rumen 751
pilier coronaire du rumen 1082
pilier crânial du rumen 1150
pilier du diaphragme 2592
pilier longitudinal du rumen
   2566
piliers du rumen 3946
pince à castrer 1445
pince de l'onglon 4587
pipérazine 3479
piqûre d'insecte 2221
pis 4721
pituitaire 3483
placebo 3486
placenta 3487
placenta diffus 1305
placenta fœtal 1672
placenta maternel 2695
placenta multiplex 2877
placenta zonaire 4998
placentome 3490
plage 3502
plagiorchiose 3491
plaie 4983
plan nasal 2942
plan naso-labial 2949
plan rostral 3921
plante médicinale 2744
plante toxique 3532
plaque incubatrice 2166
plaques de Peyer 103
plaquettaire 4548
plaquette sanguine 479
plasma sanguin 478
plasmocyte 3504
platysma 3507
plérocercoïde 3508
pleurésie 3512
pleurite 3512
pleuropneumonie 3514
pleuropneumonie contagieuse
   caprine 1042
pleuropneumonie infectieuse
   porcine 3559
plèvre 3509
plèvre costale 1104
plèvre diaphragmatique 1284
plèvre médiastinale 2738
plèvre pariétale 3322
plèvre pulmonaire 3693
plexus 3515
plexus aortique abdominal 4
plexus aortique thoracique
   4533
plexus autonome 345
plexus brachial 557

rhinotrachéite féline 1650
rhinotrachéite infectieuse bovine 2172
rhinovirus 3863
rhombencéphale 3864
riboflavine 3869
ribosome 3871
rickettsie 3874
rickettsiose 3875
rictus 3876
rifampicine 3879
rifamycine 3880
rigidité cadavérique 3897
risque sanitaire 1967
robénidine 3903
rocher 3426
ronidazole 3904
rostre 421, 3923
rostre mandibulaire 2579
rostre maxillaire 4755
rotavirus 3925
rotule 3341
rouget du porc 4384
roxarsone 3933
rumen 3935
ruménite 3938
ruménotomie 3939
rumination 3952
ruminoréticulum 3956
rupture 3958
rupture de l'aorte 241

sabot 2025
sac aérien abdominal 2
sac aérien cervical 825
sac aérien claviculaire 929
sac aérien thoracique caudal 755
sac aérien thoracique crânial 1155
sac anal 166
saccule 3959
sac dorsal du rumen 1373
sac endolymphatique 1466
sac lacrymal 2400
sacrum 3974
sacs aériens 108
sacs alvéolaires 134
sac ventral du rumen 4857
saignement du nez 1508
sain 1968
saisie des carcasses 662
salinomycine 3977
salivation 3980
salive 3978
salmonellose 3981
salmonellose des volailles 1758
salpingite 3983
sang 468
santé 1966

santé animale 189
saprolegniose 3988
sarcocystose 3989
sarcome 3990
sarcome de Rous 3931
sarcome de Sticker 4629
sarcophagide 1717
sarcopte 2678
sarcopte changeant 3996
sarcoptide 2678
sarcosporidiose 3989
satratoxine 3993
saturnisme 2471
schistosome 4003
schistosomose 4004
schizonte 4005
science vétérinaire 4917
scissure 1709
scissure interlobaire 2244
scissures cérébrales 4298
sclérose 4010
sclérotique 4009
scolex 4011
scoliose 4012
scrotum 4020
scutelle 4021
scutum 3994
séborrhée 4025
segments bronchopulmonaires 589
semence 4033
semiplume 4039
sénéciose 4042
sensibilité 4374
septicémie 4045
septicémie hémorragique à Pasteurella multocida 1939
septicémie hémorragique virale 4926
septum atrio-ventriculaire 323
septum interatrial 2224
septum interventriculaire 2304
septum lingual 2527
septum scrotal 4018
séreux 4052
seringue 4418
séroconversion 4046
sérodiagnostic 4048
sérologique 4047
séroneutralisation 4059
sérosite 4051
sérum 4056
sérum-albumine 4057
sérum antivenimeux 231
sérum hyperimmun 2073
sérum sanguin 484
service vétérinaire 4918
sétariose 4062
shigellose 4074
sillon 1904

sillon abomasal 15
sillon bicipital latéral 2440
sillon bicipital médial 2713
sillon caudal du rumen 733
sillon coronaire du rumen 1080
sillon coronal 1075
sillon coronarien 1079
sillon costal 1102
sillon crânial du rumen 1136
sillon cunéal central 794
sillon gastrique 1808
sillon infrapalpébral 2199
sillon intermammaire 2246
sillon interventriculaire 2303
sillon jugulaire 2362
sillon lacrymal 2398
sillon longitudinal du rumen 2564
sillon médian de la langue 2734
sillon mento-labial 2766
sillon omasal 3104
sillon paracunéal 988
sillon réticulaire 3836
sillon rumino-réticulaire 3954
sillon spiral externe 1594
sillon spiral interne 2279
simulie 461
sinciput 4093
sinus 4094
sinus caverneux 772
sinus coronarien 1083
sinus crâniens 1153
sinus de la dure-mère 1153
sinus frontal 1774
sinus inguinal du mouton 2213
sinus interdigital du mouton 2239
sinusite 4095
sinus lacrymal 2401
sinus lactifère 2405
sinus maxillaire 2701
sinus palatin 3231
sinus paranasaux 3290
sinus pétreux dorsal 1369
sinus pétreux ventral 4854
sinus sagittal dorsal 1374
sinus sagittal ventral 4858
sinus sous-orbitaire 2196
sinus sphénoïdal 4127
sinus sphéno-palatin 4131
sinus transverse de la dure-mère 4645
sinus urogénital 4775
sinus veineux 4840
sinus veineux de la sclère 4841
siphonaptère 1715
slafranine 4103
SNC 795
soie de porc 572
sole 4112

syndrome MMA 2692
syndrome urologique félin 1652
synéchie 4406
syngame des volailles 1803
syngamose 4407
synostose 4408
synovie 4410
synovite 4416
synovite infectieuse aviaire 354
synsacrum 4417
syrinx 4422
système endocrinien 1463
système lymphatique 2617
système majeur
    d'histocompatibilité 2642
système nerveux 2985
système nerveux autonome 344
système nerveux central 795
système nerveux
    parasympathique 3306
système nerveux périphérique
    3410
système orthosympathique
    4395
système réticulo-endothélial
    3839
système vasculaire 4826
systole 4424

T3 4671
tabanide 2045
tache jaune de la rétine 4992
tachycardie 4426
Taenia hydatigena 1348
talon 1980
talus 4574
taon 2045
tapetum lucidum 4442
tapis clair 4442
tarse 2021
tarse des paupières 4454
tarso-métatarse 4453
taux de mortalité 2863
taux de prévalence 3614
tête 1961
tête de la côte 1965
tête de l'épididyme 1963
tête fémorale 1964
tête vertébrale 1131
tectrice 499
tegmen mésencéphalique 2770
tégument commun 1010
télangiectasie 4464
télencéphale 4465
tempe 4466
température corporelle 508
température rectale 3778
temps d'attente 4976
tendinite 4478
tendon 4480

tendon calcanéen commun
    1004
tendon d'Achille 1004
tendovaginalite 4482
teneur en hémoglobine 1930
ténia 841
ténia à cénure 4430
ténia armé 3474
ténia des ovins 1765
ténia du bœuf 423
ténia du porc 3474
ténia échinocoque 1424
ténia inerme 423
ténias du côlon 4429
téniasis 4431
téniose 4431
ténosynovite 4482
tenseur du fascia lata 4484
tératogène 4485
tératogénicité 4486
tératologie 4487
tératome 4488
terminaisons nerveuses 2982
test de l'anneau 3901
testicule 4500
testostérone 4502
tétanie 4504
tétanie d'herbage 1892
tétanos 4503
tétracaïne 4505
tétracycline 4506
tétramisole 4507
thalamencéphale 4508
thalamus 4510
theilériose 4512
theilériose bovine à Theileria
    parva 1422
thélaziose 4513
thélite 2657
thèque des follicules ovariques
    4511
thérapeutique 4515
thériogénologie 4516
thermorégulation 4517
thiabendazole 4567
thiacétarsamide sodique 4518
thiamine 4519
thiopental sodique 4524
thorax 4547
thrombocytaire 4548
thrombocyte 479
thrombophlébite 4549
thrombose 4550
thymus 4553
thyrotrophine 4565
thyroxine 4566
tiabendazole 4567
tiaprost 4568
tibia 4569
tibiotarse 4578

tic aérophagique 86
tige de la plume 1636
tige du poil 1950
tige pituitaire 3484
tique 4579
tique brune du chien 592
tique de la poule 1757
tique des oreilles 1419
tissu 4583
tissu adipeux 75
tissu conjonctif 1036
tissu élastique 1438
tissu interstitiel 2297
tissu lymphoïde 2629
tissu musculaire 2883
tissu sous-cutané 4271
titrage biologique 449
titre 4585
togavirus 4589
toile choroïdienne 4463
toile sous-muqueuse 4282
toit du mésencéphale 3906
toit du quatrième ventricule
    3905
toltrazuril 4590
tonsille 4594
tonsille linguale 2528
tonsille palatine 3232
tonsille pharyngienne 3443
torsion utérine 4790
torticolis 4602
torus 3219
torus carpien 687
torus lingual 1370
toupet 1747
toupillon de la queue 4438
tourbillon des poils 1952
toux 1117
toxascariose 4605
toxémie 4603
toxémie de gestation 3599
toxémique 4604
toxicité 4607
toxicité aiguë 60
toxicité chronique 900
toxicité des médicaments 1393
toxicologie 4608
toxicose 3531
toxine 4609
toxine T2 4425
toxine trichothécène 4666
toxique 4606
toxocarose 4610
toxoplasmose 4612
trabécules charnues 2889
trabécules charnues de
    l'oreillette droite 3351
trabécules de la rate 4613
trachée 4615
trachéite 4621

Español

ácaro 2846
ácaro auricular del perro 1416
ácaro de la sarna 2678
ácaro de las patas 3996
ácaro de las plumas 1264
ácaro rojo de las gallinas 3793
ácaro subcutáneo de las
    gallinas 4270
ácido nalidíxico 2927
ácido nicotínico 3009
ácido oxolínico 3213
ácido ribonucléico 3870
ángulo de la boca 182
ángulo de la costilla 183
ángulo de la mandíbula 181
ángulo iridocorneal 2335
ángulo lateral del ojo 2439
ángulo medial del ojo 2712
árbol bronquio 582
área de incubación 2166
área vestibular inferior 2184
área vestibular superior 4332
abdomen 1
abertura caudal de la pelvis
    3367
abertura craneal de la pelvis
    3364
abertura frontomaxilar 1777
abertura maxilopalatina 2703
abertura nasomaxilar 2952
abomasitis 16
abomoso 17
aborto 18
aborto clamidial 1484
aborto contagioso 527
aborto enzoótico de la oveja
    1484
aborto vírico de la yegua 1528
absceso 19
absorción defectuosa 2645
absorción intestinal 2308
acalasia 42
acantocéfalos 20
acantocefalosis 21
acantosis 22
acardia 24
acaricida 25
acariosis de las abejas 23
acaudia 4434
acetábulo 39
acetonemia 40
acetonuria 41
acido ascórbico 4936
acidosis 47
aclaramiento 932
aclaramiento renal 3809
acondroplasia 44
acrodermatitis 49
acromion 50
ACTH 1096

actinobacilosis 51
actinomicetos 52
actinomicosis 53
acúmulo ooforo 3193
acuariosis 56
acueducto del mesencéfalo (de
    Silvio) 254
acupuntura 57
adenitis 66
adenitis contagiosa 4235
adenocarcinoma 67
adenohipófisis 68
adenoma 69
adenomatosis 70
adenomatosis pulmonar ovina
    3689
adenovirosis 72
adenovirus 71
adenovirus aviar 346
adenovirus bovino 525
adenovirus canino 638
adherencia 73
adiposidad 3053
administración oral 3136
administración parenteral 3317
adrenalina 1502
aerobia 83
aerocistitis 84
aerofagia 86
aeromonosis por Aeromonas
    salmonicida 85
afebril 88
aflatoxicosis 89
aflatoxina 90
afta 246
agalaxia 98
agalaxia contagiosa de ovejas y
    cabras 1040
agammaglobulinemia 99
agenesia 100
agente cardiovascular 676
agente del temblor enzoótico
    4014
agente de quelación 850
agente inmunosupresor 2153
agente patógeno 3345
aglutinación 101
aglutinación en placa 4104
aglutinación en tubos 4693
aglutinación lenta 4693
agnacia 104
agranulocitosis 105
agudo 58
aguja para inyección
    hipodérmica 2217
agujero 1742
agujero alar 109
agujero ciático 4006
agujero de la vena cava 4834
agujero epiploico 1506

agujero esfenopalatino 4130
agujero espinoso 4150
agujero estilomastoideo 4253
agujero etmoidal 1549
agujero infraorbitario 2193
agujero interventricular 2302
agujero intervertebral 2306
agujero magno 2423
agujero mandibular 2668
agujero mastoideo 2694
agujero mentoniano 2764
agujero nutricio 3047
agujero obturador 3057
agujero orbitorredondo 3151
agujero oval del corazón 3188
agujero palatino 3226
agujero rasgado 2392
agujero redondo 3926
agujeros alveolares 132
agujeros papilares 3266
agujeros sacros 3961
agujero supracondilar 4340
agujero supraorbitario 4350
agujero supratroclear 4366
agujero transverso 4638
agujero vertebral 4876
agujero yugular 2360
ala 4969
ala de la nariz 4973
ala del atlas 4970
ala del hueso basiesfenoides
    4971
ala del hueso ilion 4972
ala del hueso presfenoides
    4974
ala del hueso sacro 4975
alantocorión 885
alantoides 121
ala postacetabular del ílion
    3563
ala preacetabular del ílion 3587
albendazol 111
albinismo 112
albúmina sérica 4057
albuminuria 113
alcaloide 119
alcaloide ergotamina 1536
alcaloide pirrolizidínico 3727
alcalosis 120
aldosterona 114
alergénico 123
alergeno 122
alergia 125
alfaprostol 116
alfaxalona 117
algas azul-verde 495
alimento medicado 2742
almohadilla 3219
almohadilla carpiana 687
almohadilla dentaria 1256

arteria vestibular 4892
arterioesclerosis 275
arteriola 274
arteritis 277
arteritis vírica del caballo 1529
articulación 297
articulación
antebraquiocarpiana 198
articulación atlantoaxial 315
articulación atlantooccipital
316
articulación calcanocuartal 626
articulación cartilaginosa 146
articulación centrodistal 800
articulación compuesta 1020
articulación condilar 1023
articulación costocondral 1108
articulación costotransversa
1112
articulación costovertebral
1114
articulación cricoaritenoidea
1168
articulación cricotiroidea 1172
articulación cubital 1440
articulación de la cabeza de la
costilla 298
articulación de la cadera 2012
articulación de la rodilla 4222
articulación del carpo 686
articulación del codo 1440
articulación del hombro 4085
articulación del tarso 2022
articulación elipsoidal 1442
articulación en charnela 2008
articulación en pivote 3485
articulación esferoidal 403
articulación esternocoracoidea
4211
articulación esternocostal 4213
articulación femororrotuliana
1661
articulación femorotibial 1663
articulación fibrosa 4401
articulación humerocubital
2053
articulación humerorradial
2052
articulación intermandibular
2247
articulación intertransversa
lumbosacra 2601
articulación lumbosacra 2600
articulación mediocarpiana
2825
articulación plana 3494
articulación radiocarpiana
3757
articulación radiocubital 3763
articulación sacroilíaca 3971

articulación sellar 3975
articulación simple 4092
articulación sinovial 4411
articulación talocalcánea 4440
articulación tarsocrural 4450
articulación
temporomandibular 2665
articulación tibioperonea 4576
articulación tirohiodiea 4557
articulación trocoidea 3485
articulaciones
carpometacarpianas 689
articulaciones intercarpianas
2226
articulaciones interfalangianas
distales 1339
articulaciones interfalangianas
proximales 3655
articulaciones
intermetacarpianas 2258
articulaciones
intermetatarsianas 2259
articulaciones intertarsianas
2298
articulaciones intertransversas
lumbares 2588
articulaciones intracondrales
2321
articulaciones metacarpo-
falangianas 1673
articulaciones
metatarsofalangianas 2804
articulaciones
tarsometatarsianas 4451
artritis 279
artritis-encéfalitis caprina 653
artrópodo parásito 284
artrodesis 281
artrogriposis 282
artropatía 283
asa axilar 374
asa espiral del colon 4155
asas traqueales 4618
ascárido 300
ascaridiosis 301
ascaridosis 302
ascaridosis de los équidos 3295
ascitis 305
asfixia 308
aspergilosis 306
aspermia 307
asta de Ammon 2014
asta dorsal del médula espinal
1360
asta lateral del médula espinal
2447
asta rostral 3917
asta temporal 4471
asta ventral del médula espinal
4849

astenia 310
astrágalo 4574
ataxia 311
ataxia enzoótica de los
corderos 4380
atelectasia 312
aterosclerosis 313
atlas 317
atóxico 3022
atonía 318
atonía del rumen 3941
atopia 319
atresia 320
atresia anal 161
atresia anorectal 193
atresia rectal 3774
atricosis 321
atrio del corazón 324
atrio derecho 3884
atrio izquierdo 2474
atriquia 321
atrofia 326
atrofia de la retina 3846
atrofia muscular 155
aurícula 3478
aurícula del atrio 333
autoanticuerpo 340
autopsia 3569
autovacuna 341
avirulento 366
avitaminosis 367
avoparcina 368
axila 369
azaperona 381
azoospermia 382
azoturia 383

babesiosis 384
babesiosis bovina 526
babesiosis equina 1517
babesiosis ovina 3205
bacilo 386
bacilo carbuncoso 203
bacilo coliforme 983
bacilo de la tuberculosis aviar
363
bacilo de la tuberculosis bovina
550
bacilo tuberculoso 4696
bacitracina 387
bacteremia 393
bacteria 399
bacteria ácido-resistente 46
bacteria aerobia 83
bacteria anaerobia 158
bacteria del mal rojo porcino
1538
bacteria Gram-negativa 1884
bacteria Gram-positiva 1885
bacterias piogénicas 3716

bactericida 397
bacteridia carbuncosa 203
bacteriemia 393
bacteriófago 398
balanitis 400
balanopostitis 401
balantidiosis 402
bambermicina 404
baño parasiticido 4067
barba cervical 828
barba de pluma 406
barbícula 407
barbillas 4958
barbiturato 405
barra 409
barrera placentaria 3488
barrera sangre-cerebro 469
barro 1012, 4109
bartonelosis 410
base del corazón 413
base del cuerno 414
basípodo 419
basihioidea 415
basquilla 3698
bazo 4165
beatilla 3984
bencilpenicilina 431
bencilpenicilina benzatina 428
benzimidazol 429
benzocaína 430
besnoitiosis 433
betabloqueador 435
betahemolítico 436
betametasona 437
bezoar 438
bifurcación de la tráquea 4616
bilirrubina 448
bilis 445
biocenosis 450
biología molecular 2851
biopsia 452
biotecnología 453
birnavirus 455
bitionol 457
blastomicosis norteamericana 3025
blefaritis 464
bloqueo nervioso 2981
boca 2868
bocio 1873
bociogénico 1874
boldenona 509
bolsa 4020
bolsa de la cloaca 950
bolsa esternal 4202
bolsa omental 3109
bolsa ovárica 3192
bolsa podotroclear 3528
bolsa sinovial 4409
bolsa subcutánea 4268

borde coronal 1074
borde craneal de la tibia 4571
borde solear del casco 1905
borde supraorbitario 4352
borreliosis 523
botón gustativo 4455
botriocéfalo 574
botulismo 524
bradsot de los ovinos 568
branquiomicosis 567
braquignatia 564
brazo 4754
bregma 4093
brida 2574
bromocicleno 576
bromocriptina 577
bromofos 578
bromuro de homidio 2024
bromuro de piritidio 3725
broncoespasmo 581
bronconeumonía 588
bronquiectasia 584
bronquio 590
bronquios lobares 2547
bronquios segmentarios 4030
bronquitis 585
bronquitis infecciosa aviar 352
brotianida 591
brucelina 593
brucelosis 594
bruclosis bovina 527
brucelosis ovina 3206
brucelosis porcina 3552
buche 1177
buiatría 602
bulbo del pelo 1943
bulbo olfatorio 3094
bulla de la siringe 4419
bulla lacrimal 2395
bulla timpánica 4709
bunamidina 607
bunostomosis 608
bunyavirus 609
buparvacuona 610
bupivacaina 611
buquinolato 612
bursectomía 613
bursitis 614
bursitis infecciosa 2174
buserelina 615
BVD 553

cálamo 624
cálculo 628
cálculo biliar 1795
cálculo renal 3807
cálculo urinario 4768
cáliz renal 3808
cámara anterior del globo 199

cámara posterior del globo 3564
cámara vítrea del globo 4940
cánula del pezón 4458
cápsula adiposa del riñón 74
cápsula articular 2358
cápsula de la úngula 2026
cápsula del cristalino 2489
cápsula del globo (de Tenon) 4065
cápsula del glomérulo 1853
cápsula fibrosa 1685
cabello 1119
cabeza 1961
cabeza de la costilla 1965
cabeza de la vértebra 1131
cabeza del epidídimo 1963
cabeza del hueso fémur 1964
cadera 2009
calcáneo 1692
calcinosis 627
calicillo gustativo 4455
calicivirus 633
calicivirus felino 1640
califórido 493
calostro 997
cambendazol 634
campilobacteriosis 635
canal alimentario 118
canal central del epéndimo 792
canal del cuello uterino 829
canal del hipocampo 135
canal del nervio hipogloso 2086
canal del omaso 3103
canal espiral de la cóclea 4153
canales semicirculares óseos 4034
canal femoral 1656
canalículo 636
canal infraorbitario 2192
canal inguinal 2210
canal nasolacrimal 2950
canal nutricio 3046
canal pilórico 3710
canal supraorbitario 4349
canal vertebral 4874
canamicina 2367
candidiasis 637
candidosis 637
canibalismo 647
cañón de pluma 624
capilares aéreos 3521
capilariosis 648
caponización 650
capripoxvirus 655
captura de los animales 656
capuchón del espermatozoario 1962
cara 1608
carácter contagioso 1047

dermis de la pared 2416
desarrollo postnatal 3571
descargo vaginal 4803
descenso de los testículos 1273
descornación 1244
deshidratación 1245
desinfección 1337
desinfectante 1336
desintoxicación 1275
desmielinización 1250
desoxicortona 1274
dexametasona 1278
diáfisis 1285
diástole 1288
diafragma 1282
diafragma de la pelvis 3360
diagnóstico 1279
diagnóstico de laboratorio 2390
diagnóstico de la gestación 3598
diagnóstico diferencial 1304
diagnóstico ultrasónico 4736
diamfenetida 1281
diarrea 1286
diarrea de los terneros 629
diarrea vírica de los bovinos 553
diartrosis 1287
diastema 2236
diatesis exudativa 1601
diatesis hemorrágica 1937
diaveridina 1289
diazepam 1290
diazinón 1291
diclazurilo 1294
diclorofeno 1292
diclorvos 1293
dicloxacilina 1295
dicroceliosis 1297
dictiocaulosis 1298
dicumarol 1296
dieldrina 1299
diencéfalo 1300
dienestrol 1301
diente 4598
diente cortante 679
diente del eje 4599
diente del lobo 4979
dientes acústicos 329
dientes caducos 1220
dientes caninas 645
dientes de reemplazo 3416
dientes incisivos 2160
dientes molares 2850
dientes permanentes 3416
dientes premolares 3601
dietilcarbamazina 1302
dietilestilbestrol 1303
difilobotrio 574
difilobotriosis 1328

digestión de los rumiantes 3950
dihidroestreptomicina 1317
dilepididosis 1320
dimercaprol 1321
dimetridazol 1322
diminazeno 1323
dimpilato 1291
dinitolmida 1324
dinoprost 1325
dioctofimosis 1326
dipetalonemosis 1327
dipilidio canino 1388
dirofilariosis 1330
disautonomía felina 1641
disco articular 287
disco del nervio óptico 3131
disco germinal 1832
disco intervertebral 2305
discondroplasia 1407
discondroplasia tibial 4572
disentería 1408
disentería de los corderos 2407
disentería porcina 4383
disfunción 1779
displasia 1409
displasia de la cadera 2011
displasia de la retina 3848
dispnea 1410
distocia 1411
distoma hepático 1011, 2543
distoma lanceolado 2418
distoma pulmonar 2607
distomatosis hepática 1625
distrofia 1412
distrofia muscular 2885
divertículo 1344
divertículo de la nariz 2937
divertículo del estómago 1807
divertículo del la trompa auditiva 1914
divertículo de Meckel 2709
divertículo faríngeo 3438
divertículo prepucial 3605
divertículo suburetral 4294
división celular 782
dorso 388
dorso de la lengua 1380
dorso de la mano 391
dorso de la nariz 571
dorso del pie 4758
dosificación 1386
dosificación inmuno-enzimática 1485
dosificación radioinmunológica 3760
dosis 1387
dosis excesiva 3201
dosis letal 2500
dosis letal mínima 2841
dosis reforzador 519

dosis terapéutica 4514
doxiciclina 1390
duodeno 1405
duramadre del encéfalo 1453
duramadre espinal 4139
durina 1389

eccema 1431
eccema facial de la oveja 1613
ectima contagioso 1044
ectoparásito 1426
ectoparasitosis 1427
ectopia del corazón 1428
ectromelia de los ratones 1429
edema 3074
edema maligno 2652
edema mamario 2662
edema neuraxial 2990
edematoso 3075
educación veterinaria 4909
efecto adverso 81
efecto citopatogéno 1213
egipcianelosis de las aves 82
eimeriosis 1435
eje 380
eje de la pelvis 3358
eliminación de las canales 663
emasculador 1445
embadurnamiento de sangre 485
embrión 1446
embriología 1447
embrionario 1448
eminencia medial 2716
eminencia uropigial 4778
enanismo 1406
encefalitis 1454
encéfalo 1458
encefalomalacia 1456
encefalomielitis 1457
encefalomielitis aviar 348
encefalomielitis equina 1519
encefalomielitis espóradica bovina 4179
encefalomielitis hemoaglutinante porcina 3556
encefalomielitis ovina 2575
encefalomielitis vírica porcina 4495
encefalopatía 566
encefalopatía espongiforme bovina 547
encías 1910
encuesta serológica 4049
endemia 1460
endémico 1459
endocardio 1462
endocarditis 1461
endolinfa 1464

escama frontal 1775
escapo de la pluma 1636
esclerótica 4009
esclerosis 4010
escólex 4011
escoliosis 4012
escotadura 2161
escotadura cardíaca del
    pulmón 672
escotadura ciática 4008
escotadura cubital 4733
escotadura del acetábulo 37
escotadura de la escápula 2162
escotadura de la mandíbula
    2675
escotadura glenoidal 1851
escotadura peronea 1691
escotadura poplítea 3550
escotadura radial 3748
escotaduras costales 1103
escotadura supraorbitaria 4351
escotadura troclear 4682
escotadura vertebral 4878
escotadura yugular 2364
escroto 4020
escudillo 4021
escuela veterinaria 4916
esfínter del ano 167
esfingolipidosis 4136
eslafranina 4103
esmalte 1452
esófago 3082
esofagostomosis 3081
espacio inguinal 2210
espacio intercostal 2233
espacio interdigital 2240
espacio retroperitoneal 3854
espalda 4083
esparganosis 4162
espasmo de un esfínter 42
espécimen 4122
espectinomicina 4123
espermatogénesis 4125
espermatozoide 4126
espina bífida 4137
espina de la cuña 4146
espina de la escápula 4147
espina ilíaca 2116
espiramicina 4158
espirocercosis 4159
espirometrosis 4162
espiroqueta 4160
espiroquetosis 4161
espiroquetosis aviar 362
espiroquetosis del conejo 3739
espirúridos 4163
esplenectomía 4166
esplenomegalia 4173
espolón 4187
espolón del metacarpo 1535

espongiforme 4175
esporádico 4178
esporidesmina 4180
esporocisto 4181
esporoquiste 4181
esporotricosis 4182
esporozoito 4183
esqueleto 4097
esquistosoma 4003
esquistosomosis 4004
esquizonte 4005
estado portador 694
estafilococo 4197
estandarte 4819
estandarte externo 1597
estandarte interno 2283
estaquibotriotoxicosis 4192
esteatosis 1628
esteatosis hepática 1629
estefanofilariosis 4200
estefanurosis 4201
estenosis 4199
esterilidad 2186
estérnebra 4207
esternón 4220
esteroide 4221
estilesiosis 4223
estilohioideo 4248
estómago 4226
estómago de rumiante 3951
estomatitis 4228
estomatitis necrótica 2969
estomatitis papular 3274
estomatitis papular bovina 541
estomatitis vesiculosa 4889
estrabismo 4229
estradiol 3083
estrato basal 412
estrato córneo 2042
estrato espinoso 4151
estrato externo de la pared
    1583
estrato ganglionar del cerebelo
    1798
estrato granuloso 1888
estrato granuloso del cerebelo
    1887
estrato interno de la pared
    2269
estrato lúcido 933
estrato medio de la pared 2831
estrato molecular del cerebelo
    2852
estrato neuroepitelial de la
    retina 2993
estrato papilar 3267
estrato pigmentario de la
    retina 3473
estrato reticular 3837
estreñimiento 1037

estreptococo 4239
estreptomicina 4240
estreptotricosis 1269
estrías longitudinales 2567
estribo 4195
éstrido 3084
estrógeno 3085
estróngilo pulmonar 2608
estróngilos 4244
estrona 3086
estrongiloidosis 4245
estrongilosis 4246
etiología 87
etisazol 1555
etmoturbinados 1554
etomidato 1556
eutanasia 1558
examen clínico 943
exantema 1559
exantema genital equino 1518
exantema vesicular porcino
    4887
excavación pubovesical 3683
excavación rectogenital 3781
excavación vesicogenital 4885
excreta 1560
exoftalmía 1566
exostosis 1567
exostosis falangea 3900
exostosis metacarpiana 4174
exploración rectal 3775
extremidad 1600
extremidad caudal 725
extremidad craneal de la
    vértebra 1131

fármaco antialérgico 222
fármaco antibacteriano 204
fármaco anticoccidiósico 212
fármaco anticonvulsivante 213
fármaco antidiarréico 214
fármaco antifúngico 218
fármaco antiinfeccioso 223
fármaco antiinflamatorio 224
fármaco antineoplásico 225
fármaco antiparasitario 226
fármaco antiprotozoario 227
fármaco antivírico 232
fármaco contra timpanitis 208
fármaco neurotrópico 3001
fármaco quimioterápico 223
factor causal 769
factor de predisposición 3591
factor letal 2501
facultad de veterinaria 4916
fagocitario 3429
fagocito 3428
fagocitosis 3430
falange 3431
falange distal 1340

hipotermia 2097
hipotricosis 2098
hipoxia 2099
histerectomía 2100
histerotomía 622
histocompatibilidad 2015
histológico 2016
histología 2017
histomonosis 2018
histopatología 2019
histoplasmosis 2020
historia clínica 699
historia veterinaria 4910
hongo 1787
hormona 2032
hormona androgénica 177
hormona de crecimiento 4120
hormona estimulante folicular
    1734
hormona gonadotrópica 1879
hormonal 2031
hormona luteinisante 2609
hormona luteotrópica 2610
hormona paratiroides 3311
hormona tireotrópica 4565
hormona tiroidea 4562
hospedador intermediario 2253
hoz cerebral 1619
húmero 2054
huesecillos del oído 328
hueso 510
hueso accesorio del carpo 26
hueso angular 184
hueso articular 285
hueso basiesfenoides 420
hueso basioccipital 418
hueso carpiano 1° 1704
hueso carpiano 2° 4028
hueso carpiano 2 y 3 4026
hueso carpiano 3° 4525
hueso carpiano 4° 1751
hueso carpocubital 4729
hueso carporradial 3745
hueso central del tarso 797
hueso centrocuartal 801
hueso cigomático 5001
hueso compedal 1705
hueso coronal 2835
hueso corto 4077
hueso costal 3868
hueso cuadrado 3729
hueso cuboideo 1752
hueso cuneiforme intermedio
    4029
hueso cuneiforme lateral 4530
hueso cuneiforme medial 1706
hueso de la cadera 2010
hueso del pene 3377
hueso del talón 1692
hueso dentario 1252

hueso ectetmoides 1425
hueso endogloso 1480
hueso escafoideo 3745
hueso escamoso 4188
hueso esfenoides 4128
hueso esplenial 4167
hueso etmoides 1546
hueso exoccipital 1564
hueso femur 1665
hueso frontal 1767
hueso grande 4525
hueso hioideo 2067
hueso íleon 2133
hueso incisivo 2156
hueso intermedio del carpo
    2251
hueso interparietal 2289
hueso isquion 2346
hueso lacrimal 2394
hueso largo 2560
hueso mesetmoides 2775
hueso metacarpiano 3° 4527
hueso metatarsiano 3° 4528
hueso nasal 2932
hueso navicular 797
hueso naviculocuboideo 801
hueso neumático 3520
hueso ótico 3182
hueso occipital 3062
hueso orbitoesfenoides 3152
hueso palatino 3224
hueso paraesfenoides 3305
hueso parietal 3319
hueso piramidal 4729
hueso pisiforme 26
hueso plano 1713
hueso prearticular 3588
hueso prefrontal 3595
hueso presfenoides 3612
hueso pubis 3677
hueso rostral 3914
huesos de la cara 1611
huesos del carpo 683
huesos del cráneo 1126
huesos de los dedos de la mano
    1311
huesos de los dedos del pie
    1312
huesos del tarso 4444
hueso semilunar 2251
hueso sesamoideo distal 1341
huesos metacarpianos 1°-5°
    2786
huesos metatarsianos 1°-5°
    2799
huesos sesamoideos proximales
    3658
huesos supraorbitarios 4348
hueso supraangular 4337
hueso suprayugal 4343

hueso tarsiano 1° 1706
hueso tarsiano 2° 4029
hueso tarsiano 3° 4530
hueso tarsiano 4° 1752
hueso temporal 4468
hueso terigoideo 3664
hueso trapecio 1704
hueso trapezoideo 4028
hueso trapezoideo-grande 4026
hueso unciforme 1751
hueso unguicular 1340
hueso ungular 1340
huésped 2048
huésped definitivo 1701
huevo 3209
humor acuoso 255
humor vítreo 4941

íleon 2112
índice quimioterapéutico 854
íntima 2316
ictericia 2353
ictiofonosis 2102
idoxuridina 2103
Ig 2147
imidocarbo 2135
impresión cardíaca 670
impresión cólica 981
impresión esofágica 3078
impresión gástrica 1809
impresión renal 3814
inapetencia 2155
inclinación de la pelvis 2163
incontinencia urinaria 4769
incoordinación 2165
inercia uterina 4787
infarto 2168
infección 2170
infección aerógena 106
infección cruzada 1181
infección de las abejas por
    Ascosphaera apis 845
infección de prueba 846
infección experimental 1568
infección latente 2437
infección por Actinobacillus
    equuli 4074
infección por Actinobacillus
    pleuropneumoniae 3559
infección por Anaplasma spp.
    170
infección por Aspergillus spp.
    306
infección por Bacillus
    piliformis 4720
infección por Bacteroides
    nodosus 2177
infección por Bartonella spp.
    410

linfocentro inguinal profundo 2122
linfocentro inguinal superficial 2215
linfocentro inguinofemoral 2215
linfocentro lumbar 2590
linfocentro mandibular 2673
linfocentro mediastínico 2737
linfocentro mesentérico caudal 742
linfocentro mesentérico craneal 1143
linfocentro parotídeo 3328
linfocentro poplíteo 3548
linfocentro retrofaríngeo 3855
linfocentro torácico dorsal 1381
linfocentro torácico ventral 4863
linfocito 2625
linfocito B 498
linfocitosis 2627
linfocito T 4586
linfoma 2630
linfopatía 2618
linfosarcoma 2631
linguatula 3382
linguatulosis 2530
lipodistrofia 2535
lipoproteína 2536
lisis 2633
lisosoma 2634
lisotipo 3427
lisozima 2635
listeriosis 2538
litiasis 2539
lóbulo 2548
lóbulo accesorio 29
lóbulo anterior de la hipófisis 68
lóbulo caudado del hígado 762
lóbulo caudal del cerebelo 736
lóbulo craneal 247
lóbulo cuadrado del hígado 3730
lóbulo de la glándula mamaria 2550
lóbulo de la oreja 1415
lóbulo del hígado 2544
lóbulo derecho del hígado 3890
lóbulo floculonodular 1723
lóbulo frontal 1768
lóbulo hepático 2545
lóbulo izquierdo del hígado 2480
lóbulo medio 2832
lóbulo occipital 3065
lóbulo piriforme 3480

lóbulo posterior de la hipófisis 2996
lóbulo renal 2549
lóbulo rostral del cerebelo 3918
lóbulos de la glándula mamaria 2554
lóbulo temporal 4473
lobulillo 2551
lobulillo del epidídimo 2552
lobulillo del pulmón 2553
lomos 2557
loque americana 139
loque europea 1557
lordosis 2573
luteotropina 2610
lutropina 2609
luxación de la rótula 3343

mácula lútea 4992
macrófago 2637
macroscópico 2638
madeja de vermes 4982
maedi 2639
mal de altura 129
mal de coito 1389
maleina 2653
malformación congénita 1028
malófago 458
mal rojo del cerdo 4384
mama 2659
mama abdominal 8
mama inguinal 2212
mama torácica 4539
mamilitis 2657
mamilitis ulcerativa bovina 552
mandíbula 2663
mango del esternón 2682
manguito de los flexores 2679
manifestaciones clínicas 944
mano 1957
manosidosis 2680
marteilosis 2685
martillo 2654
mastadenovirus 2690
mastitis 2691
mastitis bovina 538
mastitis coliforma 984
mastitis estafilocócica 4196
mastitis estreptocócica 4238
mastitis gangrenosa 1802
mastitis micótica 2901
mastitis subclínica 4263
mastocito 2693
matriz 4794
maxilar 2696
meato 2707
meato acústico externo 1571
meato acústico interno 2261
meato nasal 2939
meato temporal 4474

mebendazol 2708
mecanismo de defensa 1237
meconio 2710
media 2711
mediastino 2741
medicación 2743
medicamento tripanocido 4688
medicina veterinaria 4913
médico veterinario 4908
medio de cultura 1191
medroxiprogesterona 2745
médula del riñón 3815
médula espinal 4138
médula ósea 513
médula ósea amarilla 4990
médula ósea roja 3791
médula oblongada 2746
médula suprarrenal 78
megestrol 2750
mejilla 848
melanocito 2751
melanoma 2753
melanosis 2754
melanostimulina 2752
melatonina 2755
melioidosis 2756
melófago de la oveja 4070
membrana 2757
membrana celular 785
membrana corio-alantoica 884
membrana del tímpano 4712
membrana espiral 4156
membrana esternocoracoclavicular 4210
membrana interósea 2286
membrana nictitante 4526
membrana pupilar 3701
membranas fetales 1671
membrana sinovial 4412
membrana vestibular (de Reissner) 3804
membrana vítrea 4942
meninges 2760
meningitis 2761
meningoencefalitis 2762
menisco articular 289
meniscos táctilos 2768
menopón de las gallinas 3580
mentón 858
merozoito 2769
mesencéfalo 2771
mesénquima 2772
mesenterio 2774
mesocolon 2776
mesoduodeno 2777
mesometrio 2778
mesonefros 2779
mesorquio 2780
mesorrecto 2781

músculo depresor de la
mandíbula 2666
músculo depresor del labio
mandibular 1265
músculo digástrico 1306
músculo dilatador 1318
músculo dilatador de la pupila
1319
músculo elevador del ano 2510
músculo elevador del párpado
2511
músculo erector de la espina
1532
músculo escaleno 3995
músculo escapulohumeral
caudal 753
músculo escapulohumeral
craneal 1152
músculo esfínter 4133
músculo esfínter de la ampolla
3073
músculo esfínter de la cloaca
952
músculo esfínter de la papila
4461
músculo esfínter de la pupila
4134
músculo esfínter del conducto
colédoco 4135
músculo esfínter del íleon 2105
músculo esfínter del píloro
3713
músculo esfínter externo del
ano 1572
músculo esfínter externo de la
uretra 4766
músculo espinoso 4141
músculo esplenio de la cabeza
4171
músculo esplenio del cuello
4172
músculo estapedio 4193
músculo esternocefálico 4208
músculo esternocoracoideo
4212
músculo esternohioideo 4214
músculo esternomandibular
4215
músculo esternomastoideo
4216
músculo esternotiroideo 4218
músculo esternotraqueal 4219
músculo estilofaríngeo caudal
754
músculo estilofaríngeo rostral
3922
músculo estilogloso 4247
músculo estilohioideo 4249
músculo estriado 4243
músculo etmomandibular 1553

músculo extensor carpocubital
4730
músculo extensor carporradial
3746
músculo extensor del dedo 1°
1569
músculo extensor del dedo 2°
1570
músculo extensor digital
común 1006
músculo extensor digital corto
4078
músculo extensor digital largo
2561
músculo extensor digital lateral
2446
músculo flexor carpocubital
4731
músculo flexor carporradial
3747
músculo flexor digital corto
4079
músculo flexor digital profundo
1227
músculo flexor digital
superficial 4322
músculo frontal 1769
músculo frontoescutular 1778
músculo fusiforme 1793
músculo gastrocnemio 1814
músculo genihioideo 1825
músculo geniogloso 1824
músculo glúteo medio 2830
músculo glúteo profundo 1228
músculo glúteo superficial 4324
músculo gluteobíceps 1865
músculo grácilis 1882
músculo hioepiglótico 2065
músculo hiogloso 2066
músculo hioideo transverso
4639
músculo ilíaco 2114
músculo iliocostal 2118
músculo iliocostal lumbar 2119
músculo iliofemoral externo
1581
músculo iliofemoral interno
2267
músculo ilioperóneo 2123
músculo iliopsóas 2129
músculo infraespinoso 2201
músculo involuntario 2333
músculo isquiocavernoso 2341
músculo isquiofemoral 2343
músculo largo de la cabeza
2568
músculo largo del cuello 2569
músculo latísimo del dorso 573
músculo levador del labio
superior 2512

músculo lingual propio 2332
músculo liso 4110
músculo longísimo 2562
músculo malar 2647
músculo masetero 2689
músculo milohioideo 2909
músculo oblicuo caudal de la
cabeza 746
músculo oblicuo craneal de la
cabeza 1146
músculo oblicuo dorsal 1366
músculo oblicuo externo del
abdomen 1586
músculo oblicuo interno del
abdomen 2270
músculo oblicuo ventral 4852
músculo obturador externo
1587
músculo obturador interno
2271
músculo occipitohioideo 3067
músculo omohioideo 3111
músculo omotransverso 3112
músculo orbicular 3142
músculo orbicular de la boca
3144
músculo orbicular del ojo 3143
músculo orbitario 3148
músculo palatino 3228
músculo palatofaríngeo 3237
músculo palatogloso 3235
músculo palmar corto 4081
músculo parotidoauricular
3329
músculo pectíneo 3353
músculo pectoral descendente
1272
músculo pectoral
profundo/ascendente 1230
músculo pectoral transverso
4643
músculo peroneo corto 4082
músculo peroneo largo 2571
músculo piriforme 3481
músculo platisma 3507
músculo pleuroesofágico 3513
músculo poplíteo 3549
músculo principal de la molleja
4520
músculo pronador cuadrado
3733
músculo pronador redondo
3929
músculo protractor 3652
músculo pseudotemporal
profundo 1233
músculo pseudotemporal
superficial 4329
músculo psóas mayor 1897
músculo psóas menor 4106

nódulos linfáticos esternales 4203
nódulos linfáticos hemales 1922
nódulos linfáticos iliofemorales 2121
nódulos linfáticos inguinales superficiales 4325
nódulos linfáticos intercostales 2230
nódulos linfáticos mamarios 2661
nódulos linfáticos mandibulares 2672
nódulos linfáticos mediastínicos 2736
nódulos linfáticos mesentéricos caudales 741
nódulos linfáticos mesentéricos craneales 1142
nódulos linfáticos parotídeos profundos 1229
nódulos linfáticos parotídeos superficiales 4327
nódulos linfáticos poplíteos 3547
nódulos linfáticos pulmonares 3692
nódulos linfáticos retrofaríngeos laterales 2459
nódulos linfáticos retrofaríngeos mediales 2724
nódulos linfáticos ruminales 3944
nódulos linfáticos sacros 3963
nódulos linfáticos subescapulares 4287
nódulos linfáticos subilíacos 4275
nódulos linfáticos traqueobronquiales 4622
no apto para la reproducción 4749
nodulillo 3018
nodulillo linfático 2615
nodulillos linfáticos agregados 103
nodulillos linfáticos solitarios 4114
no infecto 4751
no patógeno 3020
noradrenalina 3023
norepinefrina 3023
norgestomet 3024
nosemosis 3027
notario 3029
novobiocina 3032
núcleo 3041

núcleo ambiguo 137
núcleo arqueado 262
núcleo caudado 763
núcleo celular 786
núcleo central del tálamo 796
núcleo cervical lateral 2442
núcleo cuneado 1192
núcleo de la cima 1626
núcleo del cristalino 3043
núcleo fastigial 1626
núcleo grácil 3042
núcleo hipotalámico dorsomedial 1385
núcleo hipotalámico ventromedial 4867
núcleo infundibular 2204
núcleo intersticial 2296
núcleo lateral del cerebelo 2441
núcleo lenticular 2491
núcleo motor 2866
núcleo olivar 3102
núcleo parasimpático 3307
núcleo paraventricular 3313
núcleo preóptico 3603
núcleo pulposo 3044
núcleo rojo 3794
núcleos del lemnisco lateral 3037
núcleos del nervio vestibulococlear 4905
núcleos de los nervios craneales 3036
núcleos del puente 3545
núcleos del tegmento 4462
núcleos de origen 3038
núcleos habenularos 1916
núcleos interpuestos del cerebelo 2290
núcleos intralaminares del tálamo 2325
núcleos rostrales del tálamo 3920
núcleos talámicos 4509
núcleo subtálamico 4292
núcleo supraóptico 4346
núcleos ventrales del tálamo 4851
núcleos vestibulares 4897
nuca 2929
nucleótido 3040
nucleoproteína 3039
numeración celular 780

órgano de Jacobson 4950
órgano del gusto 1912
órgano del olfato 3097
órgano de vista 3157
órgano espiral 4157
órgano orobasal 3163

órganos accesorios del ojo 31
órganos de los sentidos 3159
órganos genitales femininos 1654
órganos genitales masculinos 2649
órgano subcomisural 4264
órgano subfornical 4274
órganos uropoyéticos 4770
órgano vestíbulococlear 4906
órgano vomeronasal 4950
óxido nitroso 3014
obesidad 3053
obstetricia 3054
obstrucción 3055
obstrucción del buche 1178
obstrucción del esófago 3079
occipucio 3068
oclusión intestinal 2312
ocratoxina 3071
octavo nervio craneal 4904
ofloxacino 3088
oído 1413
oído externo 1576
oído interno 2264
oído medio 2829
ojo 1602
olaquindox 3089
olecranón 3093
oligodoncia 3100
oliguria 3101
omaso 3108
ombligo 2957
omento mayor 1507
omento menor 2497
omóplato 3999
oncocercosis 3115
oncogenicidad 3117
oncogénico 3116
oncosfera 3119
oncovirus tipo C 3120
onfalitis 3113
oocisto 3122
ooquiste 3122
opacidad de la córnea 1069
operación quirúrgica 4372
opérculo nasal 2941
opistorquiosis 3127
opsonina 3128
orbita 3145
orbivirus 3153
ordeño excesivo 3203
orejilla 1415
orejuela 3478
orificio atrioventricular derecho 3882
orificio atrioventricular izquierdo 2472
orificio cecocólico 620
orificio de la aorta 239

pleuroneumonía contagiosa caprina 1042
pleuroneumonía porcina 3559
plexo 3515
plexo aórtico abdominal 4
plexo aórtico torácico 4533
plexo autónomo 345
plexo braquial 557
plexo cardíaco 673
plexo carotídeo 682
plexo celíaco 974
plexo cervical 836
plexo coroideo 892
plexo deferencial 1239
plexo de Meissner 4283
plexo dentario 1257
plexo entérico 1474
plexo esplénico 4169
plexo faríngeo 3441
plexo femoral 1658
plexo hepático 1987
plexo ilíaco 2115
plexo intermesentérico 2257
plexo linfático 2616
plexo lumbar 2593
plexo lumbosacro 2602
plexo mesentérico 2773
plexo mientérico (de Auerbach) 2907
plexo ovárico 3198
plexo pampiniforme 3251
plexo pancreático 3257
plexo parotídeo 3331
plexo pélvico 3369
plexo prostático 3641
plexo pulmonar 3694
plexo renal 3818
plexo sacro 3965
plexos gástricos 1811
plexos rectales 3776
plexos ruminales 3947
plexo subclavio 4260
plexo submucoso 4283
plexo subseroso 4291
plexo suprarrenal 4357
plexos vertebrales 4880
plexos vesicales 4882
plexo terigoideo 3668
plexo testicular 4499
plexo timpánico 4716
plexo uterovaginal 4793
plexo venoso 4839
pliegue 1730
pliegue axilar 372
pliegue del flanco 1712
pliegue frangeado 1700
pliegue ileocecal 2107
pliegue ruminorreticular 3953
pliegues de la piel 4100
pliegue semilunar 4526

pliegues glosoepiglóticos 1857
pliegue terigomandibular 3669
pliegue vestibular 4894
pliegue vocal 4943
pluma 1632
pluma de revestimiento 499
pluma de vuelo 1722
plumaje 3517
plumas de contorno 1049
plumbismo 2471
plumones 3519
pneumostrongilosis 3523
pneumovirus 3525
pododermatitis 3526
podoteca 3527
poliartritis 3533
poliartritis/poliserositis porcina 3560
polidactilia 3534
polidipsia 3535
polimastia 3536
polimelia 3537
polimixina B 3539
polimorfismo bioquímico 451
polioencefalomalacia 819
poliserositis 3541
politelia 3542
poliuria 3543
polyomavirus 3540
poples 390
porcentaje de mortalidad 2863
porción cardíaca 667
porción cartilaginosa de la trompa auditiva 697
porción ciega de la retina 466
porción dorsal de la pared 4587
porción escamosa 4189
porción fetal 1672
porción hueca del infundíbulo 774
porción inflexa 408
porción infundibular de la adenohipófisis 2205
porción intermedia de la adenohipófisis 2254
porción lumbar del diafragma 2592
porción óptica de la retina 3129
porción ósea del la trompa auditiva 517
porción orbitaria del hueso frontal 3149
porción parasimpática 3306
porción pelviana del uréter 3368
porción penácea 3380
porción petrosa del temporal 3426
porción plumácea 3516

porción proximal de la neurohipófisis 3484
porción simpática 4395
porción timpánica del temporal 4715
porción uterina 2695
porfiria 3561
poro gustativo 1913
poro sudorífero 4382
portador asintomático 693
postitis 3567
postoperativo 3573
postpatagio 3574
postvacunal 3575
potencia cancerígena 665
poxvirus 3584
poxvirus aviar 365
poxvirus porcino 4390
práctica veterinaria 4914
praziquantel 3586
precipitación radioactiva atmosférica 3755
precipitina 3590
predisposición 3592
prednisolona 3593
prednisona 3594
preinmunidad 3602
premaxilar 2156
prepucio 3604
presión sanguínea 480
prevalencia de una enfermedad 1333
primera falange 1705
primera vértebra cervical 317
principio activo 55
privado de calostro 998
próstata 3640
probiótico 3619
procaína 3620
proceso frontal 1771
proceso uretral 4765
proceso vaginal del peritoneo 4808
proctodeo 3621
profesión veterinaria 4915
profiláctico 3634
profilaxis 3635
progesterona 3622
prognatismo 3623
prolactina 3625
prolapso 3626
prolapso rectal 3777
prolapso uterino 4789
prolapso vaginal 4809
proligestona 3627
promazina 3628
prometazina 3629
prominencia laríngea 2429
promontorio de la cloaca 951
pronóstico 3624

subalimentado 4747
subclínico 4262
subletal 4276
subtálamo 4293
suero 4056
suero hiperinmune 2073
suero sanguíneo 484
sulfacloropiridazina 4299
sulfadiazina 4300
sulfadimetoxina 4301
sulfadimidina 4302
sulfadoxina 4303
sulfafurazol 4304
sulfaguanidina 4305
sulfamerazina 4306
sulfametoxazol 4307
sulfametoxipiridazina 4308
sulfamonometoxina 4309
sulfanilamida 4310
sulfapirazol 4311
sulfapiridina 4312
sulfaquinoxalina 4313
sulfatiazol 4314
sulfonamida 4315
suola 4112
superficie triturante 3069
superinfección 4331
supernumerario 4333
superovulación 4334
supuración 4336
suramina sódica 4369
surco 1904
surco bicipital lateral 2440
surco bicipital medial 2713
surco caudal del rumen 733
surco coronal 1075
surco coronario 1079
surco coronario del rumen 1080
surco costal 1102
surco craneal del rumen 1136
surco cuneal central 794
surco del abomaso 15
surco del estómago 1808
surco del omaso 3104
surco del retículo 3836
surco espiral externo 1594
surco espiral interno 2279
surco infrapalpebral 2199
surco intermamario 2246
surco interventricular 2303
surco lacrimal 2398
surco longitudinal del rumen 2564
surco medio de la lengua 2734
surco mentolabial 2766
surco paracuneal lateral/medial 988
surco ruminorreticular 3954
surcos del cerebro 4298

surco transverso del hígado 1988
surco yugular 2362
surra 4373
susceptibilidad 4374
sustancia blanca 4966
sustancia cancerígena 664
sustancia compacta del hueso 1016
sustancia cortical 1093
sustancia esponjosa 4177
sustancia gris 1901
sutura 4378
sutura escamosa 4190
sutura intermandibular 2249
sutura sagital 3976
suturas del cráneo 1154
suxametonio 4379

tábano 2045
tálamo 4510
talamoencéfalo 4508
talo 4574
talón 1980
tallo de la pluma 1636
tallo de pelo 1950
tapete lúcido 4442
taquicardia 4426
tara hereditaria 1997
tarsitis 464
tarso 2021
tarsometatarso 4453
tarso palpebral 4454
tasa de mortalidad 2863
tasa de prevalencia 3614
teca del folículo ovárico 4511
tecto del mesencéfalo 3906
tectríz 499
techo del cuarto ventrículo 3905
tegmento del mesencéfalo 2770
tegumento común 1010
teileriosis 4512
teileriosis bovina por Theileria parva 1422
tejido 4583
tejido adiposo 75
tejido conjuntivo 1036
tejido elástico 1438
tejido linfático 2629
tejido muscular 2883
tela coroidea 4463
telangiectasia 4464
tela subcutánea 4271
tela submucosa 4282
telencéfalo 4465
temblor congénito 1029
temblor enzoótico 4013
temperatura corporal 508
temperatura rectal 3778

tendinitis 4478
tendón 4480
tendón calcáneo común 1004
tendón de Aquiles 1004
tenia 841
tenia armada 3474
tenia del cenuro 4430
tenia frangeada 1765
tenia hidatigena 1348
tenia inerme 423
tenia marginada 1348
tenias del colon 4429
teniosis 4431
tenosinovitis 4482
terapia 4515
terapia antibiótica 207
terapia por fármacos 1392
teratógeno 4485
teratogenicidad 4486
teratología 4487
teratoma 4488
tercer párpado 4526
tercer ventrículo 4531
teriogenología 4516
terminaciones nerviosas 2982
termorregulación 4517
testículo 4500
testosterona 4502
teta 4456
tetania 4504
tetania de la hierba 1892
tétanos 4503
tetracaína 4505
tetraciclina 4506
tetramisol 4507
TGE 4628
tímpano 4708
título 4585
tiabendazol 4567
tiacetarsamida sódica 4518
tiamina 4519
tiaprost 4568
tibia 4569
tibiotarso 4578
tiempo de espera 4976
tiflitis 4719
tifosis aviar 1758
tilosina 4707
timo 4553
timpanitis 4718
tiña 1631
tiopental sódico 4524
tirotrofina 4565
tirotropina 4565
tiroxina 4566
tórax 4547
tórsalo 4686
tóxico 3530, 4606
togavirus 4589
toltrazurilo 4590

vena lingual 2529
vena linguofacial 2533
vena magna del cerebro 1895
vena magna del corazón 1894
vena maxilar 2702
vena mediana 2735
vena mediana del codo 2730
vena mesentérica caudal 743
vena mesentérica craneal 1144
vena musculofrénica 2893
vena obturadora 3060
vena occipital 3066
vena oftálmica externa dorsal 1359
vena oftálmica externa ventral 4848
vena oftálmica interna 2275
vena omobraquial 3110
vena ovárica derecha 3892
vena ovárica izquierda 2481
vena poplítea 3551
vena porta 3562
vena profunda 1235
vena renal 3821
vena retromandibular 3853
vena sacra media 2836
vena safena 3987
venas arqueadas 263
venas auriculares 1421
venas bronquiales 583
venas caudales 4439
venas caudales del muslo 727
venas cavernosas del pene 773
venas cecales 619
venas centrales del hígado 799
venas ciliares 916
venas circumflejas 926
venas cólicas 982
venas colaterales 990
venas comitantes 35
venas de la rodilla 1821
venas del cerebro 818
venas del corazón 674
venas del íleon 2106
venas del laberinto 4833
venas del pene 3378
venas digitales 1316
venas diploicas 1329
venas dorsales 1383
venas dorsales del cerebelo 1353
venas emisarias 1451
venas escrotales 4019
venas esofágicas 3080
venas espinales 4145
venas faríngeas 3444
venas frénicas craneales 1149
venas gástricas 1812
venas hepáticas 1989
venas intercostales 2234

venas interóseas 2288
venas intervertebrales 2307
venas labiales 2388
venas lumbares 2596
venas mediastínicas 2740
venas metacarpianas 2790
venas metatarsianas 2803
venas palatinas 3233
venas palpebrales 3250
venas pancreáticas 3259
venas pancreáticoduodenales 3256
venas perforantes 3387
venas perineales 3400
venas pudendas 3685
venas pulmonares 3696
venas radiales 3750
venas rectales 3779
venas ruminales 3949
venas sigmoideas 4091
venas suprarrenales 4358
venas tarsianas 4449
venas temporales 4477
venas tímicas 4552
venas tiroideas 4563
vena subclavia 4261
vena subcutánea del abdomen 4321
vena subescapular 4290
vena supraescapular 4361
vena supraorbitaria 4355
venas ventrales del cerebelo 4845
venas vesicales 4883
venas vestibulares 4899
venas vorticosas 4953
venas yeyunales 2355
vena testicular derecha 3895
vena testicular izquierda 2484
vena tibial caudal 759
vena tibial craneal 1159
vena torácica externa 1596
vena torácica interna 2281
vena torácica lateral 2467
vena torácica superficial 4330
vena toracodorsal 4545
vena umbilical 4742
vena uterina 4791
vena vertebral 4879
vena yugular externa 1582
vena yugular interna 2268
veneno 3530
ventana de la cóclea 3930
ventana del vestíbulo 3190
ventana oval 3190
ventana redondo 3930
ventrículo de la laringe 2431
ventrículo del corazón 4865
ventrículo derecho 3896
ventrículo izquierdo 2485

ventrículo lateral 2469
ventricular 4866
vénula 4868
verme látigo 4961
vermes capilares 1953
vermífugo 201
verminosis gástrica del cerdo 2069
vermis del cerebelo 4869
verrucosis bovina 539
vértebra 4870
vértebra caudal 761
vértebra cervical 837
vértebra coccígea 761
vértebra lumbar 2597
vértebra sacra 3969
vértebra torácica 4542
vértice de la cuña 243
vértice de la lengua 4593
vértice de la nariz 245
vértice de la oreja 1420
vértice del corazón 244
vértice del cráneo 1182
vértice del cuerno 2039
vesícula 4884
vesícula biliar 1794
vesiculovirus 4891
vestíbulo 4900
vestíbulo de la boca 4901
vestíbulo de la nariz 4902
vestíbulo de la vagina 4903
veterinario 4908
vexilo 4819
vexilo de la barba 4820
VHS 4926
vísceras 4931
vibriosis de los peces 4919
vibrisa 4920
vientre 1
villosidad 4921
villosidades del corión 888
villosidades intestinales 2314
villosidades sinoviales 4415
viremia 4922
viremia de primavera de la carpa 4184
virginiamicina 4927
viricida 232
virosis 4924
viruela aviar 1756
viruela bovina 1120
viruela caprina 1872
viruela del palomo 3469
viruela ovina 4072
viruela porcina 4389
virulencia 4928
virulento 4929
virus 4930
virus atenuado 327
virus BVD 528

virus CELO 790
virus de la anemia infecciosa
    equina 1525
virus de la arteritis vírica
    equina 1516
virus de la artritis-encefalitis
    caprina 654
virus de la bronquitis infecciosa
    aviar 2173
virus de la diarrea vírica bovina
    528
virus de la ectromelia 1430
virus de la encefalitis japonesa
    2352
virus de la encefalomielitis
    aviar 349
virus de la encefalomielitis
    equina 1520
virus de la encefalomiocarditis
    678
virus de la enfermedad de
    Aujeszky 332
virus de la enfermedad de
    Borna 522
virus de la enfermedad de
    Carré 1342
virus de la enfermedad de
    Marek 2684
virus de la enfermedad de
    Newcastle 3006
virus de la enfermedad de
    Teschen 3553
virus de la enfermedad
    vesicular porcina 4392
virus de la enteritis vírica del
    pato 1396
virus de la estomatitis papular
    bovina 542
virus de la estomatitis vesicular
    4890
virus de la fiebre catarral
    maligna 2650
virus de la fiebre efímera
    bovina 530
virus de la gastroenteritis
    transmisible del cerdo 4627
virus de la gripe 2189
virus de la hepatitis canina
    infecciosa 640
virus de la hepatitis del pato
    1394
virus de la influenza aviar 356
virus de la influenza equina
    1527
virus de la influenza porcina
    4388
virus de la inmunodepresión
    felina 1643
virus de la laringo-traqueitis
    infecciosa aviar 357

virus de la lengua azul 497
virus de la leucosis bovina
    enzoótica 535
virus de la mixomatosis 2921
virus de la necrosis pancreática
    infecciosa 2181
virus de la panleucopenia
    felina 1649
virus de la parainfluenza
    bovina 543
virus de la parainfluenza
    canina 643
virus de la peste aviar 1755
virus de la peste bovina 3899
virus de la peste de los
    pequeños ruminantes 3493
virus de la peste equina 92
virus de la peste porcina
    africana 94
virus de la peste porcina clásica
    4386
virus de la rabia 3742
virus de la rinoneumonitis
    equina 1523
virus de la rinotraqueitis del
    gato 1642
virus de la rinotraqueitis
    infecciosa bovina 2101
virus de la seudorrabia 332
virus de la viruela ovina 4073
virus del ectima contagioso
    1045
virus de leucose felina 1647
virus del exantema vesicular
    4886
virus del moquillo 1342
virus del papiloma de los
    bovinos 540
virus del polioma 3540
virus del sarcoma de Rous 3932
virus del sarcoma felino 1651
virus IBR/IPV 2101
virus IPN 2181
virus oncogénico 3118
virus sincitial bovino 548
virus sincitial respiratorio 3832
virus sincitial respiratorio
    bovino 545
virus VHS 1940
virus visna-maedi 4935
visna 4934
vitamina A 3849
vitamina B1 4519
vitamina B2 3869
vitamina B6 3723
vitamina B12 1200
vitamina C 4936
vitamina D2 1533
vitamina D3 875
vitelo 4937

vitíligo 4938
vómer 4947
vómito 4951
volumen plasmático 3506
volumen sanguíneo 492
vomitoxina 4952
vulva 4954
vulvovaginitis 4955
vulvovaginitis pustulosa
    infecciosa 2182

wohlfahrtiosis 4977

xenobiótico 4984
xenodiagnóstico 4985
xilazina 4988

yersiniosis 4994
yeyuno 2356
yunque 2167

zearalenona 4995
zónula ciliar 917
zona blanca 4968
zoonosis 4999

Deutsch

Deutsch

aviäres Adenovirus 346
aviäres Enzephalomyelitis-
Virus 349
aviäres Herpesvirus 351
aviäres Influenzavirus 356
aviäres Paramyxovirus 360
aviäres Reovirus 361
avirulent 366
Avitaminose 367
Avoparcin 368
axillare Lymphknoten 375
Axis 380
Azaperon 381
Azetonämie 40
Azetonurie 41
Azidose 47
Azoospermie 382
Azoturie 383

Babesieninfektion 384
Babesiose 384
Babesiose des Pferdes 1517
Babesiose des Rindes 526
Babesiose des Schafes 3205
Bacillus-anthracis-Infektion
202
Bacillus-larvae-Infektion 139
Bacillus-piliformis-Infektion
4720
Bacitracin 387
Backe 848
Backendrüsen 596
Backengegend 599
Backenmuskel 600
Backenpapillen 598
Backentasche des Schweins
3438
Backentasthaare 849
Bacteroides-nodosus-Infektion
2177
Badeverfahren 4067
Bakteriämie 393
bakterielle Krankheit 395
bakterielle Nierenkrankheit
der Salmoniden 396
Bakterienantigen 394
Bakteriophage 398
Bakteriostatikum 204
Bakterium 399
bakterizid 397
Balanitis 400
Balanoposthitis 401
Balantidiose 402
Ballen 3219
Bambermycin 404
Band 2514
Bandwurm 841
Bandwurmbefall 842
Barbiturat 405
Barbula 407

Barthaare 4962
Bartholinischer Gang 3617
Bartonella-Infektion 410
Bartonellose 410
Basihyoid 415
Basilarmembran des
Schneckengangs 417
Basipodium 419
Bauch 1
Bauchaorta 3
Bauchfell 3412
Bauchfellentzündung 3413
Bauchhautvene 4321
Bauchhöhle 5
Bauchluftsack 2
Bauchmuskeln 9
Bauchschnitt 2419
Bauchspalte 6
Bauchspeicheldrüse 3252
Bauchspeicheldrüsenring 3258
Bauchwassersucht 305
Baytril 1473
bazilläre Hämoglobinurie 385
Bazillus 386
Becken 3373
Beckenachse 3358
Beckenausgang 3367
Beckeneingang 3364
Beckenfaszie 3361
Beckenganglien 3362
Beckengeflecht 3369
Beckengegenden 3371
Beckengliedmaße 3365
Beckengürtel 3363
Beckenhöhle 3359
Beckenneigung 2163
Beckenstück des Harnleiters
3368
Beckensymphyse 3372
Befall 2187
Befall mit Bunostomum 608
Befall mit Cooperia 1055
Befall mit Dioctophyme renale
1326
Befall mit Gongylonema 1880
Befall mit Habronema 1917
Befall mit Haemonchus 1934
Befall mit Hyostrongylus
rubidus 2069
Befall mit Kratzern 21
Befall mit Lungenwürmer 1298
Befall mit Palisadenwürmern
4246
Befall mit Trematoden 4652
Befall mit Varroa jacobsoni
4821
Befall mit
Zwergfadenwürmern 4245
Begattungsorgan 3432
Begleitvene 35

Behandlung 4515
Bein 2517
Belastung 4241
Belastungsinfektion 846
Benzalkoniumchlorid 427
Benzathin-Benzylpenicillin 428
Benzimidazol 429
Benzocain 430
Benzylpenicillin 431
Bepheniumhydroxynaphthoat
432
Berenil 1323
Berlocke 826
Berufskrankheit 3070
Berufs- und Standesrecht 4912
Beschälseuche 1389
Besnoitiose 433
beta-hämolytisch 436
Betamethason 437
Betapropiolakton 3636
Beta-Rezeptor 434
Beta-Rezeptoren-Blocker 435
Betastung 3245
Bezoar 438
BHV-1 2101
BHV-2 537
Biegung 1720
Bienenkrankheit 422
Bienenlaus 424
Bikuspidalklappe 2473
Bilirubin 448
Bindegewebe 1036
Bindehaut 1031
Bindehautarterien 1032
Bindehaut des Augapfels 603
Bindehautentzündung 1035
Bindehautgewölbe 1033
biochemischer
Polymorphismus 451
biologische Analyse 449
biologische Prüfung 449
Biopsie 452
Biotechnik 453
Biotechnologie 453
Biozoenose 450
Birnavirus 455
Bißwunde 456
Bithionol 457
Bläschen 4884
Blasenwurm 1209
blasses, weiches, wäßriges,
exsudatives Fleisch 3238
Blastomyces-dermatitidis-
Infektion 3025
Blättchenschicht der
Hufkapsel 2269
Blättermagen 3108
blattförmige Papillen 1732
blaue Schmeißfliege 493
blaugrüne Algen 495

Ferse 1980
Fersenbein 1692
Fersenhöcker 625
Fersensehnenstrang 1004
Fesselbein 1705
Fesselbeingegend 3657
Fesselbeugesehnenscheiden 1315
Fesselgelenk 1673, 2804
Fesselgelenksgegend 2791
Fesselgelenksgegend (der Beckgliedmaße) 2805
fester Knochenmantel 1016
Festliegen 3333
fetaler Anteil der Plazenta 1672
Fettgewebe 75
fettige Degeneration 1628
Fettleber 1629
Fettpolster 3220
Fettsucht 3053
Fibrae corticoreticulares 1094
Fibrae periventriculares 3415
Fibrae pontis transversae 4636
Fibrae tectospinales laterales 2464
Fibrille 1678
Fibrin 1679
Fibrinogen 1680
Fibroblast 1681
Fibrom 1684
Fibrosa des Perikards 1686
fibröse Kapsel 1685
Fibula 1690
Fieber 1676
fieberhaft 1639
fieberlos 88
Filamentum 1693
Filarie 1694
Filariose 1695
Fimbriae tubae uterinae 1699
Fimbrie 1698
Finger 1703
Fingerglied 3431
FIP 1645
Fischbandwurm des Menschen 574
Fischkrankheit 1707
Fissur 1709
Fissura interincisiva 2243
Fissura interlobaris 2244
Fissura lateralis cerebri 4393
Fissura orbitalis 3147
Fissura palatina 3225
Fissura petrotympanica 3425
Fistel 1710
Flanke 1711
Flankenfalte 1712
Flankengegend 2438
Flaumfedern 3519

Flavivirus 1714
Flavomycin 404
Flechsigsches Bündel 4862
Fleischbeschau 2705
Fleischbeschauer 2706
Fleischfliege 1717
Fleischhaut des Hodensacks 1215
Fleischhygiene 2704
Flexur 1720
Fliege 1728
Floh 1715
Flohbefall 1716
Flotzmaul 2949
Flotzmauldrüsen 2948
Fludrocortison 1724
Flügel 4969
Flügelbein 3664
Flügelfortsatz 3667
Flügelgaumengrube 3671
Flügelgrube des Atlas 314
Flügelloch des Atlas 109
Flugeston 1725
Flumetason 1726
Fluorose 1727
Fluorvergiftung 1727
Follikel 1733
Follikelhormon 1734
follikelstimulierendes Hormon 1734
Follikelzyste 1735
Foramen 1742
Foramen ethmoidale 1549
Foramen infraorbitale 2193
Foramen interventriculare 2302
Foramen ischiadicum 4006
Foramen jugulare 2360
Foramen lacerum 2392
Foramen mandibulae 2668
Foramen mastoideum 2694
Foramen mentale 2764
Foramen orbitorotundum 3151
Foramen ovale cordis 3188
Foramen palatinum 3226
Foramen rotundum 3926
Foramen sphenopalatinum 4130
Foramen spinosum 4150
Foramen stylomastoideum 4253
Foramen supracondylare 4340
Foramen supraorbitale 4350
Foramen supratrochleare 4366
Foramen transversarium 4638
Foramen venae cavae 4834
Foramina alveolares 132
Foramina papillaria 3266
Foramina sacralia 3961
Formaldehyd 1748

Formalin 1748
Formatio reticularis mesencephali 3835
Formveränderung 1242
Fornix 1749
Fornix conjunctivae 1033
Fortpflanzungsstörung 3825
Fossa 1750
Fossa acetabuli 36
Fossa axillaris 373
Fossa clitoridis 947
Fossa condylaris 1024
Fossa coronoidea 1086
Fossa ethmoidalis 1550
Fossa hypophysialis 2091
Fossa infraspinata 2200
Fossa ischiorectalis 2345
Fossa jugularis 2361
Fossa mandibularis 2669
Fossa olecrani 3090
Fossa ovalis cordis 3189
Fossa pararectalis 3294
Fossa supracondylaris 4341
Fossa trochanterica 4675
Fötus 1675
Fovea centralis retinae 793
Fraktur 1759
Frakturbehandlung 1760
Framycetin 1761
Franse 1698
freie Körper 1762
Fremdkörper 1746
Fremdstoff 4984
Frenulum 1764
Frontalmuskel 1769
Frucht 1675
Fruchthüllen 1671
Fruchttod 1670, 4224
Frühlingsvirämie der Karpfen 4184
FSH 1734
Fundus 1780
Fundus abomasi 1782
Fundusdrüsen 1781
Fundus reticuli 1783
Fundus uteri 1785
Fungistatikum 218
Funktionsstörung 1779
Furaltadon 1788
Furazolidon 1789
Furche 1904
Furosemid 1790
Furunkulose 1791
Fusarium-Mykotoxikose 1792
Fusobacterium-necrophorum-Infektion 2965
Fuß 1739
Füßchen 3357
Fußfäule 2177
Fußrücken 4758

Kreislauf 921
Kreislaufmittel 676
Kreislaufstörung 922
Kreislaufsystem 923
Kreuzbänder 1184
Kreuzbein 3974
Kreuzbein-Darmbein-Gelenk 3971
Kreuzbeinflügel 4975
Kreuzbeinganglien 3962
Kreuzbeingegend 3966
Kreuzbein-Schwanzmuskeln 3970
Kreuzgeflecht 3965
Kreuzhöcker 3968
Kreuzimmunität 1180
Kreuzwirbel 3969
Kriebelmücke 461
Kronbein 2835
Kronbeinregion 2834
Krone 1085
Krongelenk 3655
Krongelenkgegend 3656
Krongelenkschale 3900
Kronlederhaut 1078
Kronrand 1074
Kronsaum 1075
Kronsegment 1085
Kropf 1177, 1873
Kropfentzündung 2209
kropferzeugend 1874
Kropfregion 1179
Kropfverstopfung 1178
Kruppe 3966
Kryptokokkose 1186
Kryptorchismus 1187
Kryptosporidiose 1188
Kuhpocken 1120
Kulturmedium 1191
Kümmern 4753
Kunde 2207
Kupfermangel 1056
Kupfervergiftung 1057
kurzer Knochen 4077
kurzer Wadenbeinmuskel 4082
kurzer Zehenbeuger 4079
kurzer Zehenstrecker 4078
kurznasige Rinderlaus 4080
Kutis 4098
Kyphose 2382

Lab 3823
Labmagen 17
Labmagenentzündung 16
Labmagenfurche 15
Labmagenpararauschbrand 568
Laboratoriumstier 2389
Labordiagnostik 2390
Labrum acetabulare 38

Labrum glenoidale 1850
Labyrinth 2391
Labyrinthvenen 4833
Lachgas 3014
Lachskrankheit der Hunde 3982
Lade 2236
Lahmheit 2412
Lähmung 3283
Lähmung des Nervus obturatorius 3059
Laktationstörung 2403
Lamellenkörperchen 2409
Lamina 2413
Lamina arcus vertebrae 2414
Lamina basalis 411
Lamina choroidocapillaris 891
Lamina medullaris 2749
Lamina nuchae 2411
Lamina parietalis pericardii 3320
Lamina pr(a)evertebralis 3615
Lamina spiralis ossea 518
Lamina superficialis 4326
Lamina suprachoroidea 4338
Lamina visceralis pericardii 4932
Lämmerdysenterie 2407
langer Brustkorbnerv 2572
langer Halsmuskel 2569
Langerhanssche Inseln 3254
langer Knochen 2560
langer Rückenmuskel 2562
langer Seitwärtszieher des Schwanzes 966
langer Wadenbeinmuskel 2571
langer Zehenstrecker 2561
langnasige Rinderlaus 2570
Längsband der Wirbelsäule 2565
Längsfurche der Zunge 2734
Längsfurche des Rumens 2564
Längspfeiler des Rumens 2566
Längsspalt des Gehirns 2563
Lanzettegel 2418
Laparotomie 2419
Läppchen 2551
Lappen 2548
Lappenbronchen 2547
Laryngitis 2432
Laryngotracheitis 2434
Larynx 2435
Lasalocid 2436
Läsion 2496
latente Infektion 2437
lateral Brustkorbvene 2467
laterale Gehirnkammer 2469
laterale Halsregion 2458
laterale Kreuzbeinarterie 2461

laterale Retropharyngeallymphknoten 2459
lateraler Gelenkknorren 2444
lateraler Strahlschenkel 2460
laterale Sohlenarterie 2453
Lateralhorn der grauen Substanz 2447
Laus 2576, 4295
Läusebefall 2578
Lausfliege 2577
Lebendvakzine 2546
Lebensmittelhygiene 1736
Lebensmittelüberwachung 1737
Lebensmittelvergiftung 1738
Leber 2540
Leberarterie 1986
Leberegel 2543
Leberegelkrankheit 1625
Leberentzündung 1990
Leberkrankheit 2542
Leberläppchen 2545
Leberlappen 2544
Leberpforte 1988
Lebersteatose 1629
Lebervenen 1989
Lebervergrößerung 1994
Leberzelle 1991
Leberzirrhose 2541
Lecksucht 3466
Lederhaut 1065
Lederhautblättchen 2408
Lederzecke 265
Leerdarm 2356
Leerdarmarterien 2354
Legeabfall-Syndrom 1433
Leichenstarre 3897
Leishmaniose 2486
Leiste 1903
Leistenbruch 2211
Leistengegend 2214
Leistenkanal 2210
Leistenlymphknoten 4325
Leitungsanästhesie 2981
Lende 2557
Lendenanschwellung 2586
Lendenarterien 2585
Lendenganglien 2587
Lendengeflecht 2593
Lendengegend 2594
Lendennerven 2591
Lendenvenen 2596
Lendenwirbel 2597
Lentivirus 2492
Leporipoxvirus 2493
Leptospira 2494
Leptospirose 2495
letal 2499
Letalfaktor 2501

Lordose 2573
lösliches Antigen 4116
Louping-ill 2575
LTH 2610
Luftkapillaren 3521
Luftröhre 4615
Luftröhrenast 590
Luftröhrendrüsen 4619
Luftröhrenentzündung 4621
Luftröhrengabelung 4616
Luftröhrenmuskel 4620
Luftröhrenwindungen 4618
Luftröhrenwurm 1803
Luftsack (des Pferdes) 1914
Luftsäcke 108
Luftsackentzündung 107
Luftschlucken des Pferdes 86
Lumbalnerven 2591
Lumbalregion 2594
Lunge 2606
Lungenadenomatose des
    Schafes 3689
Lungenalveolen 3690
Lungenegel 2607
Lungenemphysem 3691
Lungenentzündung 3522
Lungenfell 3693
Lungenhilus 2005
Lungenläppchen 2553
Lungenlymphknoten 3692
Lungenpfeife 3276
Lungensegmente 589
Lungenseuche der Ziege 1042
Lungenseuche des Rindes 1041
Lungenvenen 3696
Lungenwurm 2608
Lungenwurzel 3908
Luteinisierungshormon 2609
luteotropes Hormon 2610
Luteotropin 2610
Lutropin 2609
Lyme-Krankheit 2611
Lymphadenitis 2613
Lymphangitis 2614
Lymphe 2612
Lymphgefäß 2632
Lymphgefäßentzündung 2614
Lymphgefäßgeflecht 2616
Lymphgefäßklappe 2619
Lymphgewebe 2629
Lymphherz 2621
Lymphknötchen 2615, 2620
Lymphknoten 2622
Lymphknotenentzündung 2613
Lymphocentrum coeliacum 973
Lymphocentrum iliofemorale
    2122
Lymphocentrum iliosacrale
    2130

Lymphocentrum
    inguinofemorale 2215
Lymphocentrum ischiadicum
    2340
Lymphocentrum lumbale 2590
Lymphocentrum mandibulare
    2673
Lymphocentrum mediastinale
    2737
Lymphocentrum mesentericum
    caudale 742
Lymphocentrum mesentericum
    craniale 1143
Lymphocentrum parotideum
    3328
Lymphocentrum popliteum
    3548
Lymphocentrum
    retropharyngeum 3855
Lymphocentrum thoracicum
    dorsale 1381
Lymphocentrum thoracicum
    ventrale 4863
Lymphocystis-Krankheit 2624
Lymphoidzelle 2628
Lymphom 2630
Lymphonodi iliofemorales
    2121
Lymphonodi poplitei 3547
Lymphonodi subscapulares
    4287
Lymphonodi
    tracheobronchales 4622
Lymphopathie 2618
Lymphosarkom 2631
Lymphozyt 2625
lymphozytäres
    Choriomeningitisvirus 2626
Lymphozytose 2627
Lymphsystem 2617
Lymphzentrum 2623
Lyse 2633
Lysis 2633
Lysosom 2634
Lysotyp 3427
Lysozym 2635

Maedi 2639
Magen 4226
Magen- 1806
Magenbremse 2043
Magen-Darm- 1818
Magendarmentzündung 1817
Magendasselfliege 2043
Mageneindruck 1809
Magenentzündung 1813
Magenfundus 1784
Magengeschwür 4227
Magengrübchen 1810
Magengrund 1784

Magenkrümmung 1194
Magenpförtner 3714
Magenrinne 1808
Magenvenen 1812
Magnesiummangel 2640
Mähne 2676
Makrolidantibiotikum 2636
Makrophage 2637
makroskopisch 2638
Malabsorption 2645
maligne Hyperthermie des
    Schweins 2651
malignes Ödem 2652
Mallein 2653
Mamillitis 2657
Mamillitis-Virus 537
Mandel 4594
Mandelentzündung 4597
Mandelgrübchen 4596
Mandelgrube 4595
Mandelkörper 153
Mangel 1240
Mangelkrankheit 1241
Manica flexoria 2679
männliche Geschlechtsorgane
    2649
männliches Glied 3379
männliches Sexualhormon 177
Mannosidose 2680
Manubrium 2682
Mareksche Krankheit 2683
Marek-Virus 2684
Margo supraorbitalis 4352
Markhöhle 2747
Markpyramide 3819
Marteilia-Infektion 2685
Massetergegend 2688
Mastadenovirus 2690
Mastdarm 3787
Mastdarmarterien 3773
Mastdarmgekröse 2781
Mastdarm-Harnröhrenfistel
    3782
Mastdarm-Scheidenfistel 3784
Mastdarmvenen 3779
Mastdarmvorfall 3777
Mastitis 2691
Mastitis gangraenosa 1802
Mastitis-Metritis-Agalaktie-
    Syndrom 2692
Mastozyt 2693
Mastzelle 2693
Maul- und Klauenseuche 1740
Mäusepocken 1429
Mäusepockenvirus 1430
Maxilla 2696
maxillare Backendrüse 5006
Meatus 2707
Mebendazol 2708
Meckelsches Divertikel 2709

Media 2711
mediale Palmararterien 3240
medialer Darmbeinwinkel 3968
medialer Gelenkknorren 2714
medialer Strahlschenkel 2725
mediale Sohlenarterie 2720
mediane Euterfurche 2246
mediane Kreuzbeinarterie
    2726
Medianspalt des Rückenmarks
    2731
Mediastinallymphknoten 2736
Mediastinum 2741
Medikation 2743
Medizinalfutter 2742
Medroxyprogesteron 2745
Medulla oblongata 2746
Megestrol 2750
Mehrfachovulation 4334
Mehrfachvakzine 1000
Meibomsche Drüsen 4445
Mekonium 2710
Melanom 2753
Melanose 2754
Melanozyt 2751
melanozytstimulierendes
    Hormon 2752
Melatonin 2755
meldepflichtige Krankheit
    3030
Melioidose 2756
Melissococcus-pluton-
    Infektion 1557
Melkerknoten 3660
Membran 2757
Membrana interossea 2286
Membrana pupillaris 3701
Membrana
    sternocoracoclavicularis
    4210
Membrana vitrea 4942
Meningen 2760
Meningitis 2761
Meningoenzephalitis 2762
Menopon gallinae 3580
Mentalbüschel 859
Merkelsche Tastzellen 2768
Merozoit 2769
Mesencephalon 2771
Mesenchym 2772
Mesenterium 2774
Mesometrium 2778
Mesonephros 2779
Mesorchium 2780
Mesosalpinx 2782
Mesovarium 2783
Metacarpus 2792
Metakarpalarterien 2785
Metakarpalballen 2788
metakarpale Exostose 4174

Metakarpalien I-V 2786
Metakarpalregionen 2789
Metakarpalvenen 2790
Metallibur 2794
Metapatagium 2795
Metaphyse 2796
Metastrongylose 2797
Metatarsalarterien 2798
Metatarsalballen 2801
Metatarsalien I-V 2799
Metatarsalregionen 2802
Metatarsalvenen 2803
Metatarsus 2806
Metazerkarie 2793
Methadon 2808
Methämoglobin 2809
Metomidat 2810
Metrifonat 2811
Metritis 2812
Metronidazol 2813
Meynert-Bündel 2814
Mikrobe 2821
Mikrobenzahl 994
mikrobiell 2815
Mikrobiologie 2816
Mikrogamet 2819
Mikrogamont 2820
Mikrokokk 2818
Mikroorganismus 2821
Mikrophthalmie 2822
Mikroskopie 2823
Mikrosporidien 2824
Mikrozephalie 2817
Milbe 2846
Milchader 4321
Milchbrustgang 4536
Milchdrüse 2659
Milchdrüsenläppchen 2554
Milchdrüsenlappen 2550
Milchgang 2404
Milchhygiene 2837
Milchmangel 98
Milchzähne 1220
Milchzisterne 2405
Milz 4165
Milzarterie 4168
Milzbrand 202
Milzbrandbazillus 203
Milzhilus 2007
Milztrabekeln 4613
Milzvene 4170
Milzvergrößerung 4173
Mineralokortikoid 2840
Mineralstoffmangel 2838
minimal letale Dosis 2841
Mirazidium 2845
Mißbildung 1028, 1242
Mitochondrien 2847
Mitralklappe 2473
Mittelfell 2738

Mittelfellvenen 2740
Mittelfuß 2806
Mittel gegen Protozoen 227
Mittel gegen Trypanosomen
    4688
Mittel gegen Tympanie 208
Mittelhand 2792
Mittelhirn 2771
Mittelhirnhaube 2770
Mittelhirnkanal 254
Mittellappen der Lunge 2832
Mittelohr 2829
Mittelohrentzündung 3186
Mittelschicht der Hufplatte
    2831
Mittelstück der Rippe 4064
Mittelstück des
    Oberschenkelbeins 4063
Mittelteil des Nebenhodens
    502
mittlere Augenhaut 4827
mittlere Bauchregion 2826
mittlere Kruppenmuskel 2830
mittlere Nasenmuschel 2833
mittlere
    Retropharyngeallymphknoten
    2724
mittlerer sympathischer
    Halsnervenknoten 2827
mittlere Schicht der
    Gefäßwand 2711
mittlere Strahlfurche 794
MKS 1740
MMA-Komplex 2692
Moderhinke 2177
Modiolus 2848
Molaren 2850
Molekularbiologie 2851
Molekularschicht der
    Kleinhirnrinde 2852
Molybdänose 2853
Molybdän-Vergiftung 2853
Monensin 2854
Monieziose 2855
monoklonale Gammopathie
    2857
monoklonaler Antikörper 2856
Morantel 2858
Moraxella-bovis-Infektion 533
Morbillivirus 2859
Morphologie 2861
morphologisch 2860
Mortalität 2862
Mortalitätsziffer 2863
motorischer Kern 2866
motorischer Nerv 2865
motorische Wurzel des Nervus
    trigeminus 2867
MSH 2752

Orbitalflügel des Keilbeins 4974
Orbivirus 3153
Orchitis 3154
Orchitomie 704
organische Chlorverbindung 3156
organische Phosphorverbindung 3158
Organkrankheit 3155
Ormetoprim 3161
Ornithose 3162
Orthomyxovirus 3165
Orthopoxvirus 3166
Orthopoxvirus der Lumpy skin disease 2605
örtliche Betäubung 2555
Os angulare 184
Os articulare 285
Os basioccipital 418
Os carpale I 1704
Os carpale II 4028
Os carpale III 4525
Os carpale IV 1751
Os carpi accessorium 26
Os carpi intermedium 2251
Os carpi radiale 3745
Os carpi ulnare 4729
Os centroquartale 801
Os dentale 1252
Os dorsale 3029
Os ectethmoidale 1425
Os entoglossum 1480
Os exoccipitale 1564
Os mesethmoidale 2775
Os orbitosphenoidale 3152
Os otica 3182
Os parasphenoidale 3305
Os prearticulare 3588
Os prefrontale 3595
Os scaphoideum 3745
Ossifikation 3168
Os spleniale 4167
Os squamosum 4188
Os supraangulare 4337
Os suprajugale 4343
Os supraorbitalia 4348
Os tarsi centrale 797
Osteoblast 3170
Osteochondritis 3171
Osteodystrophie 3174
Osteogenese 512
Osteogenesis imperfecta 3175
Osteoklast 3172
Osteomalazie 3176
Osteomyelitis 3177
Osteopathie 511
Osteopetrose 3178
Osteophyt 3179
Osteoporose 3180

Osteosynthese 1760
Osteozyt 3173
Ostertagiose 3181
Ostitis 3169
Ostium pyloricum 3712
Ostium trunci pulmonalis 3124
Ostium venae cavae 3125
Ostküstenfieber 1422
Otitis 3184
Otitis externa 3185
Otitis media 3186
Otobius 1419
Otodektesräude 3187
Ovar 3200
Ovarektomie 3199
Ovidukt 3204
ovine Pestivirose 520
ovines und caprines Morbillivirus 3493
Ovulation 3208
Oxacillin 3210
Oxfendazol 3211
Oxibendazol 3212
Oxolinsäure 3213
Oxyclozanid 3214
Oxytetracyclin 3216
Oxytocin 3217
Oxyuriose 3218

Pacchionische Granulationen 256
Paecilomykose 3221
Palatoschisis 934
Palisadenwürmer 4244
Pallium 3239
Palpation 3245
Panacur 1666
Pankreas 3252
Pankreaserkrankung 3253
Pankreasinseln 3254
Pankreassaft 3255
Pankreatitis 3260
Pansen 3935
Pansenatonie 3941
Pansenegel 3287
Pansenegelbefall 3288
Pansenfistel 3936
Pansenflora 3937
Pansengeflechte 3947
Pansenhaubenöffnung 3955
Panseninhalt 3942
Panseninsel 3943
Pansenlymphknoten 3944
Pansenpapillen 3945
Pansenpfeiler 3946
Pansenprotozoen 3940
Pansenschnitt 3939
Pansenstillstand 3941
Pansen und Haube 3956
Pansenvenen 3949

Panzootie 3261
Papel 3275
Papilla duodeni 1404
Papilla ilealis 2109
Papilla incisiva 2159
Papillarkörper 3267
Papillarmuskeln 3268
Papille 3262
Papillen der Drüsenmagen 3263
Papillom 3270
Papillomavirus 3271
Papovavirus 3272
papulös 3273
Paradidymis 3278
Paragonimose 3279
Parainfluenza 3281
Parakeratose 3282
Paralyse 3283
paralytisch 3284
paralytische Myoglobinurie des Pferdes 3285
Paramphistomose 3288
Paramunisierung 3021
Paramyxovirus 3289
Paramyxovirus der Newcastle-Krankheit 3006
Paraplegie 3291
Parapoxvirus 3292
Parapoxvirusinfektion 3660
Paraproteinämie 3293
Pararauschbrand 2652
Parasit 3297
Parasitämie 3296
parasitär 3298
parasitäre Krankheit 3304
Parasitenbefall 3300
parasitischer Arthropode 284
Parasitismus 3301
Parasitologie 3303
parasitologisch 3302
Parasitose 3304
Parasit-Wirt-Beziehung 2049
Paraskaridose 3295
parasympathischer Kern 3307
parasympathisches Nervensystem 3306
Parasympatholytikum 3308
Parasympathomimetikum 3309
Parathormon 3311
Parathyreoidea 3310
Paratuberkulose 3312
Parazentese 3277
Parazitizid 3299
Parbendazol 3314
Pärchenegel 4003
Parenchym 3315
parenteral 3316
parenterale Verabreichung 3317

plötzlicher Tod 4296
pneumatischer Knochen 3520
Pneumonie 3522
Pneumostrongylose 3523
Pneumothorax 3524
Pneumovirus 3525
Pododermatitis 3526
Podotheca 3527
Podotrochlose 2958
Polioenzephalomalazie 819
Pollex 4551
Polyarthritis 3533
Polydaktylia 3534
Polydipsie 3535
Polymastie 3536
Polymelie 3537
polymorphkerniger Granulozyt
    3538
Polymyxin B 3539
Polyomavirus 3540
Polyserositis 3541
Polyserositis und Polyarthritis
    des Schweines 3560
Polythelie 3542
Polyurie 3543
Pons 3544
porcine Gastroenteritis-
    Coronavirus 4627
porcines Herpesvirus 3557
porcines Parvovirus 3558
Porphyrie 3561
porzines Bläschenkrankheit-
    Enterovirus 4392
porzines Enzephalomyelitis-
    Enterovirus 3553
Posphorsäureester 3158
Posthitis 3567
postmortale Veränderung 3568
postnatale Entwicklung 3571
postnatale Periode 3572
postoperativ 3573
Postpatagium 3574
postvakzinal 3575
Poxvirus 3584
prädisponierender Faktor 3591
Prädisposition 3592
Pralidoxim-Jodid 3585
Prämolaren 3601
Prämunität 3602
Präputialbeutel 3605
Präputialkatarrh 401
Präputialöffnung 3609
Präputium 3604
Prävalenz einer Krankheit 1333
Präzipitationsreaktion 3589
Präzipitin 3590
Praziquantel 3586
Prednisolon 3593
Prednison 3594
Primärinfektion 3616

Probe 4122
Probiotikum 3619
Procain 3620
Processus anconeus 174
Processus coronoideus 1087
Processus jugularis 2366
Processus muscularis 2888
Processus papillaris hepatis
    3269
Processus xiphoideus 4986
Processus zygomaticus 5004
Proctodeum 3621
Progesteron 3622
Prognathie 3623
Prognose 3624
Prolaktin 3625
Prolapsus ani 3777
Proligeston 3627
Promazin 3628
Promethazin 3629
Prominentia laryngea 2429
Properdin 3633
prophylaktisch 3634
Prophylaxe 3635
Propiolakton 3636
Prosenzephalon 3637
Prostaglandin 3639
Prostata 3640
Prostazyklin 3638
Prosthogonimose 3642
Protease 3643
Protein 3644
Proteinurie 3646
proteolytisches Enzym 3643
Prothrombin 3647
Protostrongylose 3648
protozoal 3650
protozoäre Infektion 3651
Protozoen 3649
Protozoeninfektion 3651
Protuberantia 3653
Protuberantia occipitalis
    externa 1588
Protuberantia occipitalis
    interna 2272
Pruritus 3659
Psalter 3108
Psalterblatt 3105
Psalterhals 2964
Psalterkanal 3103
Psalterlabmagenöffnung 3107
Psalterpapille 3106
Psalterrinne 3104
PSE-Fleisch 3238
Pseudogeflügelpest 3005
Pseudomonas-mallei-Infektion
    1841
Pseudomonas-pseudomallei-
    Infektion 2756
Pseudopocken 3660

Pseudorotz 1514
Pseudoträchtigkeit 3661
Pseudotuberkulose 3662
Pseudotuberkulose des Schafes
    700
Pseudowut 331
Psittakose 3162
Psoroptesräude 3663
Puerperalstörung 3687
Puerperium 3688
Pulmonalklappe 4818
Pulsfrequenz 3699
Pulvipluma 3583
punktförmig 3424
punktförmige Blutung 3423
Pupille 3700
Purkinje-Zellschicht 1798
Pustel 3705
pustulös 3704
Putamen 3706
Puten-Herpesvirus 4706
Pyelitis 3707
Pyga 3957
Pygostyl 3708
Pylorus 3714
Pylorusdrüsen 3711
Pylorusvorhof 3709
Pyodermie 3715
Pyramidenbahn 3718
Pyramidenkreuzung 1222
Pyramis medullae oblongatae
    3719
Pyrantel 3720
Pyrethrin 3721
Pyrexie 3722
Pyridoxin 3723
Pyrimethamin 3724
Pyritidium-Bromid 3725
Pyrogen 3726
Pyrrolizidinalkaloid 3727

Q-Fieber 3728
Quadratlappen der Leber 3730
Quadratum 3729
Quarantäne 3735
Quecksilbervergiftung 2767
querer Bauchmuskel 4630
querer Brustkorbmuskel 4646
querer Brustmuskel 4643
Querfortsatz 4644
quergestreifter Muskul 4243
Querkolon 4635
Querspalt des Gehirns 4637
querverlaufende Arterien 4632
Quesenbandwurm 4430

Rabenschnabel-Armmuskel
    1059
Rabenschnabelbein 1061
Rabenschnabelfortsatz 1062

Latine

catarrhus 707
cauda 4432
cauda epididymidis 4436
cavitas glenoidalis 1849
cavum abdominis 5
cavum articulare 2359
cavum cranii 1127
cavum dentis 3697
cavum epidurale 1498
cavum hypophysis 2090
cavum laryngis 2426
cavum medullare 2747
cavum nasi 2934
cavum oris 3137
cavum pelvis 3359
cavum pericardii 3389
cavum peritonei 3411
cavum pharyngis 3437
cavum pleurae 3510
cavum subarachnoideale 4256
cavum subdurale 4272
cavum thoracis 4535
cavum tympani 4711
cavum uteri 4783
cecum 621
cefalexinum 775
cefaloridinum 776
cefalotinum 777
cefotaximum 778
cellula 779
cellulae ethmoidales osseae
    1547
cellulae reticuli 787
cementum 1254
centrum tendineum perinei 798
cera 806
ceratohyoideum 804
Ceratophyllus gallinae 857
cerebellum 811
cerebrum 823
cervix 2584
cervix uteri 840
cervix vesicae 2961
Cestoda 841
cetrimidum 843
cetylpyridinii chloridum 844
chiasma opticum 3130
Chlamydiales 861
chloramphenicolum 865
chlorhexidinum 866
chlormadinonum 867
chloroquinum 869
chloroxylenolum 870
chlorpromazinum 871
chlortetracyclinum 872
choanae 3566
chordae tendinae 4479
chorioidea 890
chorion 886
choroidea 890

chylus 901
chymotrypsinum 906
chymus 904
Ciliata 918
cilium 1605
cingulum membri pelvini 3363
cingulum membri thoracici
    4084
ciprofloxacinum 920
cirrus capitis 1747
cirrus caudae 4438
cirrus metacarpeus 1674
cirrus metatarseus 1674
cisterna cerebellomedullaris
    810
cisterna chyli 902
cisternae subarachnoideales
    4255
cisterna magna 810
claudicatio 927
clavicula 928
clazurilum 931
clenbuterolum 941
clioquinolum 945
clioxanidum 946
clitoris 948
cloaca 949
clofenotanum 1216
clofenvinfosum 954
clopidolum 955
cloprostenolum 956
clorsulonum 957
closantelum 958
cloxacillinum 961
clunis 616
coccyx 4432
cochlea 967
cochlea tibiae 4570
Cochliomyia hominivorax 4015
Coenurus cerebralis 976
colecalciferolum 875
colistinum 985
colliculus caudalis 721
colliculus rostralis 3915
collum 2959
collum humeri 2963
collum omasi 2964
collum ossis femoris 2962
colon 993
colon ascendens 304
colon crassum 2421
colon descendens 1271
colon dorsale 1354
colon sigmoideum 4089
colon tenue 4105
colon transversum 4635
colon ventrale 4846
columna vertebralis 4875
commissura alba 4963

commissurae supraopticae
    4345
commissura grisea 1900
commissura labiorum oris 1002
commissura palpebrarum 1001
concha auriculae 335
concha nasalis 2935
concha nasalis dorsalis 1363
concha nasalis media 2833
concha nasalis ventralis 4850
condylus 1026
condylus humeri 1027
condylus lateralis 2444
condylus medialis 2714
condylus occipitalis 3063
conus arteriosus 271
conus epididymidis 2552
conus medullaris 2748
coprodeum 1058
cor 1969
coracoideum 1061
corium 1065
corium coronae 1078
corium cunei 1066
corium limbi 3404
corium parietis 2416
corium soleae 1067
cor lymphaticum 2621
cornea 1068
cornu 2033
cornu ammonis 2014
cornua uteri 2038
cornu dorsale 1360
cornu laterale 2447
cornu rostrale ventriculi
    lateralis 3917
cornu temporale ventriculi
    lateralis 4471
cornu ventrale 4849
corona 1085
corona dentis 1183
corona radiata 1076
corpora libera 1762
corpora nigra 1886
corpora quadrigemina 721,
    3915
corpus adiposum 3220
corpus albicans 1088
corpus alienum 1746
corpus amygdaloideum 153
corpus callosum 1089
corpus cavernosum penis 771
corpus ciliare 911
corpus costae 4064
corpuscula nervosa terminalia
    4493
corpusculum bicellulare 440
corpusculum bulboideum 605
corpusculum lamellosum 2409

glandulae linguales 2524
glandulae molares 2849
glandulae nasales 2938
glandulae odoriferae 4002
glandulae (o)esophageae 3076
glandulae olfactoriae 3095
glandulae oris 3139
glandulae palatinae 3227
glandulae pelvis renalis 1842
glandulae pharyngea 3439
glandulae plani nasolabiales 2948
glandulae pr(a)eputiales 3607
glandulae pyloricae 3711
glandulae sebaceae 4024
glandulae sine ductibus 1399
glandulae sudoriferae 4381
glandulae tarsales 4445
glandulae tracheales 4619
glandulae uretericae 4762
glandulae urethrales 4764
glandulae uterinae 4786
glandulae vesicales 1844
glandulae vestibulares minores 2844
glandula lacrimalis 2397
glandula mammaria 2659
glandula mandibularis 2671
glandula maxillaris 2698
glandula mentalis 859
glandula mucosa 2873
glandula nasalis lateralis 2452
glandula orbitalis 5006
glandula parathyroidea 3310
glandula parotis 3327
glandula pinealis 3476
glandula pituitaria 2092
glandula profunda palpebrae tertiae 1959
glandula seromucosa 4050
glandula serosa 4053
glandula sublingualis 4280
glandula superficialis 4323
glandula suprarenalis 77
glandula thyr(e)oidea 4561
glandula uropygialis 4779
glandula ventis 4843
glandula vesicularis 4888
glandula vestibularis major 2644
glandula zygomatica 5006
glans penis 1846
glomerulus 1855
glomus caroticum 681
Glossina 4692
glottis 1860
gloxazonum 1861
glucagonum 1862
gomphosis 1875
gonadorelinum 1877

gonadotrophinum chorionicum 887
gonadotrophinum sericum 4058
Goniocotes gallinae 3578
Goniodes gigas 2420
granula iridica 1886
granulationes arachnoideales 256
griseofulvinum 1902
guaifenesinum 1908
gubernaculum testis 1909
gyri cerebri 813
gyrus cinguli 919
gyrus dentatus 1258
gyrus parahippocampalis 3280

habenula 1915
Haematobia irritans 2035
Haematopinus asini 2047
Haematopinus eurysternus 4080
Haematopinus suis 3471
Haematopinus tuberculatus 601
halofuginonum 1954
halothanum 1955
haloxonum 1956
hamulus 2028
hamulus pterygoideus 3665
haustra c(a)eci 618
helicotrema 1981
helix 1982
hemiplegia laryngis 3902
hemispherium 815
hepar 2540
hepatitis contagiosa canis 646
hepatitis distomatosa 1625
hepatitis infectiosa necrotica 460
hernia abdominalis 7
hernia diaphragmatica 1283
hernia incarcerata 4236
hernia inguinalis 2211
hernia perinealis 3396
hernia scrotalis 4016
hernia umbilicalis 4739
hexachlorphenum 2002
hiatus 2004
hiatus aorticus 237
hiatus (o)esophageus 3077
hilus lienis 2007
hilus ovarii 2006
hilus pulmonis 2005
hilus renalis 3813
Hippobosca equina 2046
Hippoboscus 2577
hippocampus 2014
homidii bromidum 2024
humerus 2054

humor aquosus 255
humor vitreus 4941
hyaluronidasum 2058
hydrocortisonum 2060
Hypoderma bovis 1012
Hypoderma lineatum 4109
hypopenna 97
hypophysis (cerebri) 2092
hypothalamus 2096

idoxuridinum 2103
ileum 2112
imidocarbum 2135
impressio cardiaca 670
impressio colica 981
impressio gastrica 1809
impressio (o)esophagea 3078
impressio renalis 3814
incisura 2161
incisura acetabuli 37
incisura cardiaca pulmonis 672
incisurae costales sterni 1103
incisura fibularis tibiae 1691
incisura glenoidalis 1851
incisura ischiadica 4008
incisura jugularis 2364
incisura mandibulae 2675
incisura poplitea 3550
incisura radialis ulnae 3748
incisura scapulae 2162
incisura supraorbitalis 4351
incisura trochlearis 4682
incisura ulnaris radii 4733
incisura vertebralis 4878
inclinatio pelvis 2163
incontinentia urinae 4769
incus 2167
inertia uteri 4787
infundibulum 2206
infundibulum dentis 2207
infundibulum hypothalami 3484
infundibulum tubae uterinae 2208
ingluvies 1177
inguen 1903
Insecta 2220
insulae pancreaticae 3254
insula ruminis 3943
integumentum commune 1010
interferonum 2241
interstitium 2297
intestinum 2315
intestinum crassum 2422
intestinum tenue 4107
intumescentia cervicalis 831
intumescentia lumbalis 2586
involutio uteri 4788
iridocyclitis recidivans 3789
iris 2337

medulla glandulae suprarenalis 78
medulla oblongata 2746
medulla ossium 513
medulla ossium flava 4990
medulla ossium rubrum 3791
medulla renis 3815
medulla spinalis 4138
megestrolum 2750
Melophagus ovinus 4070
membrana 2757
membranae fetales 1671
membrana interossea 2286
membrana pupillaris 3701
membrana spiralis 4156
membrana sternocoracoclavicularis 4210
membrana synovialis 4412
membrana tympani 4712
membrana vestibularis 3804
membrana vitrea 4942
membrum 2517
membrum pelvinum 3365
membrum thoracicum 4538
meninges 2760
menisci tactus 2768
meniscus articularis 289
Menopon gallinae 3580
mentum 858
mesencephalon 2771
mesenterium 2774
mesocolon 2776
mesoduodenum 2777
mesometrium 2778
mesonephros 2779
mesorchium 2780
mesorectum 2781
mesosalpinx 2782
mesovarium 2783
metacarpus 2792
metalliburum 2794
metapatagium 2795
metaphysis 2796
metatarsus 2806
metencephalon 2807
methadonum 2808
metomidatum 2810
metrifonatum 2811
metronidazolum 2813
Micrococcus 2818
Microsporidia 2824
modiolus 2848
monensinum 2854
morantelum 2858
morbus vesicularis suum 4391
mucus 2874
Multiceps multiceps 4430
Musca autumnalis 1609
Musca domestica 2050

musculi abdominis 9
musculi apteriales 253
musculi arrectores pilorum 268
musculi auriculares 1417
musculi bulbi 1599
musculi capitis 2882
musculi caudae/coccygis 4435
musculi colli 2960
musculi cutanei 1197
musculi dorsi 389
musculi femorotibiales 1664
musculi gemelli 1819
musculi hyoidei 2068
musculi iliotibiales 2131
musculi iliotrochanterici 2132
musculi intercostales 2231
musculi interflexorii 2242
musculi interossei 2287
musculi interspinales 2294
musculi intertransversarii 2300
musculi laryngis 2428
musculi linguae 4592
musculi lumbricales 2603
musculi multifidi 2876
musculi papillares 3268
musculi pectinati 3351
musculi pennales 1635
musculi perinei 3397
musculi pr(a)eputiales 3608
musculi rotatores 3924
musculi sacrocaudales/sacrococcygis 3970
musculi skeleti 4096
musculi subcoracoscapulares 4265
musculi subcostales 4266
musculi syringeales 4421
musculi thoracis 4540
musculus 2879
musculus abductor cruris caudalis 718
musculus abductor cruris cranialis 1125
musculus abductor digiti I brevis 4075
musculus abductor digiti II 12
musculus abductor digiti I longus 2558
musculus adductor 61
musculus adductor brevis 4076
musculus adductor digiti I 63
musculus adductor digiti II 65
musculus adductor digiti V 62
musculus adductor longus 2559
musculus adductor magnus 1893
musculus adductor mandibulae caudalis 737

musculus adductor mandibulae externus 1584
musculus adductor rectricius 64
musculus ancon(a)eus 175
musculus articularis 290
musculus articularis coxae 291
musculus articularis genus 293
musculus articularis humeri 292
musculus arytenoideus transversus 4634
musculus biceps brachii 441
musculus biceps femoris 442
musculus bipennatus 454
musculus brachialis 556
musculus brachiocephalicus 560
musculus brachioradialis 563
musculus bronchoesophageus 587
musculus buccinator 600
musculus bulbi rectricium 3786
musculus bulbocavernosus 604
musculus bulbospongiosus 604
musculus caninus 642
musculus caudofemoralis 766
musculus caudo-ilio-femoralis 767
musculus ceratohyoideus 805
musculus ciliaris 914
musculus cleidobrachialis 935
musculus cleidocephalicus 936
musculus cleidocervicalis 937
musculus cleidomastoideus 938
musculus cleidooccipitalis 939
musculus cleidotrachealis 940
musculus coccygeus 966
musculus constrictor 1038
musculus constrictor colli 1039
musculus coracobrachialis 1059
musculus costoseptalis 1110
musculus costosternalis 1111
musculus crassus 4520
musculus cremaster 1163
musculus cricoarytenoideus 1169
musculus cricopharyngeus 1171
musculus cricothyroideus 1174
musculus cucullaris 1189
musculus cutaneus colli 835
musculus cutaneus trunci 10
musculus deltoideus 1247
musculus depressor labii inferioris 1265
musculus depressor mandibulae 2666
musculus digastricus 1306
musculus dilatator 1318

musculus serratus dorsalis 1377
musculus serratus ventralis cervicis 4859
musculus serratus ventralis thoracis 4860
musculus soleus 4113
musculus sphincter 4133
musculus sphincter ampullae hepatopancreaticae 3073
musculus sphincter ani 167
musculus sphincter ani externus 1572
musculus sphincter cloacae 952
musculus sphincter ductus choledochi 4135
musculus sphincter ilei 2105
musculus sphincter papillae 4461
musculus sphincter pupillae 4134
musculus sphincter pylori 3713
musculus sphincter urethrae 4766
musculus spinalis 4141
musculus splenius capitis 4171
musculus splenius cervicis 4172
musculus stapedius 4193
musculus sternocephalicus 4208
musculus sternocoracoideus 4212
musculus sternohyoideus 4214
musculus sternomandibularis 4215
musculus sternomastoideus 4216
musculus sternothyroideus 4218
musculus sternotrachealis 4219
musculus styloglossus 4247
musculus stylohyoideus 4249
musculus stylopharyngeus caudalis 754
musculus stylopharyngeus rostralis 3922
musculus subclavius 4258
musculus subscapularis 4288
musculus supinator 4335
musculus supracoracoideus 4342
musculus supramammarius 4344
musculus supraspinatus 4364
musculus tarsalis 4447
musculus temporalis 4475
musculus tensor fasciae antebrachii 4483
musculus tensor fasciae latae 4484

musculus tensor propatagialis 3631
musculus tenuis 4523
musculus teres major 2424
musculus teres minor 4108
musculus thyroarytenoideus 4554
musculus thyrohyoideus 4558
musculus thyropharyngeus 4564
musculus tibialis caudalis 758
musculus tibialis cranialis 1158
musculus trachealis 4620
musculus tracheolateralis 4624
musculus transversospinalis 4647
musculus transversus abdominis 4630
musculus transversus thoracis 4646
musculus trapezius 4648
musculus triceps brachii 4656
musculus triceps surae 4657
musculus ulnaris lateralis 4730
musculus vastus intermedius 2256
musculus vastus lateralis 2468
musculus vastus medialis 2728
musculus zygomaticus 5002
Mycobacterium 2895
Mycobacterium avium 363
Mycobacterium bovis 550
Mycobacterium tuberculosis 4696
Mycoplasma 2898
myelencephalon 2904
myocardium 2912
myometrium 2917
myxomatosis 2920

naftalofosum 2924
nandrolonum 2928
narasinum 2930
naris 3028
nasus 3026
nasus externus 1585
nates 616
Nematoda 2970
neomycinum 2972
neopallium 2974
nequinatum 2979
nervi auriculares 337
nervi cardiaci cervicales 830
nervi cardiaci thoracici 4534
nervi carotici externi 1574
nervi caroticotympanici 680
nervi caudales/coccygei 745
nervi cervicales 834
nervi ciliares 915
nervi clunium caudales 720

nervi clunium craniales 1129
nervi clunium medii 2828
nervi craniales 1145
nervi digitales dorsales 1358
nervi digitales palmares 3241
nervi digitales plantares 3497
nervi intercostales 2232
nervi labiales 2386
nervi lumbales 2591
nervi mentales 2765
nervi metacarpei palmares 3243
nervi metatarsei dorsales 1362
nervi metatarsei plantares 3499
nervi olfactorii 3096
nervi palatini 3229
nervi pectorales 3355
nervi pelvini 3366
nervi perineales 3398
nervi rectales caudales 752
nervi sacrales 3964
nervi spinales 4142
nervi splanchnici lumbales 2595
nervi splanchnici sacrales 3967
nervi supraclaviculares 4339
nervi temporales profundi 1234
nervi thoracici 4541
nervi vaginales 4807
nervus 2980
nervus abducens 11
nervus accessorius 30
nervus alveolaris mandibularis 2664
nervus ampullaris 152
nervus auriculopalpebralis 338
nervus auriculotemporalis 339
nervus axillaris 377
nervus buccalis 597
nervus caroticus internus 2263
nervus costoabdominalis 1105
nervus cutaneus antebrachii medialis 2715
nervus cutaneus femoris caudalis 723
nervus cutaneus femoris lateralis 2445
nervus depressor 1266
nervus dorsalis penis 1364
nervus dorsalis scapulae 1365
nervus ethmoidalis 1552
nervus facialis 1614
nervus femoralis 1657
nervus frontalis 1770
nervus genitofemoralis 1828
nervus glossopharyngeus 1858
nervus gluteus caudalis 731
nervus gluteus cranialis 1134
nervus hypogastricus 2085
nervus hypoglossus 2087

os carpi accessorium 26
os carpi intermedium 2251
os carpi radiale 3745
os carpi ulnare 4729
os centroquartale 801
os compedale 1705
os coronale 2835
os costale 3868
os coxae 2010
os cuboideum 1752
os cuneiforme intermedium
    4029
os cuneiforme laterale 4530
os cuneiforme mediale 1706
os dentale 1252
os dorsale 3029
os ectethmoidale 1425
os entoglossum 1480
os ethmoidale 1546
os exoccipitale 1564
os femoris 1665
os frontale 1767
os hamatum 1751
os hyoideum 2067
os ilium 2133
os incisivum 2156
os interparietale 2289
os ischii 2346
os lacrimale 2394
os longum 2560
os lunatum 2251
os mesethmoidale 2775
os metacarpale III 4527
os metatarsale III 4528
os nasale 2932
os naviculare 797
os naviculocuboideum 801
os occipitale 3062
os orbitosphenoidale 3152
os otica 3182
os palatinum 3224
os paraglossum 1480
os parasphenoidale 3305
os parietale 3319
os penis 3377
os pisiforme 26
os planum 1713
os pneumaticum 3520
os pr(a)esphenoidale 3612
os prearticulare 3588
os prefrontale 3595
os pterygoideum 3664
os pubis 3677
os rostrale 3914
ossa carpi 683
ossa cranii 1126
os sacrum 3974
ossa digitorum manus 1311
ossa digitorum pedis 1312
ossa faciei 1611

ossa metacarpalia I-V 2786
ossa metatarsalia I-V 2799
ossa sesamoidea proximalia
    3658
ossa tarsi 4444
os scaphoideum 3745
os sesamoideum distale 1341
ossicula auditus 328
ossis 510
os sphenoidalis 4128
os spleniale 4167
os squamosum 4188
os supraangulare 4337
os suprajugale 4343
os supraorbitalia 4348
os tarsale I 1706
os tarsale II 4029
os tarsale III 4530
os tarsale IV 1752
os tarsi centrale 797
os temporal 4468
osteogenesis imperfecta 3175
ostium aortae 239
ostium atrioventriculare
    dextrum 3882
ostium atrioventricularis
    sinistrum 2472
ostium c(a)ecocolicum 620
ostium ileoc(a)ecale 2108
ostium omasoabomasicum
    3107
ostium papillare 4460
ostium pharyngeum tubae
    auditivae 3440
ostium pr(a)eputiale 3609
ostium pyloricum 3712
ostium reticuloomasicum 3841
ostium ruminoreticulare 3955
ostium trunci pulmonalis 3124
ostium tympanicum tubae
    auditivae 4714
ostium urethrae externum 1591
ostium urethrae internum 2276
ostium uteri externum 1589
ostium uteri internum 2273
ostium vaginae 3160
ostium venae cavae 3125
os trapezium 1704
os trapezoideum 4028
os triquetrum 4729
os ungulare 1340
os zygomaticum 5001
otitis externa 3185
otitis media 3186
Otobius megnini 1419
Otodectes cynotis 1416
ovarium 3200
ovocytus 3123
ovum 3209
oxacillinum 3210

oxfendazolum 3211
oxibendazolum 3212
oxyclozanidum 3214
oxytetracyclinum 3216
oxytocinum 3217

palatum 3222
palatum durum 1960
palatum fissum 934
palatum molle 4111
palatum osseum 516
palea 4958
palear 1277
pallium 3239
palma manus 3244
palpebra 1606
palpebra III 4526
palpebra inferior 2580
palpebra superior 4756
pancreas 3252
panniculus adiposus 4267
papilla 3262
papilla duodeni 1404
papillae buccales 598
papillae conicae 1030
papillae filiformes 1696
papillae foliatae 1732
papillae fungiformes 1786
papillae labiales 2387
papillae linguales 2526
papillae proventriculares 3263
papillae reticuli 3264
papillae ruminis 3945
papillae vallatae 4814
papilla ilealis 2109
papilla incisiva 2159
papilla mammae 4456
papilla omasi 3106
papilla parotidea 3330
papilla pili 1946
papilla renis 3816
papula 3275
parabronchus 3276
paradidymis 3278
paradigitus 1276
Paragonimus 2607
paralysis bulbaris infectiosa
    331
Paramphistomum 3287
paraplexia enzootica ovium
    4013
paraungula 1276
parbendazolum 3314
parenchyma 3315
paries corneus 4956
paries tympanicus ductus
    cochlearis 4156
paries vestibularis ductus
    cochlearis 3804
paroophoron 3324

vallum 2518
valva aortae 4816
valva atrioventricularis dextra
    3883
valva atrioventricularis sinistra
    2473
valva bicuspidalis 2473
valva cordis 1974
valva ileoc(a)ecalis 2109
valva mitralis 2473
valva tricuspidalis 3883
valva trunci pulmonalis 4818
valvula cordis 1974
valvulae semilunares 4036
valvula lymphatica 2619
valvula venosa 4817
variola avium 1756
variola ovina 4072
vas anastomoticum 172
vasa vasorum 4822
vas capillare 649
vas collaterale 991
vas lymphaticum 2632
vas sanguina 491
velum medullare caudale 738
velum medullare rostrale 3919
velum palatinum 4111
vena 4832
vena abdominalis caudalis 717
vena abdominalis cranialis
    1124
vena angularis oculi 185
vena axillaris 379
vena azygos dextra 3885
vena azygos sinistra 2475
vena bicipitalis 444
vena brachialis 559
vena brachiocephalica 562
vena cava caudalis 760
vena cava cranialis 1160
vena cephalica 802
vena cephalica accessoria 27
vena cerebri magna 1895
vena cervicalis profunda 1226
vena cervicalis superficialis
    4320
vena cordis magna 1894
vena costoabdominalis dorsalis
    1357
vena costocervicalis 1107
vena cremasterica 1162
vena cutanea 1198
venae arcuatae 263
venae auriculares 1421
venae bronchales 583
venae c(a)ecales 619
venae caudales 4439
venae caudales femoris 727
venae cavernosae penis 773
venae centrales hepatis 799

venae cerebelli dorsales 1353
venae cerebelli ventrales 4845
venae cerebri 818
venae ciliares 916
venae circumflexae 926
venae colicae 982
venae collaterales 990
venae comitantes 35
venae cordis 674
venae digitales 1316
venae diploicae 1329
venae dorsales 1383
venae emissariae 1451
venae gastricae 1812
venae genus 1821
venae hepaticae 1989
venae ilei 2106
venae intercostales 2234
venae interosseae 2288
venae intervertebrales 2307
venae jejunales 2355
venae labiales 2388
venae labyrinthi 4833
venae lumbales 2596
venae mediastinales 2740
venae metacarpeae 2790
venae metatarseae 2803
venae (o)esophageae 3080
venae palatinae 3233
venae palpebrales 3250
venae pancreaticae 3259
venae pancreaticoduodenales
    3256
venae penis 3378
venae perforantes 3387
venae perineales 3400
venae pharyngeae 3444
venae phrenicae craniales 1149
vena epigastrica caudalis 724
vena epigastrica cranialis 1130
vena epigastrica cranialis
    superficialis 4321
venae pudendae 3685
venae pulmonales 3696
venae radiales 3750
venae rectales 3779
venae ruminales 3949
venae scrotales 4019
venae sigmoideae 4091
venae spinales 4145
venae suprarenales 4358
venae tarseae 4449
venae temporales 4477
vena ethmoidalis externa 1577
venae thymicae 4552
venae thyr(e)oideae 4563
venae vesicales 4883
venae vestibulares 4899
venae vorticosae 4953
vena facialis 1616

vena femoralis 1660
vena gastroduodenalis 1816
vena gastroepiploica dextra
    3888
vena gastroepiploica sinistra
    2478
vena glutea caudalis 732
vena glutea cranialis 1135
vena ileocolica 2111
vena iliaca communis 1009
vena iliaca externa 1580
vena iliaca interna 2266
vena iliolumbalis 2128
vena infraorbitalis 2198
vena jugularis externa 1582
vena jugularis interna 2268
vena laryngea cranialis 1139
vena lienalis 4170
vena lingualis 2529
vena linguofacialis 2533
vena maxillaris 2702
vena mediana 2735
vena mediana cubiti 2730
vena mesenterica caudalis 743
vena mesenterica cranialis 1144
vena musculophrenica 2893
vena obturatoria 3060
vena occipitalis 3066
vena omobrachialis 3110
vena ophthalmica externa
    dorsalis 1359
vena ophthalmica externa
    ventralis 4848
vena ophthalmica interna 2275
vena ovarica dextra 3892
vena ovarica sinistra 2481
vena phrenica caudalis 750
vena poplitea 3551
vena portae 3562
vena profunda 1235
vena renalis 3821
vena retromandibularis 3853
vena sacralis mediana 2836
vena saphena 3987
vena scapularis dorsalis 1376
vena sternocleidomastoidea
    4209
vena stylomastoidea 4254
vena subclavia 4261
vena subcutanea abdominis
    4321
vena subscapularis 4290
vena supraorbitalis 4355
vena suprascapularis 4361
vena testicularis dextra 3895
vena testicularis sinistra 2484
vena thoracica externa 1596
vena thoracica interna 2281
vena thoracica lateralis 2467

Printed and bound by CPI Group (UK) Ltd, Croydon, CR0 4YY

03/10/2024

01040330-0003